This book is an authoritative and comprehensive account of the principles and practice of modern NMR spectroscopy of solids as applied to polymeric materials.

NMR spectroscopy has been applied to the characterisation of polymers in the solid state for over 40 years. The past two decades have seen the development of many new NMR capabilities, including high-resolution techniques for solids, multidimensional methods, deuterium NMR and others. All of these developments have contributed to a dramatic increase in the power and applicability of NMR for the characterisation, at a molecular level, of the dynamics and structural organisation of polymeric solids.

This book is intended for polymer physicists, chemists and materials scientists. The emphasis of the applications chapters (5–8) is on polymer types and properties to make it more accessible to this audience. To help those with little knowledge of NMR the authors have included an introduction to the main principles of the technique involved in its application to solid polymers. Those more knowledgeable on the subject will find that rigorous and detailed analytical treatments are also available, often in appendices.

All research workers, whether graduate students beginning their studies or established professionals, with a concern for polymer characterisation and the relationship between structure, dynamics and function for these materials will find this book of value in their work.

NUCLEAR MAGNETIC RESONANCE IN SOLID POLYMERS

Cambridge Solid State Science Series

EDITORS

Professor E. A. Davis
Department of Physics, University of Leicester

Professor I. M. Ward, FRS
IRC in Polymer Science and Technology, University of Leeds

Titles in print in this series

NUCLEAR MAGNETIC RESONANCE IN SOLID POLYMERS

VINCENT J. McBRIERTY
Professor of Polymer Physics, Trinity College, Dublin

and KENNETH J. PACKER
Chief Scientist, British Petroleum Co. p.l.c.

CAMBRIDGE
UNIVERSITY PRESS

CAMBRIDGE UNIVERSITY PRESS
Cambridge, New York, Melbourne, Madrid, Cape Town, Singapore, São Paulo

Cambridge University Press
The Edinburgh Building, Cambridge CB2 2RU, UK

Published in the United States of America by Cambridge University Press, New York

www.cambridge.org
Information on this title: www.cambridge.org/9780521301404

First published 1993
This digitally printed first paperback version 2006

A catalogue record for this publication is available from the British Library

Library of Congress Cataloguing in Publication data
McBrierty, Vincent J.
Nuclear magnetic resonance in solid polymers/by Vincent J.
McBrierty and Kenneth J. Packer.
p. cm. – (Cambridge solid state science series)
Includes bibliographical references and indexes.
ISBN 0-521-30140-8 (hc)
1. Polymers – Analysis. 2. Nuclear magnetic resonance
spectroscopy. I. Packer, K. J. II. Title. III. Series.
QD139.P6M38   1993
668.4′2–dc20   92-45158   CIP

ISBN-13 978-0-521-30140-4 hardback
ISBN-10 0-521-30140-8 hardback

ISBN-13 978-0-521-03172-1 paperback
ISBN-10 0-521-03172-9 paperback

Dedicated to Kay and Christine for their
patience and support throughout

Contents

Preface

Since its inception, nuclear magnetic resonance (NMR) has been used with remarkable success to investigate polymeric materials. However, application to solid polymers was for many years largely the province of physicists and physical chemists because of the need for specialised spectrometers to gain access to the broad spectra (usually ^{1}H) typical of solids, and because interpretation of these spectra and associated relaxation times required theoretical models of a strongly physical nature. The chemist, meanwhile, was more than satisfied to exploit the enormous potential provided by the increasing power of liquid-state NMR spectroscopy which had benefited considerably from the introduction of Fourier transform (FT) methods, the availability of higher fields generated by superconducting magnets with concomitant enhanced sensitivity, formidable on-line computing capabilities, and the added flexibility of multidimensional NMR. The rich site-specific information in high-resolution liquid-state NMR remained undetected in early solid-state spectra because of the dominant dipolar contribution. Sustained efforts to achieve comparable results for solids led to procedures to suppress dipolar contributions using high-power decoupling techniques, sample spinning and the application of ingenious pulse sequences. Today the full power of high-resolution one-, two- and three-dimensional NMR is available for solid materials, albeit requiring more sophisticated experimentation and analysis. Specifically, multidimensional NMR permits different spin interactions to be correlated or separated, exchange between different states of a resonant nucleus to be monitored over selected timeframes and the intricacies of complex molecular motions to be elucidated.

These developments initially centred on ^{13}C NMR, having circumvented difficulties of low natural abundance. Refinements in ^{2}H NMR have since added rich site-specific detail. The latest generation of spectrometers have access to a wide range of resonant nuclei and experiments on a relatively routine basis. Aside from constraints of time, high-resolution NMR of both liquids and solids may now be carried out on essentially the same spectrometer system.

Our intention is to provide the polymer scientist of whatever discipline with a view of what NMR is about, why and how it can be applied usefully to solid polymers and what practical information may be extracted from NMR experiments. The method has many unique attributes beyond merely complementing and supporting other methods in examining polymer properties such as chain chemistry, conformation, packing, orientation and dynamics; but NMR is a form of coherent spectroscopy and, as such, requires somewhat more investment of effort on the part of the newcomer to appreciate fully its scope and capability. Such effort will reveal a remarkable flexibility of application in diverse areas of science, not least those of polymer physics and chemistry.

The book primarily addresses the needs of the polymer scientist who is not a specialist in NMR and, consequently, the approach is more pedagogical than definitive. It does not focus exclusively on the most informative NMR experiments since they are often the most demanding and sophisticated and so are not readily accessible to the routine user. On the one hand, the various transitions in the polymer may be established using fairly conventional NMR methods to give much useful information of a practical nature. On the other, the maximum available information on the underlying molecular dynamics can require state-of-the-art procedures available only in the most advanced and specialist NMR laboratories.

With this in mind, Chapter 1 presents a broad overview of the versatility of the technique in the context of solid polymers while Chapters 2 and 4 introduce the reader to the prerequisite concepts and experimental procedures. For those who wish to delve more deeply into the fundamentals of NMR, Chapter 3 examines in more detail the key spin interactions used to probe the local structure and dynamics in solid polymers. The approach exploits the elegance of rotation matrices but in a way that does not compromise the needs of those merely interested in practical application. Expressions essential for data interpretation and analysis are cast in Cartesian representation in the main text and may be used directly without recourse to the formal derivations which are treated separately in the appendices. The remaining chapters are organised around polymer properties rather than the isotopes generally used in NMR experiments which often serve to delineate specific competences and interests in the published literature. This approach is more in keeping with the intended spirit of the book. These chapters cover structure and motion (Chapter 5), heterogeneity (Chapter 6), and orientation (Chapter 7). The final chapter 8 selects a number of specialist topics which demonstrate further the diversity of NMR applications. A glossary of NMR terms along with useful information on polymers referred to in the book are included.

Acknowledgements

It is a pleasure to acknowledge Dr Dean Douglass's sustained advice and inspiration. We are grateful also to Heather Browne, Elaine Kavanagh and Michelle Gallagher who typed the manuscript. Thanks are due to John Kelly and Vincent Weldon for the artwork and photography. Finally, the support and assistance of Edward and Gwen Morgan during the preparation of the book are gratefully acknowledged.

Glossary of terms

$90_{x'}, 180_{y'} \ldots$	rf pulses producing rotations of 90°, 180° etc., about the x', y' ... axes of the rotating reference frame
ADRF	Adiabatic demagnetisation in the rotating frame
ADC	Analogue-to-digital convertor
$\mathbf{B}_0$	Large, static, polarising magnetic field
$\mathbf{B}_{1i'}$	radiofrequency (rf) field along the i' axis of the rotating reference frame
$\mathbf{b}_{z'}$	effective z' magnetic field in the rotating reference frame
$\mathbf{B}_e$	effective rf field in the rotating reference frame
$\mathbf{B}_{loc}$	local magnetic field, for example, dipolar
β_0	angle between principal symmetry axis of spin coupling tensor and the external field $\mathbf{B}_0$
B_{1I}	amplitude of B_1 field applied to spins I
CP	cross polarisation: double irradiation method for magnetisation transfer between unlike spins
C	Curie constant
CW	Continuous wave: NMR detected by sweeping B_0/ω_0
CPMAS	Cross polarisation/magic-angle spinning experiment for obtaining high-resolution spectra from solids
CORD	Three-dimensional NMR experiment in solids which Correlates ORder and Dynamics
CRAMPS	Combined Rotation and Multiple Pulse experiments to determine (usually) isotropic chemical shift NMR spectra of homonuclear dipolar-coupled spins in solids
CSA	Chemical shift anisotropy
DD	Dipolar decoupling: strong resonant irradiation of one species of spin to remove its dipolar coupling effects from NMR spectra
$D(\Omega)$	Wigner operator
$D_{mn}^{(l)}(\Omega)$	Wigner rotation matrix elements
D_s	Spin diffusion coefficient
DRSE	Dipolar rotational spin echo experiment for determining heteronuclear dipolar lineshapes in solids
DSC	Differential scanning calorimetry
ΔE	Activation energy
eQ	nuclear electric quadrupole moment

eq	electric field gradient at nucleus
EFG	electric field gradient
η	asymmetry parameter of a tensor spin interaction
$\bar{\eta}$	effective asymmetry parameter after averaging over specific motions
FID	free induction decay signal
$F_n(t)$	time dependent function of lattice variables (n labels the component of the tensor)
FT	Fourier transformation
γ	nuclear magnetogyric ratio
$G(t)$	FID signal
$G^n(t)$	time correlation function of spin coupling interaction tensor (n labels the tensor component)
$\mathcal{H}_D$	dipole–dipole spin coupling Hamiltonian
$\mathcal{H}_{CS}$	magnetic shielding (chemical shift) Hamiltonian
$\mathcal{H}_Q$	nuclear electric quadrupole coupling Hamiltonian
$\mathcal{H}_{rf}$	rf field Zeeman coupling Hamiltonian
$\mathcal{H}_J$	indirect (scalar) spin coupling Hamiltonian
$\mathcal{H}_{SR}$	spin–rotation coupling Hamiltonian
$\mathcal{H}_z$	Zeeman interaction Hamiltonian for $\mathbf{B}_0$
$\mathcal{H}'(t)$	zero-average, time-varying coupling Hamiltonian
I/I_z	nuclear spin angular momentum quantum numbers
$\mathbf{I}/\mathbf{I}_z$	nuclear spin angular momentum operators
$I(\omega)$	NMR spectrum
$\mathbf{I}^{+/-}$	raising/lowering spin angular momentum operators
I_{MN}	spinning sideband intensities in two-dimensional NMR spectra of solids
$I(\tau_c)$	distribution of correlation times τ_c
$J(\omega)$	power spectrum of a time-dependent spin coupling
$J^n(n\omega_0)$	power spectrum of the nth component of a fluctuating tensor coupling at frequency $n\omega_0$
LFSLR	laboratory frame spin–lattice relaxation
$\bar{\mathbf{M}}_n$	number average molecular weight
$\bar{\mathbf{M}}_w$	weight average molecular weight
MAS	magic-angle spinning
μ	nuclear magnetic moment
$\mathbf{M}$	nuclear magnetisation vector
M_0^I	equilibrium value of nuclear magnetisation of ensemble of I spins
$M_i^{eq}(i = x, y, z)$	laboratory frame components of $\mathbf{M}_0$
$M_{i'}(i = x', y', z')$	components of nuclear magnetisation in the rotating reference frame
m	quantum number for the z-component of spin (see I_z)
$M_n(\omega)$	nth moment of a spectrum
$M(t)$	time-dependent nuclear magnetisation
MREV-8	an eight-pulse cycle sequence for obtaining high-resolution spectra from homonuclear dipolar-coupled spin systems in solids
ν_c	correlation frequency of a motion ($= 2\pi\tau_c)^{-1}$
NOE (F)	nuclear Overhauser enhancement (factor)
NQS	non-quaternary suppression (dipolar dephasing) experiment for obtaining NMR spectra in solids of dilute spins with weak

	dipolar couplings to abundant spins (for example, quaternary ^{13}C spins)
ω_0^I	precession frequency of nuclear spin I in field $\mathbf{B}_0$
Ω	apparent precession frequency of spins in the rotating reference frame
ω_e	precession frequency of spins about $\mathbf{B}_e$ in the rotating reference frame
ω_Q	quadrupole coupling energy expressed as a frequency
$\Omega_i(\alpha\beta\gamma)$	Euler rotation angles
ω_r	angular frequency of sample rotation
$\omega_1(=\gamma B_1)$	precession frequency of spins about on-resonance $B_1(\omega)$ field in the rotating reference frame
ω_{MAS}	angular frequency of sample rotation about an axis making an angle θ_m with respect to $\mathbf{B}_0$
PAS	principal axis system
$P_l(\cos\beta)$	Legendre polynomial of order l
$P(\Omega)$	orientation distribution in an oriented polymer
Q	quality factor of rf tuned circuit or component
r_{CH}	carbon–hydrogen internuclear distance
R_1	spin–lattice relaxation rate in the laboratory frame
$R_{1\rho}$	spin–lattice relaxation rate in the rotating reference frame
R_2	transverse spin relaxation rate
$R_1^{II/SS}$	homonuclear relaxation rates arising from dipole–dipole coupling
$R_1^{IS/SI}$	heteronuclear relaxation rates arising from dipole–dipole coupling
RFSLR	rotating frame spin–lattice relaxation (see $T_{1\rho}$, $R_{1\rho}$)
RAD	rotational angular distribution function for motions in solids
σ	magnetic shielding tensor
$\sigma_{ii}(i=1,2,3)$	principal components of σ
$\sigma_{\parallel/\perp}$	principal components of σ for axial symmetry
$\tilde{\sigma}$	isotropic shielding constant
σ_{zz}	component of σ along the field $\mathbf{B}_0$
$\bar{\bar{\sigma}}$	shielding tensor in rotor frame
$\sigma^{(R)}$	shielding tensor averaged over fast motions about axis R
$S(\omega)$	NMR spectrum
$S(t)$	NMR signal in the time domain: FID (see $G(t)$)
$S(\Omega_1, \Omega_2)$	Two-dimensional NMR spectrum
$S(t_1, t_2)$	time domain signal which Fourier transforms to $S(\Omega_1, \Omega_2)$
SPE	single-pulse excitation: experiment which relies on dilute spin T_1 to produce magnetisation
SLR	spin–lattice relaxation (see T_1, R_1 etc.)
SANS	small-angle neutron scattering
SAXS	small-angle X-ray scattering
SSB	spinning sideband
τ_c	correlation time
T_g	glass transition temperature
θ	angle of $\mathbf{B}_e$ with respect to z' axis in rotating reference frame
θ_p	rotation produced by an rf pulse
θ_m	magic angle $(\cos^{-1}(3)^{-\frac{1}{2}}) = 54.7°$
t_p	duration of rf pulse
T_2	transverse or spin–spin relaxation time

Glossary of terms

T_1	laboratory frame spin–lattice relaxation time
$T_{1\rho}$	rotating frame spin–lattice relaxation time
T_L	lattice temperature
T_S	spin temperature
T_{CR}	cross-relaxation time
$T_{CH}^{(SL)}$	cross-relaxation time between ^{13}C and ^1H spins under matched Hartmann–Hahn spin-locked cross polarisation
$T_{2lt/ht}$	transverse relaxation times at temperatures below/above a linewidth transition
T_R	recycle (repetition) time of a pulse sequence
t_D	dwell time: interval between successive digitisation of a time-varying signal
T_{AQ}	acquisition time
τ	general time interval in an NMR pulse sequence
τ_D	dipolar dephasing time in an NQS experiment
TOSS	total sideband suppression experiment to eliminate spinning sidebands from MAS spectra
t_m	mixing time – often in a multidimensional NMR experiment or in studies of spin diffusion
$\langle T_j(^{13}C)\rangle$	average spin–lattice relaxation time ($j = 1, 1\rho$) for ^{13}C spins
X, Y, Z	axis labelling for the laboratory frame of reference
X_0, Y_0, Z_0	axis labelling for the sample-fixed coordinate system
x, y, z	axis labelling for the molecule-fixed coordinate system

1

The NMR of solid polymers: an overview

1.1 The nature of polymers

One of the remarkable features of current progress in materials technology is the ever-increasing ability to control and design the physical and chemical responses of synthetic polymers. An unprecedented range of properties is now routinely accessible through a variety of processing techniques which include chemical synthesis and substitution, thermal and electrical treatment, addition of fillers, blending of dissimilar polymers and mechanical deformation to name but a few.

Polymers are long chain molecules comprising large numbers of basic repeat units: the chemical structures of some of the more common types are furnished in table A1.1 (Appendix 1). For the most part, they are organic materials of high molecular weight in crystalline, glassy or rubbery states, one or more of which may be simultaneously present. Attempts to understand more fully the underlying reasons behind the properties of polymers are bedevilled by a characteristically complex morphology which defies precise description even in the chemically simplest cases. There is concomitant complexity too in the way in which polymer molecules move. Motions of flexible backbone chains, for example, derive from many coupled degrees of freedom and there is often a broad distribution of spectral frequencies associated with a given motional event. Two or more discrete motions may be active at the same time.

It is important to understand structure and motion in polymers because of the manner in which they influence properties of practical interest. Changes in crystallinity or tacticity, for example, can induce marked differences in overall performance and potential applications of a polymer; chain orientation can confer remarkable strength as in ultradrawn polymers; plasticisers and fillers alter mechanical properties; molecular motion underpins thermal distortions, creep and impact strength; the motion of molecules with dipole moments influences electrical loss; and control of morphology and material defects is the key to improving the performance of semiconducting polymers and so on.

In discussing the application of NMR to the study of polymer systems, the somewhat crude though pragmatic approach of distinguishing between *structure* and *motion* is adopted even though the two are inevitably mutually interdependent. Indeed it is often the case that different structural regions in a polymer are identified precisely on the basis of their characteristically different motions. It is a matter of convenience, in the interests of clarity of presentation, to retain this arbitrary dichotomy between structure and motion, at least for the time being, subject to the proviso that the strong inter-relationship between the two is constantly borne in mind.

1.2 The role of NMR

NMR spectroscopy has featured prominently among the host of investigative techniques used in polymer research. One consequence of the short-range nature of NMR interactions is the inherent sensitivity to different molecular environments as typically reflected in NMR lineshapes and decays (fig. 1.1(a)); but, as with most forms of experimental data, the translation of raw NMR results into meaningful information on structure and molecular relaxation is not wholly unambiguous. In particular, interpretation of low-resolution NMR spectra where lineshapes are dominated by dipolar contributions inevitably appeals to intuitive models which lead to conclusions that are not always definitive. While the onset of motion can be readily detected, the precise mechanism is often obscure. This contrasts with the identification of individual chemical sites in high-resolution spectra for liquids in which the usually dominant dipolar interactions are averaged to zero as a gift of nature. Consider, for example, the high-resolution proton spectrum of a moderately low molecular weight ($M_n = 2600$) polybutadiene terminated at both ends with phenyl groups (fig. 1.1(b)). CH_2, CH and phenyl protons, for which experimental intensities agree rather well with theoretical estimates, are readily discernable. That the relaxation behaviour of the resolved moieties can be individually explored greatly facilitates the study of chain dynamics.

Selective suppression of the dominant dipolar contribution to the Hamiltonian in the solid state reveals a wealth of information associated with the weaker chemical shift and *J*-coupling contributions. This has been achieved by means of cross polarisation (CP), dipolar decoupling (DD)

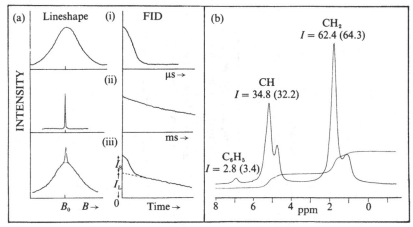

Figure 1.1. (a) Schematic lineshapes and corresponding free induction decays (FIDs) for (i) a solid, (ii) a liquid and (iii) a material such as a partially crystalline polymer that contains both rigid and mobile molecules. The timescales for the solid and liquid FIDs are of the order of microseconds and milliseconds, respectively. The initial component intensities of the composite FID and, correspondingly, the areas under the resolved lineshapes, denote the number of participating resonant nuclei contributing to each phase. (b) Proton spectrum (300 MHz) of a moderately low molecular weight ($M_n = 2600$) polybutadiene terminated at each end with a phenyl group (25 % *trans* 1,4, 50 % vinyl, 99 % unsaturated). Experimental intensities, I, are computed automatically by the spectrometer from the areas under the component peaks. The theoretical components (in parenthesis) are computed as follows: $10/(6n+10)$ phenyl protons; $2n/(6n+10)$ CH protons; $4n/(6n+10)$ CH$_2$ protons where $n = 45.3$ is the number of butadiene units in the chain. Proton chemical shifts are relative to TMS.

and magic-angle spinning (MAS) techniques. Table 1.1 illustrates the progression towards ^{13}C spectra in solids (cf fig. 5.1). However, because MAS averages out anisotropic spin interactions there is a loss of important information on the structural and motional complexity in the solid state, information which is vital to the complete characterisation of solid polymers. Separated local field and rotational spin echo experiments (Opella and Waugh, 1977; Munowitz and Griffin, 1982; Schaefer *et al.*, 1983) were among the early successes in retrieving this suppressed information. More recent developments in multidimensional NMR allow different spin interactions to be correlated or separated, exchange between different states of a resonant nucleus to be monitored over selected time periods, and the details of complex motions to be elucidated (Ernst *et al.*, 1987; Nakai *et al.*, 1988b; Spiess, 1991). The term *nuclear magnetic resonance crystallography* has been coined to describe one form of two-dimensional NMR which can elucidate the relative orientations and

Table 1.1. *Solution state versus high-resolution solid-state ^{13}C NMR spectroscopy*

Total interaction =	Zeeman (MHz)	+	Dipolar[a] (kHz)	+	Scalar[a] (Hz)	+	Chemical shift
Solution state	50		0		200		isotropic, single frequency
Solid state	50		50		200		200 ppm-wide chemical shift anisotropy
Solid-state NMR			high-power (dipolar) proton decoupling		high-power decoupling also removes J coupling		magic-angle spinning gives the isotropic line

[a] Couplings are to ^1H.

internuclear distances of neighbouring molecules in polycrystalline and non-crystalline solids (Tycko and Dabbagh, 1991; Raleigh *et al.*, 1989). Selective isotope labelling and the ability to control spin diffusion (which tends to average out much detailed information) have also resulted in more meaningful data for comparison with theoretical models.

Efforts to unravel the complexities of polymer behaviour are also eased by correlating the relaxation results of NMR with those of related techniques which include dielectric relaxation, dynamic mechanical thermal analysis (DMTA) and quasi-static experiments such as differential scanning calorimetry (DSC). Each technique responds in its own characteristic manner. For example, NMR and DMTA are sensitive to the motion of CH_3 and CH_2 groups, whereas dielectric relaxation is not since groups such as these do not have a dielectric moment. DMTA probes a wide range of low frequencies typical of those which influence impact strength and which are not systematically accessible by other methods. In general, it transpires that many diverse macroscopic properties of a polymer reflect the same relaxation mechanism which, in turn, can be identified with a specific molecular motion (cf Section 1.5).

1.3 Structural considerations: a model for the polymer

Despite the inherent structural complexities of polymers, the use of idealised models that are generally consistent with the known morphology has facilitated the interpretation of experimental data. A two-phase model was first proposed to account for the readily discernible behaviour of crystalline and amorphous regions in a partially crystalline polymer (Wilson and Pake, 1953). Spectra consisted of superimposed broad and narrow resonances, assigned respectively to crystalline and amorphous regions (fig. 1.1(a)). The implicit assumption of sharp boundaries between the two phases has since been refined in those cases where interfacial material is present in such proportions as to constitute a third phase (fig. 1.2). This general approach carries over naturally to a treatment of composites such as block copolymers and heterogeneous blends.

However, the interpretation of experimental data in terms of non-interacting phases is compromised by evidence of transport between crystalline and amorphous phases (Mansfield, 1987; Schmidt-Rohr and Spiess, 1991b) and by a number of factors that are inherent features of NMR. Spin diffusion, for example, can affect the magnitudes and intensities of component spin–lattice and rotating frame relaxation times in

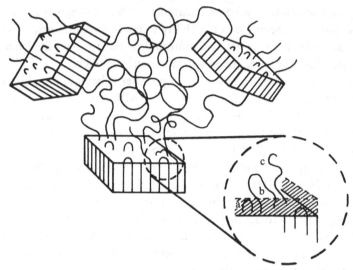

Figure 1.2. Visualisation of crystalline and amorphous material in a partially crystalline polymer. (a) tight fold; (b) loose fold; (c) cilium or part of a tie molecule. It is possible, on occasion, to resolve an interfacial phase on the surface of the crystalline component as defined by the dashed region in the insert.

a way that complicates the quantitative analysis of separate phases (cf Section 3.8.3). The observation of complex relaxation decay may be an inherent feature of NMR relaxation behaviour itself rather than a manifestation of distinct morphological regions or structural entities in the polymer. The observed decay may also arise from a broad distribution of structural environments (or a broad distribution of motional correlation frequencies) and the extraction of only two or three components may simply reflect the limitations of the decomposition procedure rather than a definitive indication of the presence of a specific number of discrete regions or molecular entities as such.

Although there are obvious limitations in the approach, the description of a polymer in terms of one or more non-interacting phases forms the basis of a rational and consistent interpretation of experimental NMR data. To a first approximation, it is presumed that each phase is made up of characteristic structural units in the form of isotropic amorphous material, single crystals, crystallites, chain segments or other appropriate structural entity. The properties of units with well-defined crystallographic structure tend to be spatially anisotropic and in this respect the NMR response is no exception. Lineshapes for single crystals, for example, often

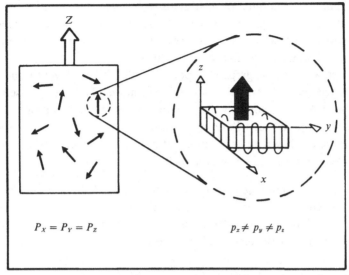

$$P_X = P_Y = P_Z \qquad\qquad p_z \neq p_y \neq p_z$$

Figure 1.3. Schematic diagram showing the spatially random orientation of structural crystal units in an isotropic bulk polymer. The arrows denote the polar axes of the units. At the macroscopic level, the property, P, under examination does not depend on direction because of spatial randomisation in the bulk. At a molecular level, the tensor property is generally anisotropic, that is, $p_x \neq p_y \neq p_z$.

depend sensitively upon sample orientation in the laboratory magnetic field. In bulk materials, on the other hand, the constituent structural units are randomly distributed in space and, as such, the macroscopic measurement represents an isotropic spatial average for the system as a whole (fig. 1.3). Consequently, much useful information contained in the molecular anisotropy is lost in macroscopic measurements on isotropic polymers. The situation is redressed somewhat by mechanically deforming the sample, which has the effect of generating at least partial alignment of the constituent polymer chains, thereby recovering some of the molecular anisotropy at a macroscopic level (McBrierty and Ward, 1968, 1971b; Ward, 1982; Nomura, 1989). This macroscopic anisotropy, which is revealed in a wide range of experimental NMR and other data, can be used to characterise the spatial distribution of molecules in deformed polymers as described in Chapter 7.

The principal ingredients of a model for a polymer are the following.

- The intrinsic response of the structural unit which, in the NMR context, requires information on the positions and motions of contributing nuclei.

- The relative amounts of distinguishable phases present since the macroscopic response involves contributions from the different phases. In many cases this information is contained in component intensities of resolved contributions to the NMR linewidth.
- In calculating the aggregate response of all the constituent structural units, account must be taken of their orientation relative to a chosen axis in the polymer sample, which in turn requires information on the statistical spatial distribution of units in the sample. Conversely, experimental NMR data for oriented polymers may be used to extract information on the distribution itself.

The practical implementation of these ideas will become evident in succeeding chapters. Where appropriate, the basic model is refined to take account of additional considerations such as distributions of correlation frequencies and/or the effects of spin diffusion.

1.4 Molecular motion: an overview

The sensitivity of NMR to molecular motion of appropriate frequency and amplitude underpins the usefulness and, on occasion, indispensability of the method in unravelling the complexities of molecular relaxation. An initial, somewhat general, consideration of molecular motion in polymers will help to clarify and put into perspective the analytical methodology and interpretations encountered in subsequent chapters.

Complexity arises in several ways.

- Cooperativity, anisotropy, and the effects of geometrical constraints are typical earmarks of molecular motion in polymers.
- Several discrete motions may be active simultaneously.
- Broad distributions of correlation frequencies, routinely encountered in practice, can only be dealt with semi-quantitatively.
- The complex motion of *inter*nuclear vectors generated from the *relative* motions of nuclei is difficult to deal with analytically other than in a few special cases.
- The way in which defects promote motion within their sphere of influence, particularly in polymers that are dilute in impurity defects, can be difficult to quantify.
- Both timescale *and* spatial dimension of the motion are important.
- Whereas the detection of specific molecular motions can be reasonably straightforward, the nature of the motional mechanism is often obscure.

Table 1.2. *Typical motions observed in polymers*

Motion	Region[a]	Nature	Typical activation energy (kJ mol^{-1})	Remarks
Primary main chain	C	Hindered rotations, oscillations or translations	> 125	May be activated by defects
Primary main chain	A	Large-scale rotations and translations	< 400	Associated with glass transition and breakdown of long-range order: ΔE is usually temperature-dependent
Secondary chain motion	C/A	Localised motion of interfacial material: cilia, folds, tie molecules, chains of low molecular weight	40–60	Characteristic of linear polymers: some or all chains may be involved
Sidegroup	C/A	Highly localised motion of specific moiety	8–80	Usually in the low-temperature regime below T_g
Impurity motions	C/A	Motions induced by trace solvents or low molecular weight material which can act as a plasticiser	—	Dominates relaxation at low temperature

[a] C = crystalline; A = amorphous.

Much effort has been expended in seeking a better overall understanding of molecular motion in polymers. Many models based upon a range of fundamentally different premises have been developed, some principally to describe dielectric and dynamic mechanical relaxation while others have addressed the problem from the specific standpoint of NMR. Detailed comparison of a number of the more frequently used models has shown gratifying consistency (McBrierty and Douglass, 1980), which supports the contention that the approach adopted reasonably predicts the behaviour of real systems and that the correlation of diverse experimental data is not strongly model-dependent. Significant progress has also been achieved in computer generation of realistic polymer models (Spiess, 1983; Hirschinger *et al.*, 1990). At the microscopic level, multidimensional NMR is progressively removing the ambiguities implicit in molecular motional assignments. This is rather graphically illustrated for the α-relaxation in poly(vinylidene fluoride) as described in Section 5.4.5 (Hirschinger *et al.*, 1991).

Having placed the problem in context, procedurally it is usual to adopt a somewhat pragmatic interpretive approach whereby initially assumed idealised models for specific molecular motions are subsequently refined, intuitively or otherwise, to take account of the range of complexities discussed above. In the spirit of this approach, McCall (1969) classified molecular motion in polymers under a number of broad headings (table 1.2).

Generally, models which describe the Brownian motion of a chain of elements reasonably account for polymer behaviour in solution and can be extended in modified form to describe adequately the rubbery state (Cohen-Addad and Guillermo, 1984; Brereton, 1990). They are wholly inadequate when dealing with the glassy or crystalline states where more local motions involving activated motion over barriers between conformational isomers are encountered. In such cases one must resort to models that are based, for example, on the motion of defects, originally devised by Glarum (1960) and adapted subsequently by several groups. Unlike the defect diffusion model, the graphically appealing reptation model developed by Edwards (1967), de Gennes (1971, 1976) and Doi (1975) was designed specifically to describe polymer relaxation involving motion over long distances and times. This model visualises the random movement of defects along a polymer chain which is geometrically confined within a 'tube' whose axis assumes random coil conformation (fig. 1.4). In the polymer, the tube may be thought of in terms of the topological constraints imposed by neighbouring molecules.

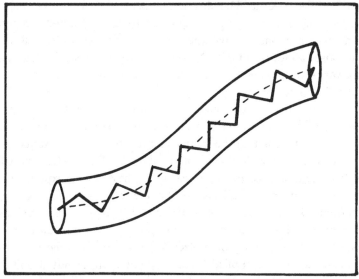

Figure 1.4. Schematic diagram of a polymer chain confined within a hypothetical tube as visualised in the reptation model. Within the tube, the polymer can undergo motion activated by defect propagation while the tube itself is presumed to behave like a random coil.

In developing further these introductory ideas, it is pertinent to review the way in which NMR relates to other experiments. The ensuing discussion emphasises the advantages of correlating the results of a number of different relaxation experiments in arriving at a more complete description of polymer behaviour and alludes to the various levels of approximation involved in drawing comparisons between different methods.

1.5 Comparison with other relaxation experiments: transition maps

In considering the relationship between NMR, dielectric and dynamic mechanical relaxation it is appropriate first to define the terms used to describe relaxation and molecular motion in polymers. The *correlation frequency* v_c is $(2\pi\tau_c)^{-1}$ where τ_c is an average time between motional events. These motions need not necessarily be sudden reorientations of large amplitude; they may, for example, involve diffusive jumps. v_c and τ_c refer to motional events on a molecular scale whereas the *relaxation time* is a characteristic time associated with the macroscopic response of the polymer. Typically it denotes the reciprocal of the frequency of maximum loss in DMTA spectra. The distinction between correlation time and relaxation time is numerically unimportant for dielectric and mechanical experiments since macroscopic and microscopic times are for all intents and purposes about equal. For NMR there are established relationships linking various experimentally measured relaxation times and v_c or τ_c (cf Chapters 2 and 3).

Now consider the way in which the experiment couples to the molecular motion. In NMR and dielectric measurements the imposed force field is nearly uniform throughout the sample and it is possible to describe events at a molecular level with reasonable confidence. For NMR this will be readily apparent in subsequent discussion. In the dielectric experiment one of the dominant contributions to the dielectric response is the re-orientational time dependence of permanent electric dipoles, which again is reasonably well understood. It is conceptually more difficult to achieve comparable understanding of molecular events in dynamic mechanical experiments where the applied force may not be transmitted uniformly throughout the sample volume. In partially crystalline polymers, for example, there may be orders of magnitude difference in the response of crystalline and amorphous regions and, unlike NMR and the dielectric experiment, the nature of the interface between different regions may exert a disproportionate influence on the overall response of the polymer. The difference in dielectric response of crystalline and amorphous regions in contrast may only be slight. The ability to distinguish between different regions of a polymer is one of the strengths of NMR over its dielectric and dynamic mechanical counterparts for which an average response for the specimen as a whole is observed.

A fundamental implication of the way in which external forces couple to the spin system relates to the ease with which the nuclear spins can be driven by external forces which often exceed by several orders of magnitude the forces exerted on the spins by their molecular and electronic environments. Both the laboratory field $\mathbf{B}_0$ and the perturbing resonant radiofrequency field $\mathbf{B}_1(t)$ impose forces of this type, which is reflected in the generation of sharp absorption NMR resonances, in contrast, for example, to the over-damped motions characteristic of dielectric relaxation where the motion of dipoles is tightly coupled to the molecular framework. This feature of NMR stems from an ability to render the spin Hamiltonian rapidly time-dependent as required in high-resolution NMR.

There are subtle implications in comparing NMR and dielectric data when molecular motions exhibit a broad distribution of correlation times. It is often tacitly assumed that both sets of experimental data can be described by the same distribution function and it is common practice to use models developed for the dielectric experiment to interpret NMR data (cf Section 3.7.2). This approach should be viewed with some caution for a number of reasons: (i) NMR and dielectric measurements assign different weights to the various molecular motions depending upon the relative orientation of internuclear vectors and electric dipoles in the molecule; (ii) different correlation functions describe NMR and dielectric relaxation (first- and second-order Legendre polynomials respectively); (iii) the distribution may be modified in certain circumstances by spin diffusion which has no counterpart in dielectric relaxation; and (iv) it is assumed implicity that all correlation times in the distribution have the same activation energy which is valid in many but not all cases.

In evaluating diverse experimental relaxation data, care is necessary when collating the results of techniques which probe motion on a very local dimensional scale with those that sample motional events on a more macroscopic scale. This is particularly true for the glass transition. In practice these complications may not be particularly serious in view of the level of approximation involved in modelling distributions in τ_c (Connor, 1963). Despite these and other problems associated, for example, with overlapping transitions and variation in sample characteristics, the correlation of relaxation data from diverse experiments shows remarkable consistency and forms a useful starting point in understanding overall relaxation behaviour for polymers. Indeed, this is another indication that the correlation of data from different techniques is not overly model-dependent.

Correlation frequencies which characterise molecular motions in polymers generally fall within a range that typically spans 13 decades of frequency, $10^{-4} < v_c < 10^9$ Hz. More specialised techniques permit extension of the range to higher frequencies, as in quasi-elastic neutron scattering measurements for which $v_c = 2 \times 10^{10}$ Hz (Duplessix *et al.*, 1980). Even then the bounds are dictated by experimental limitations rather than by the spectrum of motions as such. It may be noted in passing that the ease with which temperature can be varied over a wide range in conventional low-resolution experiments has yet to be matched in the more sophisticated methodology.

Having extracted meaningful correlation frequencies as a function of temperature from the available body of experimental relaxation data, the results are compiled as plots of $\log v_c$ (or τ_c) versus inverse temperature. The family of plots for a given polymer is referred to as a *transition map*, a typical example of which is portrayed in fig. 1.5. The data points tend to

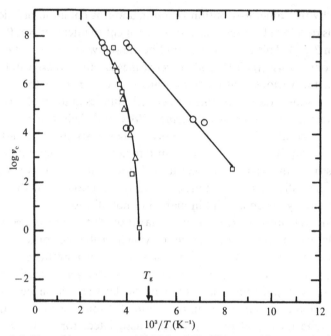

Figure 1.5. Plot of $\log v_c$ versus $10^3/T$ for PIB. This transition map includes data for NMR ($\bigcirc$), mechanical ($\square$) and dielectric ($\triangle$) relaxation experiments. T_g denotes the quasi-static glass transition temperature.

lie on well-defined loci, indicating that diverse experimental observations are assignable to the same molecular motion. In the data shown for polyisobutene (PIB), the low-temperature transition is due to the rotation of methyl groups, of which there are two on every other carbon atom on the chain. This transition exhibits a linear dependence of $\log v_c$ on reciprocal temperature, manifesting an Arrhénius process described by

$$\tau_c = \tau_0 \exp(\Delta E/kT) \qquad (1.1a)$$

or equivalently

$$v_c = v_0 \exp(-\Delta E/kT) \qquad (1.1b)$$

ΔE is the activation energy which may be obtained directly from the slope of the straight line in fig. 1.5. Many motions in polymers, particularly side group and local main chain motions, exhibit Arrhénius behaviour and obey eq 1.1. This is clearly not the case for the higher temperature, glass-to-rubber transition. Emperical theories, first developed to account for the temperature dependence of viscosity (Vogel, 1921; Fulcher, 1925; Tamman and Hesse, 1926), later complemented by models based on free-volume considerations (Fox and Flory, 1950, 1951, 1954; Doolittle, 1951, 1952;

Williams *et al.*, 1955) formed the basis of a description of the glass transition. The empirical expression

$$\ln\frac{\tau_{c}}{\tau_{0}} = \frac{-A(T-T_{0})}{B+(T-T_{0})}$$ (1.2)

devised by Williams, Landel and Ferry (WLF) has gained widespread appeal. T_{0} is a reference temperature related to the glass transition temperature T_{g} and A and B are constants. It is evident from fig. 1.5 that the temperature at which the glass transition is detected is frequency-dependent, a point which must be borne in mind when comparing results from different experiments.

Transition maps are limited in a number of respects. First, they do not parametrise the strength or intensity of a given relaxation, apart from the trivial case when the transition has zero intensity. It is not uncommon for a point corresponding to a weak dielectric loss to lie on the same locus as a point arising from a strong mechanical loss. Second, unequivocal assignments of diverse relaxation data to one phase or another are not always correct. For example, dielectric and NMR methods assign the α relaxation in polyethylene to the crystalline regions whereas the mechanically active α process, while requiring the presence of a crystal phase, is assigned to the amorphous regions (Boyd, 1985). Third molecular weight dependence, relaxation due to residual solvent or low molecular weight fractions, the presence of water, or impurity relaxation constitute typical additional refinements that must be taken into account in specific situations.

It is remarkable and perhaps a blessing in disguise, that transition maps, albeit in the form of semilog plots, are so insensitive to sample history and morphology. Although transition maps provide incomplete information, they do identify the gross features of relaxation which are important in practical applications.

2

Basic concepts in NMR

2.1 Nuclear magnetisation

Many atomic nuclei have a spin angular momentum, I, and, hence, a magnetic moment, μ, related by

$$\boldsymbol{\mu} = \gamma \boldsymbol{I} \qquad (2.1)$$

where γ is the magnetogyric ratio characteristic of the isotope. Table 2.1 summarises some of the properties of magnetic nuclei most often used in NMR studies of polymers.

NMR in bulk matter involves probing, and sometimes modifying, the nuclear magnetic properties of samples containing large numbers (typically $\geqslant 10^{19}$) of magnetic nuclei, using one or more frequencies of radiation which are close to the natural frequencies of motion of the spins. Normally, the spins are subjected to a dominant, static interaction such as a coupling with an externally applied magnetic field $\mathbf{B}_0$. This coupling is known as the Zeeman interaction and, to a first approximation, generally determines the energy states of the system. Local spin couplings slightly modify these energy states and, in reality, it is these local interactions on which the whole utility of NMR in physical science rests. In this chapter we introduce and discuss some of the concepts required to understand the experiments described later.

The interaction of a nuclear spin with the static field $\mathbf{B}_0$, taken as defining the laboratory Z axis, is described by the Hamiltonian

$$\mathscr{H}_z = \gamma_I B_0 \mathbf{I}_z = -\omega_0^I \mathbf{I}_z \qquad (2.2)$$

which has eigenvalues, E_{I_z}, given by

$$E_{I_z} = -\hbar \omega_0^I I_z \qquad (2.3)$$

$\mathbf{I}_z$ and I_z are the operator and quantum number, respectively, for the z

Table 2.1. *Nuclei typically encountered in the NMR of polymers*
$(B_0 = 1\ T)$

Isotope	Spin	Relative[a] sensitivity	Abundance (%)	μ[b]	$\gamma \times 10^{-8}$[c]	ν_0 (MHz)
^{1}H	1/2	1.0000	99.980	2.7927	2.6752	42.577
^{2}H	1	0.0096	0.016	0.8574	0.4107	6.536
^{13}C	1/2	0.0159	1.108	0.7022	0.6726	10.705
^{14}N	1	0.0010	99.635	0.4036	0.1933	3.076
^{15}N	1/2	0.0010	0.365	−0.2830	−0.2711	4.315
^{19}F	1/2	0.8340	100.000	2.6273	2.5167	40.055
^{29}Si	1/2	0.0785	4.700	−0.5548	−0.5316	8.460
^{31}P	1/2	0.0664	100.000	1.1305	1.0829	17.235

(a) For equal numbers of nuclei at constant field.
(b) Magnetic moment in units of the nuclear magneton.
(c) Magnetogyric ratio (rad s^{-1} T^{-1}).

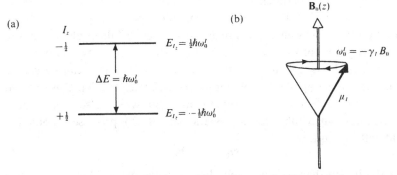

Figure 2.1. (a) Energy levels of spin $I = \frac{1}{2}$ in a magnetic field $\mathbf{B}_0$. ω_0 is the precession frequency. (b) Precession of an isolated spin with magnetic moment μ_I about $\mathbf{B}_0(z)$.

component of spin. For the purposes of this introduction, we restrict our attention to spins with $I = \frac{1}{2}$ and, hence, $I_z = \pm\frac{1}{2}$, in which case there are two energy states with $E_{\pm\frac{1}{2}} = \mp \frac{1}{2}\hbar\omega_0^I$ as illustrated in fig. 2.1(a).

In a system comprising a large number, N_T, of such uncoupled spins, at thermal equilibrium they will be distributed between these two states according to a Boltzmann distribution of populations $N_{+\frac{1}{2}}$ and $N_{-\frac{1}{2}}$ where

$$N_{-\frac{1}{2}} = N_{+\frac{1}{2}}\exp\left(-\hbar\omega_0^I/kT_L\right) \approx N_{+\frac{1}{2}}(1 - \hbar\omega_0^I/kT_L) \qquad (2.4)$$

T_L is the temperature of the 'lattice', the term used to describe the atomic and molecular degrees of freedom of the system which contains the spins and with which they are in thermal contact. The second equality in eq 2.4, known as the high-temperature approximation, is valid for all nuclear spins in normally available fields except at temperatures $\leqslant 0.1$ K.

The small excess of population, Δn, in the lower energy state gives rise to a resultant nuclear magnetisation M_0 for the whole sample, given by

$$
\begin{aligned}
M_0 = \Delta n \mu_{I_z = -\frac{1}{2}} &= N_{+\frac{1}{2}}(\hbar\omega_0^I/kT_{\mathrm{L}})(\gamma\hbar/2) \\
&\approx N_{\mathrm{T}}(\gamma^2\hbar^2 B_0/4kT_{\mathrm{L}}) \\
&= CB_0/kT_{\mathrm{L}}
\end{aligned}
\tag{2.5}
$$

This is the Curie law of temperature-dependent paramagnetism and illustrates that M_0^I is proportional to N_{T} and B_0 and inversely proportional to temperature. Since the signals observed in NMR experiments are ultimately derived from M_0^I, sensitivity is generally greater for large numbers of spins with large intrinsic magnetic moments, high-B_0 fields and low temperatures.

The description given above in terms of energy states and populations is incomplete and is inadequate to account for the response of spin systems to typical NMR experiments. The main reason for this inadequacy, as we shall see, is that NMR is a form of coherent spectroscopy in which the radiation used has well-defined phase properties. While a full description of the properties of spin systems is best carried out using density operator (matrix) methods (Slichter, 1990), a great deal of insight can be gained from the simpler approach involving a semi-classical model. This is based on the simple fact that a magnetic moment which also has angular momentum, when subjected to a static field $B_0(Z)$, precesses about that field with a frequency

$$
\boldsymbol{\omega}_0 = -\gamma_I \mathbf{B}_0
\tag{2.6}
$$

It is significant that ω_0 is precisely the frequency obtained when the usual spectroscopic relationship $\Delta E = \hbar\omega$ is applied to the energy levels of fig. 2.1(a). Figure 2.1(b) illustrates the precession described by eq 2.6. Note that the sense of precession illustrated is for a positive γ_I, the negative sign in eq 2.6 arising from the use of a right-handed corkscrew convention. The precession behaviour can be readily deduced by application of Newton's second law as follows

$$
(d\boldsymbol{\mu}/dt) = \gamma(d\mathbf{I}/dt) = \gamma\mathbf{T} = \gamma\boldsymbol{\mu} \times \mathbf{B}_0
\tag{2.7}
$$

$\mathbf{T}$ is the torque acting on the magnetic moment $\boldsymbol{\mu}$ arising from $\mathbf{B}_0$.

This classical view of the behaviour of $\boldsymbol{\mu}$ can be merged with the simple quantum model to give a picture of the many-spin system as illustrated for $I = \frac{1}{2}$ in fig. 2.2(a). The two eigenstates are represented by the two cones in which the spins precess. Of course, since the spins are distributed more or

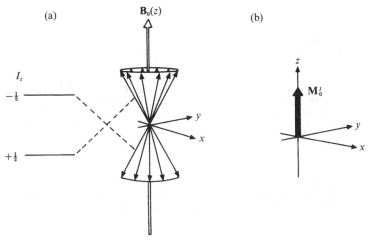

Figure 2.2. (a) Representation of the two eigenstates corresponding to $I_z = \pm\frac{1}{2}$. The cones define the precessing spins. The upper cone corresponds to the lower energy level. (b) Thermal equilibrium state of the magnetisation vector $\mathbf{M}_0^I$.

less uniformly throughout the sample, their magnetic moment vectors do not all coincide at a point as in fig. 2.2(a); this is purely for diagrammatic convenience. At thermal equilibrium there are more spins in the upper cone (lower energy level) leading to the nett magnetisation $\mathbf{M}_0^I$. The new feature is that we must now say something about the components of magnetisation in the xy plane perpendicular to $\mathbf{B}_0$. At thermal equilibrium, where $\mathbf{B}_0$ is the only field experienced by the spins, there is no energetically preferred direction in the xy plane and, hence, we would expect the phase angles ϕ defining the instantaneous xy coordinates of the spins to be distributed evenly between 0 and 2π. This implies that the thermal equilibrium state of the macroscopic magnetisation vector is

$$M_z^{eq} = M_0; \qquad M_x^{eq} = M_y^{eq} = 0 \qquad (2.8)$$

This is illustrated in fig. 2.2(b).

2.2 The radiofrequency (rf) field

In order to determine or modify the spectrum or other properties of the nuclear spin system, one or more frequencies of electromagnetic radiation are applied to the sample. The nuclei, being magnetic dipolar in nature, interact with the magnetic component of the radiation and this can be represented as a field $\mathbf{B}_1(\omega)$. Since, for readily available strengths of the $\mathbf{B}_0$ field, typical precession frequencies $\nu_0 (= \omega_0/2\pi)$ are of the order 1–500 MHz, $\mathbf{B}_1(\omega)$ is often referred to as a 'radiofrequency' or rf field.

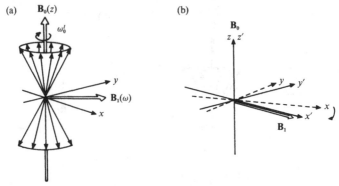

Figure 2.3. (a) Orientation and direction of perturbing field $\mathbf{B}_1(\omega)$ relative to $\mathbf{B}_0(z)$. (b) Coordinate frame $(x'y'z')$ rotating at angular frequency ω about z.

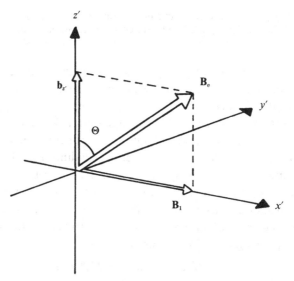

Figure 2.4. Resultant field $\mathbf{B}_e$ experienced by the spins in the resonantly rotating frame $(x'y'z')$.

The effect of an rf field may be illustrated by incorporating it into the semi-classical model of fig. 2.2 as in fig. 2.3(a). $\mathbf{B}_1(\omega)$ must lie in the xy plane to be most effective in causing spins to undergo transitions between the two states (cones). In addition, the frequency ω at which $\mathbf{B}_1(\omega)$ rotates in the xy plane must be close to ω_0, the nuclear spin precession frequency, in order to produce any marked perturbation of the spins' motion. This can be appreciated qualitatively in the semi-classical model in that the response of an individual magnetic moment, $\boldsymbol{\mu}$, to the field $\mathbf{B}_1(\omega)$ will be to attempt to precess about the instantaneous direction of $\mathbf{B}_1(\omega)$. If this direction is changing rapidly, as it will if ω is very different from ω_0, then there will be no significant perturbation of the precession of μ about $\mathbf{B}_0$ by $\mathbf{B}_1(\omega)$. Because of the requirement for $\omega \approx \omega_0$ (which, of course, is the same as the $\Delta E = \hbar\omega_0$ condition) it is helpful to consider the behaviour of both spins and rf field in a so-called 'rotating frame' of reference which removes the common high-frequency motion ω. This is illustrated in fig. 2.3(b). The simplest rotating frame is that in which the x and y axes are rotated about the z axis with the frequency ω, that is, at the same rate and with the same sense as $\mathbf{B}_1(\omega)$. The axes in this frame are usually designated $(x'y'z')$. This choice of rotating frame ensures that $\mathbf{B}_1(\omega)$ appears stationary which, as we shall see, can be helpful in considering the effects of $\mathbf{B}_1(\omega)$ on the nuclear magnetisation.

In the ω-rotating frame, magnetic moments with precession frequency ω_0 appear to precess at the frequency Ω where

$$\Omega = (\omega_0 - \omega) \tag{2.9}$$

which can be thought of as being due to a static field $\mathbf{b}_{z'}$ in the z' direction (the same as z) given by

$$\mathbf{b}_{z'} = \Omega/\gamma = (\omega/\gamma - \mathbf{B}_0) \tag{2.10}$$

This is illustrated in fig. 2.4 for $\omega \approx \omega_0$ where it is apparent that the resultant field experienced by the spins, viewed in this rotating frame, is static and of amplitude B_e where

$$B_e = (B_1^2 + b_{z'}^2)^{\frac{1}{2}} \tag{2.11}$$

and makes an angle Θ with the z' axis given by

$$\Theta = \tan^{-1}(B_1/b_{z'}) \tag{2.12}$$

For convenience, $\mathbf{B}_1(\omega)$ has been chosen quite arbitrarily to lie along the x' axis. The advantage of this device is that the effects of a 'near-resonant' rf field, $\mathbf{B}_1(\omega)$, can be described and visualised independently of the high-frequency precession ω_0.

Now suppose that a field $\mathbf{B}_1(\omega)$, with $\omega \approx \omega_0$, is applied to a spin system at thermal equilibrium (eq 2.8). We expect that, in the conventional way, this field will induce transitions of the spins between their energy states and will thus, probably, change M_z. Such transitions are referred to as 'magnetic dipole transitions' in optical spectroscopy. If that were the only effect, however, there would be no need for the rotating frame and precessional motion description. The resonant rf field, in addition, destroys the cylindrical symmetry and forces the spins to develop a phase coherence in the xy plane which produces transverse components of magnetisation; that is, after $\mathbf{B}_1(\omega)$ has been on for some time

$$M_{z'} < M_0 \qquad \text{and} \qquad M_{x',y'} \neq 0$$

It is the transverse magnetisation that is generally responsible for the NMR signals observed and it is the coherent nature of the exciting field $\mathbf{B}_1(\omega)$ that leads to its creation.

2.3 Rf pulses and nuclear magnetic relaxation

Most NMR experiments are carried out using bursts or pulses of the field $\mathbf{B}_1(\omega)$. Thus, the amplitude B_1 is time-dependent. In considering the response of a nuclear spin system to such a pulse we ignore for the moment the fact that it will have a spectrum which may contain a number or a continuous range of frequencies. The effect of a near-resonant rf pulse on the equilibrium magnetisation, $\mathbf{M}_0$, is illustrated in fig. 2.5(a). The magnetisation precesses at frequency $\omega_e = -\gamma \mathbf{B}_e$, in a cone around $\mathbf{B}_e$, periodically returning to the z' axis. If the spin system is irradiated exactly at resonance ($\omega = \omega_0$) then $\mathbf{B}_e = \mathbf{B}_1$ and $\mathbf{M}_0$ precesses or 'nutates' in the $z'y'$ plane as illustrated in fig. 2.5(b). If $\mathbf{B}_1(\omega)$ is switched on for a time t_p such that

$$\gamma B_1 t_p = \omega_1 t_p = \theta_p = \pi/2 \tag{2.13}$$

then $\mathbf{M}_0$ is tipped, unchanged in magnitude (if $t_p \ll T_1$), into the y' direction. This is referred to as a 90_x or $(\pi/2)_x$ pulse and creates the conditions

$$\mathbf{M}_{z'} = 0; \qquad \mathbf{M}_{x'} = 0; \qquad \mathbf{M}_{y'} = \mathbf{M}_0$$

Similarly, a pulse which inverts $\mathbf{M}_0$ so that

$$\mathbf{M}_{z'} = -\mathbf{M}_0; \qquad \mathbf{M}_{x'} = 0; \qquad \mathbf{M}_{y'} = 0$$

is referred to as a 180_x or $(\pi)_x$ pulse.

These are non-equilibrium states for the spin system which, if left to themselves, will return towards equilibrium. The processes bringing about this return to equilibrium are referred to as relaxation, of which there are two essentially different types. The first, known as spin–lattice or longitudinal relaxation, is the recovery of the component of magnetisation along the main field $\mathbf{B}_0$. It is generally represented by one or more time constants given the symbol T_1 and, if it is a simple first-order process, may be represented as

$$[M_0 - M_z(t)] = [M_0 - M_z(0)] \exp(-t/T_1) \tag{2.14}$$

where $[M_0 - M_z(t)]$ is the deviation of M_z from equilibrium at time t. This

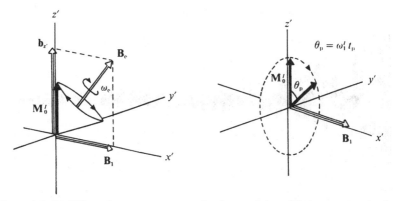

Figure 2.5. (a) Effect of a near-resonant rf pulse on the equilibrium magnetisation $\mathbf{M}_0^I$. Precession at ω_e about B_e is induced. (b) Precession of $\mathbf{M}_0^I$ about $\mathbf{B}_e = \mathbf{B}_1$ (at resonance $\omega = \omega_0$) in the $y'z'$ plane.

process is one of re-establishing the Boltzmann population ratio between the energy states of the spin system and is thus an energy transfer process between the magnetic energy of the spins in $\mathbf{B}_0$ and the energy of the lattice which was defined earlier.

The second type of relaxation is the decay of transverse magnetisation components to zero which is often represented by

$$M_{xy}(t) = M_{xy}(0)\exp\left(-t/T_2\right) \tag{2.15}$$

where T_2 is the transverse relaxation time. This process is one whereby the xy components of the magnetisation are dephased and is fundamentally different in nature from spin–lattice relaxation. When brought about by static nuclear magnetic dipole–dipole coupling, T_2 is described as the spin–spin relaxation time. The timescales of T_1 and T_2 vary considerably depending on the nuclear isotope involved and the physical properties of the system containing the spins. Magnitudes can range from microseconds to minutes. We should also note that the exponential form of eq 2.15 is by no means always an accurate description of the transverse relaxation process, particularly in solids, and that there are also a number of different spin–lattice relaxation processes in which the rf or internal fields replace $\mathbf{B}_0$. These are dealt with elsewhere.

2.4 NMR signals: the free induction decay (FID)

The application of one or more rf pulses may leave the spin system with a non-zero transverse magnetisation. Left to itself this magnetisation

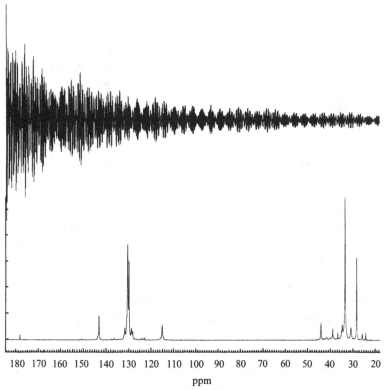

180 170 160 150 140 130 120 110 100 90 80 70 60 50 40 30 20

ppm

Figure 2.6. A typical high-resolution FID and its spectrum.

precesses about $\mathbf{B}_0$ and will induce a corresponding rf voltage in a coil surrounding the sample and tuned to the appropriate frequency range. This induced signal is then amplified and detected by mixing or 'demodulating' with a reference frequency which has a similar effect to observing the precession of the magnetisation in the ω-rotating frame. The NMR spectrum excited by the pulses, denoted $I(\omega)$, is related to the transient signal – the free induction decay $G(t)$ – by the relationships

$$\left.\begin{aligned} G(t) &= \frac{1}{2\pi} \int_{-\infty}^{\infty} I(\omega) \exp\left(-\mathrm{i}\omega t\right) \mathrm{d}\omega \\ I(\omega) &= \int_{-\infty}^{\infty} G(t) \exp\left(\mathrm{i}\omega t\right) \mathrm{d}t \end{aligned}\right\} \tag{2.16}$$

The spectrum and free induction decay (FID) form a Fourier transform pair and contain equivalent information. In some experiments the spectrum is determined either directly or *via* Fourier transformation of the

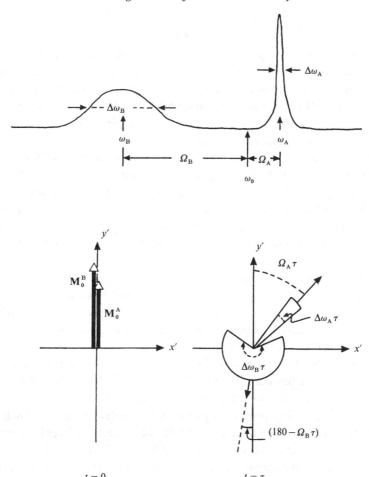

Figure 2.7. Evolution of the different spectral magnetisations in the rotating ω_0 frame for a two-line spectrum.

FID, whilst in others, only the amplitude of the FID is measured as a function of some parameter(s) of the pulse sequence applied to the spin system as, for example, in the measurement of relaxation behaviour.

A typical 'high-resolution' FID and its corresponding spectrum are shown in fig. 2.6. It should be noted that a short-lived FID corresponds to a broad, low-resolution spectrum (cf fig. 1.1(a)) whereas a long-lived FID with complex beat pattern yields a spectrum with many sharp lines. This is readily appreciated by relating the FID to the evolution of the different spectral magnetisations viewed in the rotating frame as illustrated for a two-line spectrum in fig. 2.7.

2.5 Spin temperature

In Section 2.3 the concept of spin–lattice relaxation was introduced. It was described as the process governing the return to their thermal equilibrium values of the populations of the spin energy states, which are generally dominated by the Zeeman interaction with the field B_0. This is an energy transfer process since the magnetic energy of the spin system, in general, is $-MB_0$ whereas, at thermal equilibrium, it is $-M_0 B_0$. Following a 180° rf pulse, for example ($M_z = -M_0$), the spin system has an energy $M_0 B_0$ and, to achieve equilibrium, an energy of $2M_0 B_0$ must be transferred to the surroundings or 'lattice'. This process is illustrated schematically in fig. 2.8. For a pair of energy levels, as appropriate for $I = \frac{1}{2}$, or for any equally spaced set of levels, we can think of the ratio of the populations of adjacent levels in terms of a spin temperature, T_s, different from that of the lattice, T_L. Thus, in qualitative terms, a $\theta°$ rf pulse (other than 2π) 'heats' the spin system to a temperature greater than T_L and the subsequent spin–lattice relaxation can be thought of as a cooling of the spin system by transfer of heat to the lattice. Note, however, that $\pi/2 < \theta < \pi$ gives rise to negative values of T_s, which correspond to a higher energy than that specified by $T_s = \infty$, which is achieved when $\theta = \pi/2$!

This concept of a temperature for the spin system which is different to that of the lattice has been developed much further than indicated by the above discussion. For solids in particular the concept of a properly defined spin temperature has proved essential to the understanding of nuclear magnetisation behaviour. For more details the interested reader is referred to the books by Goldman (1970), Wolf (1979) and Abragam and Goldman (1982). Here, we restrict ourselves to the extension of the concept from the laboratory frame Zeeman interaction to the rotating frame Zeeman and dipolar interactions.

Generalising eq 2.5, and assuming that a temperature T_s may be defined,

$$M = CB/kT_s \tag{2.17}$$

where M is the magnetisation along a field B which corresponds to a spin temperature T_s. The object of writing eq 2.17 in this way is to emphasise the point that the spin temperature concept may be extended to situations in which the field is not the equilibrium field B_0. For example, a common situation encountered in pulsed NMR experiments is that in which the magnetisation M_0 is subjected to a 'spin-locking' pulse sequence. This is illustrated in fig. 2.9 for rf irradiation applied at exact resonance ($\omega = \omega_0$).

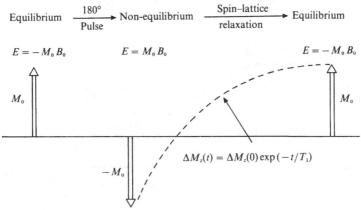

Figure 2.8. Recovery of magnetisation $\mathbf{M}_0$ following inversion by a 180° pulse.

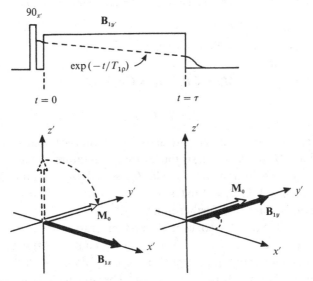

Figure 2.9. Pulse sequence which spin-locks $\mathbf{M}_0$ along B'_{1y} in the rotating frame where it decays at a rate given by $\exp(-t/T_{1\rho})$.

The sequence consists of a 90° pulse applied along the x axis, denoted 90_x, followed immediately by application of a $\mathbf{B}_1$ field phase-shifted by 90°, for example, $\mathbf{B}_{1y}$ as shown. This aligns $\mathbf{M}_0$ and $\mathbf{B}_1$ in the rotating frame. The description 'spin-locked' arises because, if $B_1 \gg (\Delta\omega/\gamma)$, where $\Delta\omega$ is the frequency spread of the spectrum (for example, the linewidth), then any tendency for the magnetisation to dephase is suppressed by the rapid precession about B_1 ($\omega_1 = \gamma B_1$). If it is assumed that this situation may be

Laboratory frame   Rotating frame
                                    (spin-locking)

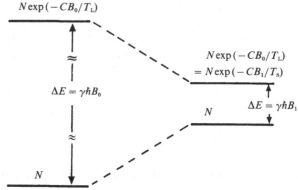

Figure 2.10. Comparison of populations in the laboratory and rotating frames (see text).

thought of as a quasi-equilibrium state, characterised by a temperature, then we can write

$$M_0 = CB_0/kT_L = CB_1/kT_s \qquad (2.18a)$$

whence

$$T_s = T_L(B_1/B_0) \qquad (2.18b)$$

Thus, the effective spin temperature associated with the initial spin-locked state in which M_0 lies along B_1, is the lattice temperature at which M_0 was generated by polarisation in the field B_0, scaled by the ratio (B_1/B_0). Typically, if $B_0 = 1$ T, $B_1 = 1$ mT and $T_L = 300$ K, then $T_s = 0.3$ K! This phenomenon is illustrated further in fig. 2.10.

As mentioned above, the concept of spin temperature has been extended considerably and can be used, for example, in the context of spin couplings. Consider a system of dipolar coupled spins in a solid in which the nuclear magnetisation M_0 has been spin-locked. The so-called 'adiabatic' reduction of the amplitude of the spin-locking B_1 field to a value less than the dipolar linewidth of the spins transfers spin order (of which magnetisation is only one type) into the dipolar couplings. To be adiabatic, the instantaneous precession frequency of the spins in their effective fields in the rotating frame must always be greater than the rate of change of the field. Following such an adiabatic reduction of $\mathbf{B}_1$ to less than the dipolar local fields, $\mathbf{B}_{loc}$, the order which was present as Zeeman polarisation with respect to $\mathbf{B}_1$ is converted to ordering of the spins with respect to their local fields. Under these conditions the dipolar energy states can be ascribed a temperature which will be of the order $T_L B_{loc}/B_0$. Such a dipolar ordered state will be characterised by a low spin temperature because, usually, $B_{loc}/B_0 \ll 1$.

Both the spin-locked and dipolar ordered states, when characterised by such low spin temperatures, are non-equilibrium states and will relax towards the lattice temperature. These spin–lattice relaxation processes will thus reduce the magnitude of the spin-locked magnetisation or the dipolar order and the timescales are usually denoted $T_{1\rho}$ and T_{1D}, that is, spin–lattice relaxation in the rotating frame and dipolar spin–lattice relaxation, respectively.

2.6 Spin couplings, lineshapes and motional averaging

As mentioned in Section 2.1, the utility of NMR lies in the existence of spin interactions which produce breadth and/or fine structure in the spectra as well as inducing relaxation. These are dealt with theoretically in Chapter 3 and illustrated experimentally in subsequent chapters. They typically include magnetic shielding (chemical shift), dipolar, indirect (scalar) and quadrupolar couplings and all, to a lesser or greater extent, produce characteristic splittings and broadening of the NMR spectra obtained from solids because of their anisotropic, tensorial character (Gerstein, 1983). In polycrystalline materials and isotropic bulk polymers, the continuous distribution of structural entities and associated molecular orientations usually renders the spectra broad and often somewhat featureless (cf Section 1.3). The contributions of these various interactions to spectral shapes and extents depends on such factors as the isotopes concerned, the field strength B_0 and the presence of any time dependence in the interaction.

Typical proton NMR spectra in polycrystalline organic solids are symmetric and may be 10–100 kHz wide, due predominantly to homonuclear proton–proton dipolar couplings. ^{13}C spins in the same materials may also have dipolar dominated spectra with linewidths in the range 1–10 kHz, due in this case to ^{13}C–^1H dipolar couplings. These dipolar lineshapes and widths are independent of B_0. Magnetic shielding, on the other hand, gives rise to asymmetric lineshapes in powder samples with extents proportional to B_0 and which can be as small as a few parts per million (ppm) for protons to of the order 10^3 ppm for heavy nuclei such as ^{59}Co and ^{205}Tl (Duncan, 1990). Quadrupole interactions can vary considerably in magnitude giving linewidths from 100 kHz to 100 MHz or more. For polymers, the principal spins available for study are ^1H, ^2H, ^{13}C,

^{14}N, ^{15}N, ^{19}F, ^{29}Si and ^{31}P (table 2.1). Of these, only ^{2}H and ^{14}N are quadrupolar and both have $I = 1$. ^{2}H in a C–^{2}H bond has a quadrupolar coupling of the order of 170 kHz whereas ^{14}N quadrupole couplings can range up to 5 MHz.

2.6.1 Spin couplings

We review here, briefly, the nature of the more important spin couplings. They are dealt with in more detail in Chapter 3 and the reader already familiar with their general characteristics should proceed there directly.

2.6.1.1 Magnetic dipole–dipole coupling

Typical local magnetic fields produced by nuclear magnetic moments at the sites of neighbouring spins are of the order of 1 mT (10 gauss) or less. As such, the dipolar interaction energies are much less than the usual Zeeman couplings of the spins with the external field $\mathbf{B}_0$, but can be of a similar order to those with the perturbing $\mathbf{B}_1$ fields. This means that, usually, the magnetic dipole–dipole coupling is a small perturbation on the large, static, Zeeman interaction. For a qualitative introductory discussion of the effects of dipolar couplings, we can imagine that the spins are quantised along $\mathbf{B}_0$ and then consider the effects arising only from mutual coupling of their z components. For an isolated pair of spins ($I = \frac{1}{2}$), the presence of each spin produces a local field $\mathbf{B}_{loc}$ which is additional to the field $\mathbf{B}_0$ and thus influences the transition frequency of the other spin. Since the one spin can be either 'up' or 'down' with regard to $\mathbf{B}_0$ there will be two equally probable values of the local field, the magnitude of which will depend on both the distance, r, between the spins (as r^{-3}) and on the orientation of $\mathbf{r}$ with respect to $\mathbf{B}_0$. Thus the other spin can undergo a transition with this local field either adding to or subtracting from $\mathbf{B}_0$, giving rise to a doublet spectrum whose splitting is dependent on the orientation and magnitude of the internuclear vector and the size of the two magnetic moments. The spectrum of a single crystal of such 'isolated' spin pairs will be a doublet with a separation which alters with orientation of the crystal with respect to $\mathbf{B}_0$. A powder or polycrystalline sample will give rise to a superposition of spectral doublets with a weighting determined by the probability distribution of particular crystallite orientations in the sample. A completely random set of orientations covering all space gives rise to the Pake powder lineshape (Pake, 1956) illustrated in fig. 2.11 and discussed in more detail in the next chapter.

In general, spins are not distributed in simple, isolated, geometric groupings but have couplings spanning a wide range of values arising from

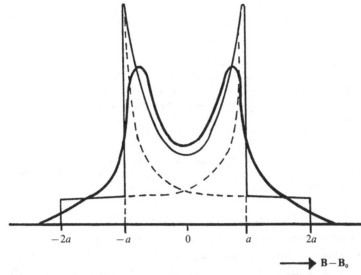

Figure 2.11. Pake doublet powder lineshape for a pair of identical nuclei. The dashed lines refer to the individual components, the thin solid line indicates their sum and the thick solid line denotes the broadened spectrum. For the dipolar interaction, $a = 3\gamma\hbar/4r_k^3$.

the more or less complex spatial disposition of the nuclear moments. The consequent distribution of local fields and the more usual investigation of polycrystalline or powder materials typical of polymers often leads to broad, monotonic spectra, sometimes approximating to a Gaussian shape (cf fig. 1.1 and Section 3.5). The width of these spectra – or equivalently the decay time of the FID – reflects the rms value of the local dipolar magnetic field, thereby providing an important probe of the local molecular environment. Recall that there may be two or more regions with distinguishably different molecular environments in polymers, leading to superimposed characteristic spectra.

In more formal terms, the Hamiltonian operator for the dipolar coupling is

$$\mathscr{H}_D = \frac{\mu_0}{4\pi} \sum_{j<k} \frac{\hbar\gamma_j\gamma_k}{r_{jk}^3}[A+B+C+D+E+F] \qquad (2.19)$$

The functional forms of A–F are given in table 2.2. θ_{jk} and ϕ_{jk} are the polar and azimuthal angles of the vector $\mathbf{r}_{jk}$ with respect to a right-handed set of coordinates with $\mathbf{B}_0$ defining the polar z-axis. In the presence of large Zeeman coupling, only those terms which commute with the Zeeman term are retained to describe the effects of $\mathscr{H}_D$ on the spectrum of the spins

Table 2.2. *Terms that contribute to the dipolar Hamiltonian $\mathscr{H}_{\mathrm{D}}$*

Term	$m^{(a)}$	Expression
A	0	$(1-3\cos^2\theta_{jk})[I_{z_j}I_{z_k}]$
B	0	$-\frac{1}{4}(1-3\cos^2\theta_{jk})[I_j^+I_k^-+I_j^-I_k^+]$
C	$+1$	$-\frac{3}{2}\sin\theta_{jk}\cos\theta_{jk}\exp(-i\phi_{jk})[I_j^+I_{z_k}+I_{z_j}I_k^+]$
D	-1	$-\frac{3}{2}\sin\theta_{jk}\cos\theta_{jk}\exp(i\phi_{jk})[I_j^-I_{z_k}+I_{z_j}I_k^-]$
E	$+2$	$-\frac{3}{4}\sin^2\theta_{jk}\exp(-2i\phi_{jk})[I_j^+I_k^+]$
F	-2	$-\frac{3}{4}\sin^2\theta_{jk}\exp(2i\phi_{jk})[I_j^-I_k^-]$

$^{(a)}$ The significance of m will be apparent in Section 3.3.3. Note that $C = D^*$ and $E = F^*$ where $*$ denotes complex conjugate.

around ω_0. For coupling between like spins (for example, ^{1}H–^{1}H) this implies that only the A and B terms are retained. For unlike spins only the A term remains. This term is the one which corresponds most closely to the simple local field model just described, since it involves only the $\mathbf{B}_0(z)$ components of the magnetic moments. For unlike spins this is a reasonable, though pictorial, model. For like spins however, there is also the B term. This can be thought of in terms of a local field model in the following way. The effect of $\mathbf{B}_0$ on the moments is to cause them to precess at $\omega_0 = -\gamma\mathbf{B}_0$. In addition to the static local fields there will, therefore, also be rotating local fields, arising from the transverse components of the precessing spins. If a neighbouring spin is of the same type as those producing the rotating local fields, these time-dependent fields will be 'resonant' with the precession of the spin and, hence, will be able to cause it to undergo a transition. However, this must be a mutual process and, to be energy-conserving, one spin must 'flip' up whilst the other 'flops' down. That is precisely the nature of the spin operators in term B which couple an 'up–down' state with a 'down–up' state. Thus, for like spins, in addition to the static local field picture, we must allow for the fact that the spins are able to communicate with each other in a dynamic sense by undergoing these energy-conserving flip–flop transitions at a rate which is of the order of the strength of their dipolar couplings. This extra contribution to the secular dipolar coupling for like spins is the source of the process referred to as spin diffusion. This describes the tendency for such a spin system to even out any spatial gradients in longitudinal magnetisation, which can be of particular relevance to complex and heterogeneous solids such as polymers.

2.6.1.2 Magnetic shielding

It is widely recognised that one of the fundamental strengths of NMR as a spectroscopic structural/analytical method rests in the fact that the same nuclei in different chemical environments give rise, in solutions or liquids, to resonance lines which show characteristic frequency shifts – the *chemical shift*. It is less widely appreciated that this magnetic shielding of the nuclear spin by the surrounding electrons and nuclei is a second-rank tensor property and, hence, has a dependence on the orientation of the local molecular axis system with respect to the external field direction. The form of the Hamiltonian $\mathcal{H}_{cs}$ is

$$\mathcal{H}_{cs} = \gamma \mathbf{I} \cdot \boldsymbol{\sigma} \cdot \mathbf{B}_0 \qquad (2.20)$$

where the shielding tensor $\boldsymbol{\sigma}$ is a 3×3 matrix which can be made diagonal by choosing an appropriate set of molecule-fixed axes. Under these conditions, the number of independent elements depends on the site symmetry, having a maximum number of three for symmetries less than axial. These so-called principal shielding tensor components are often labelled σ_{11}, σ_{22} and σ_{33}. For axial symmetry of the nuclear site, only two independent components $\sigma_{\parallel}$ and $\sigma_{\perp}$ are required, corresponding, respectively, to shielding when $\mathbf{B}_0$ is parallel and perpendicular to the principal symmetry axis. In isotropic solutions, the average or isotropic shielding tensor $\tilde{\sigma}$, is what gives rise to the observed chemical shift as described by

$$\tilde{\sigma} = \tfrac{1}{3}(\sigma_{11} + \sigma_{22} + \sigma_{33}) \quad \text{for less than axial symmetry}$$

$$\tilde{\sigma} = \tfrac{1}{3}(\sigma_{\parallel} + 2\sigma_{\perp}) \qquad \text{for axial symmetry}$$

The significance of the anisotropic character of the chemical shift for solids is that its contribution to the spectral position, shape and extent depends on the nature of the sample. A single crystal will exhibit sharp resonances for each magnetically distinguishable site (including crystallographic inequivalences) and the frequencies will change as the orientation of the crystal in $\mathbf{B}_0$ is changed. In a powder, on the other hand, each inequivalent site will give rise to a so-called powder lineshape, the exact details of which will depend on site symmetry and the precise relative values of the shielding tensor's principal components. Figure 2.12 illustrates the features of powder shielding lineshapes for both axial and asymmetric sites. The principal components of the chemical shift tensor

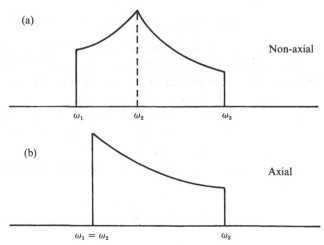

Figure 2.12. Powder shielding lineshapes for axial and non-axial symmetry.

can, in principle, be extracted from the frequency positions ω_1, ω_2 and ω_3 at the discontinuities.

Of course in practice both dipolar and shielding contributions to lineshapes are present simultaneously and, unless suitable experimental devices are used, it is not always easy to separate them unless one or other is dominant. For protons, dipolar couplings usually predominate even at high fields whereas for ^{19}F nuclei and other heavier nuclei, in materials dilute in protons, the shielding can be the more important contribution. For example, dipolar splitting in the CF_2 moiety is 33 kHz whereas ^{19}F chemical shift widths are typically 54 kHz (150 ppm) in a field of 9 T (9×10^4 gauss; $\nu_0 = 360$ MHz).

Early experiments attempted to delineate dipolar and chemical shift contributions by exploiting the linear dependence of the spectral width arising from shielding on B_0 and the independence of dipolar broadening from B_0. Difficulties arising largely from morphological complexity were only partially circumvented in this approach, with the result that early chemical shift data for polymers are generally unreliable. More recent experiments such as the coherent narrowing techniques described in Chapter 4 have been designed to delineate these effects and to reduce the breadth due to the anisotropic character of these interactions. Such line-narrowing techniques are essential, for example, in the detection of proton chemical shifts through suppression of the dominant dipolar contribution.

2.6.1.3 *Nuclear electric quadrupole coupling*

For spins $I > \frac{1}{2}$, the ellipsoidal electric charge distribution in the nucleus results in an electric quadrupole moment. The nuclear moment is collinear with the major axis of the nuclear charge. In the same way that the magnetic dipole moment of the nucleus has an energy which depends on its orientation in the field $\mathbf{B}_0$, so does the electric quadrupole moment have an orientation-dependent energy when in an electric field gradient (EFG). Such gradients arise from the electron distribution around the quadrupolar nucleus which, in solids, tends to quantise the spins in this locally fixed axis system defined by the gradient.

In an NMR experiment the effects produced by the quadrupole coupling will depend on the relative magnitudes of the Zeeman and quadrupole coupling energies. In practice, electric quadrupole effects are categorised according to their relative strength: interactions that are significantly stronger than nuclear Zeeman couplings constitute a related branch of spectroscopy known as pure quadrupole resonance spectroscopy. At the other end of the scale, the weaker quadrupole interactions are treated as a small perturbation of the Zeeman levels. Here we are concerned exclusively with the latter, which includes interactions involving deuterium (^{2}H), the principal quadrupolar nucleus of interest in polymers.

The general form of the quadrupole Hamiltonian $\mathcal{H}_Q$ is

$$\mathcal{H}_Q = \mathbf{I} \cdot \mathbf{V} \cdot \mathbf{I} \tag{2.21}$$

$\mathbf{V}$, the quadrupole coupling tensor whose elements are the second derivative of the electric potential at the nucleus, may also be diagonalised in an appropriately chosen local axis system such that, in general, only three independent quantities are required to describe the interaction fully. For axial symmetry which describes deuterium in C–^{2}H or O–^{2}H bonds, only two independent components of the tensor survive and the effective Hamiltonian is

$$\mathcal{H}_Q = \frac{\omega_Q}{8I(2I-1)}(3\cos^2\beta_0 - 1)(3I_z^2 - I^2) \tag{2.22}$$

where $\omega_Q = (1/4\pi\varepsilon_0)\,e^2qQ/\hbar$ is the quadrupole coupling constant, eQ is the magnitude of the nuclear electric quadrupole moment, eq is the magnitude of the electric field gradient (EFG) and β_0 is the angle between the symmetry axis of the EFG and $\mathbf{B}_0$. As with the dipolar coupling between a pair of spins, this Hamiltonian for spin $I = 1$ (for example ^{2}H and ^{14}N) leads to a doublet spectrum whose splitting varies as $(3\cos^2\beta_0 - 1)$ for a single crystal sample. A powder gives rise to a Pake powder lineshape which is identical in form to that from a pair of dipolar-coupled $I = \frac{1}{2}$ spins (fig. 2.11) with, in this case, $a = 3\omega_Q/8$, $\Delta\nu_Q = 2a = 128$ kHz for the $C-^2H$ interaction and $-a$ and $2a$ denoting the orientation of the $C-^2H$ bond perpendicular and parallel to $\mathbf{B}_0$, respectively. For perdeuterated materials, each site with differing quadrupole coupling strengths gives rise to a distinct powder lineshape and the total spectrum is a superposition of all of these.

The power of ^2H spectroscopy for investigating solid polymers is best realised when selective deuteration of specific sites is carried out, thus avoiding the problem of unscrambling overlapping lineshapes. This is illustrated in succeeding chapters.

2.6.2 *Motional averaging*

As mentioned in Chapter 1, thermal energy in a solid can produce motions ranging from lattice vibrations to restricted rotations of parts or of whole molecules and even translational diffusion. All of these motions modulate the anisotropic spin interactions and the effect they produce depends on their frequency spectrum, spatial amplitude and the relationship of the motional axes with respect to those defining the spin interactions.

The fundamental principle underlying motional averaging is that motion only affects the spectrum arising from a particular spin coupling when the motional frequency becomes comparable to or greater than the spectral width arising from the interaction. Once this condition is attained or exceeded, the effective interaction which contributes to the spectral shape and width is the average over this motion. This is why dipolar couplings, which have a zero isotropic average, vanish in NMR spectra of isotropic fluids since, in such systems, motions about all directions are fast in the sense described above, thus leading to dipolar contributions to the spectral width which average to zero.

The process of spectral averaging, whilst complicated, can be exceedingly valuable in following the details of motions. At this stage we content ourselves with noting that materials which may be structurally heterogeneous, such as polymers in the solid state, may often be characterised by spectra representing superpositions of component spectra from different regions which show distinguishably different degrees of motional averaging at a given temperature, as described and illustrated in Chapter 1. The separation of these contributions can often be achieved by a variety of

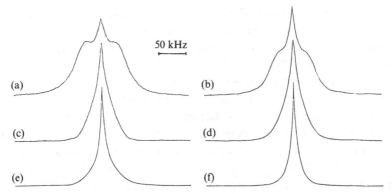

Figure 2.13. Typical ^{1}H spectra of some solid polymers: (a) solution crystallised high-density PE (HDPE); (b) melt-crystallised HDPE; (c) and (d) annealed and quenched isotactic polypropene; (e) and (f) annealed and quenched polybut-1-ene. (Reprinted with permission from Packer *et al.* (1984). Copyright (1984), John Wiley and Sons, Inc.)

experiments which exploit their differences in relaxation and spectral properties. Typical ^{1}H NMR spectra of some solid polymers are shown in fig. 2.13.

This thermally driven averaging, or motional narrowing as it is often called, occurs though the imposition of a time dependence on the local molecular axis systems defining the spin couplings. The Hamiltonians for such couplings have the general form

$$\mathscr{H} = F(I)f(\Omega)$$

where $f(\Omega)$ describes the orientation/internuclear vector dependence and $F(I)$ the dependence on the nuclear spin (magnetic moment components). It is $f(\Omega)$ which is modulated by the thermally activated motions described above. Another approach to introducing time dependence into $\mathscr{H}$, and hence spectral averaging/narrowing, is to perturb the system through externally generated effects which can operate on either $F(I)$ or $f(\Omega)$.

Macroscopic rotation of the sample at frequency ω_R about an axis making an angle θ with respect to $\mathbf{B}_0(Z)$ renders $f(\Omega)$ time-dependent. For second-rank tensor interactions such as shielding, dipolar coupling and, to first order, quadrupole coupling, the averaging which occurs when ω_R is larger than the spectral spread due to these interactions scales the spectrum by the factor $(1 - 3\cos^2\theta)/2$. For $\theta = \cos^{-1}(\sqrt{3})^{-1} = 54.7°$, this is zero and, in principle, the anisotropic parts of the interactions are removed from the spectrum. This is the basis of the technique called magic-angle spinning (MAS) which is essential for achieving high-resolution spectra in solids, first introduced by Andrew and coworkers (1958) and Lowe (1959). Its main use is in averaging chemical shift anisotropy to yield spectra determined only by the isotropic shielding constant, $\tilde{\sigma}$, once the dominant dipolar interactions, if present, have been removed by other methods. This is discussed in more detail in Section 3.6.1.

Averaging may also be introduced through the effects of rf fields on the $F(I)$ term. A simple example is that of spin-echo formation (cf Section 4.4.1.3) which proceeds as follows. A 90° pulse generates a free induction decay (FID) where the decay is considered to arise from a spread of static magnetic fields across the sample dimensions. A 180° pulse applied at a time τ after the 90° pulse refocusses the transverse magnetisation, forming an echo with a maximum at 2τ. The echo has the shape of two FIDs back-to-back. The 180° pulse has caused the signal to last longer than the FID and has therefore 'narrowed' the spectrum, thus averaging (or removing) the effect of the broadening of the spectrum arising from the spread of the static field. For spectra dominated by homonuclear dipole–dipole or quadrupole couplings, the pulse sequences required to effect the necessary spin–space averaging are more complex. However, the general principle is that a sequence of pulses, usually called a pulse cycle, designed to average the relevant interactions, is applied repetitively to the spin system (Haeberlen, 1976). Such multiple-pulse sequences average the relevant interactions as long as the time for the individual pulse cycle can be made short compared with the timescale of the unperturbed FID, that is, the T_2 timescale associated with the interaction. This, of course, is the same condition that was introduced in discussing the requirements for thermally driven motions to average interactions.

Averaging can thus be brought about in a number of ways. All that is required is that the Hamiltonian is made time-dependent with rates that exceed the spread of frequency in the spectrum arising from the static interactions. A further important example is the removal of heteronuclear dipolar couplings in the generation of high-resolution spectra of spins such as ^{13}C (cf Chapter 4). Further details are given elsewhere, as required, and accounts of these effects can be found in many texts, for example, Haeberlen (1976) and Mehring (1976).

2.7 Spin–lattice relaxation and motion

In the previous section it was noted that thermally driven movements of the atoms and molecules containing the nuclear spins could produce a time dependence in their anisotropic spin couplings and, hence, an averaging of the contributions of these couplings to the spectra. These time-dependent couplings are also vital in determining the spin–lattice relaxation characteristics of the nuclear magnetisation.

2.7.1 Dipolar relaxation between like spins

Spin–lattice relaxation involves transitions induced between nuclear magnetic states of different energy. The process is broadly similar to that of irradiating the spins with a resonant rf field where the role of the rf field is mimicked by internal, time-dependent spin interactions, the time variation arising from the thermally driven motions of the molecules.

Consider the proton spins on a polymer chain which can undergo a variety of motions as in a rubber. On average, a given proton experiences the distribution of near-neighbour protons, each of which contributes to the local dipolar magnetic field. If this polymer chain were at a very low temperature such that all the motions were very slow then the ^{1}H NMR spectrum would cover about 100 kHz, corresponding to a rms local field of the order of 10 gauss (10^{-3} T). However, the corresponding spectrum of the polymer in its rubbery state at higher temperatures above T_g may be 10^4 times narrower, having linewidths of less than 10 Hz. As has been seen, this arises by averaging the local fields through 'fast' relative movements of the magnetic dipoles. The *instantaneous* local dipolar magnetic field experienced by a proton in a rubbery polymer may, however, still be of the same order as in the low-temperature rigid solid. The difference is that this local field is now fluctuating chaotically, the fluctuations being characteristic of the detailed dynamics of the 'lattice', that is, the translational, rotational and vibrational motions of the molecules containing the nuclear spins. These fluctuating local fields can drive nuclear spin transitions – the requirement for spin–lattice relaxation – and indeed, any spin interaction which can be made time-dependent by the thermal energy of the system can act as a mechanism for spin–lattice relaxation. Detailed discussions of relaxation mechanisms and their mathematical formulation may be found in a number of standard texts (Abragam, 1961; Carrington and McLachlan, 1967; Farrar and Becker, 1971; Harris, 1983a, b; Slichter, 1990). We shall restrict ourselves to a general overview of the description of spin–lattice relaxation processes and their characteristic behaviour.

Since the source of the relaxation transitions is incoherent, chaotic fluctuations in the spin couplings driven by the thermal energy of the system, we must use a statistical average method to describe the process. The fluctuating spin couplings, represented by a time-dependent local field,

$\mathbf{b}(t)$, may be characterised for our purposes by a time correlation function, $G(t)$, which can be written as

$$G(t) = \langle \mathbf{b}(t) \cdot \mathbf{b}(t+\tau) \rangle_{eq} \qquad (2.23)$$

where $\langle \ \rangle_{eq}$ denotes an average over an equilibrium ensemble.

As its name implies, this function describes how, on average, the values of this local field at one time, t, are correlated with its values at time τ later. In principle, $G(\tau)$ could be a function of absolute time, t, but, for systems at equilibrium and which are therefore statistically stationary, it is only a function of τ. The nature of $G(\tau)$ can be appreciated by noting that for a given proton, in our example of a mobile polymer chain, the local field at any instant will depend on the exact details of the local environment and may change with a wide range of behaviours reflecting the complexity of the system. However, on average – that is, if we observe all protons in the system and average their behaviour, or observe one chosen proton many times and similarly average – the local field will show some characteristic memory or persistence before it has changed substantially either in magnitude or direction. Thus the correlation function, $G(\tau)$, will have the general characteristic that it eventually decays to zero. It may show some oscillatory behaviour but, often, for the purposes of describing spin–lattice relaxation, it is assumed to decay exponentially:

$$G(\tau) = \langle \mathbf{b}(t) \cdot \mathbf{b}(t+\tau) \rangle_{eq} = \langle b^2 \rangle_{eq} \exp\left(-|\tau/\tau_c|\right) \qquad (2.24)$$

where $\langle b^2 \rangle_{eq}$ is the mean squared value of the local field (a measure of the power in the fluctuating field) and τ_c is the correlation time (cf Section 1.5) which defines the time scale for the decay of the correlations in the fluctuations of b.

Calculation of the rate of spin–lattice relaxation involves analysis of the fluctuating local fields to determine the power available at the relevant nuclear spin transition frequencies, typically ω_0 and $2\omega_0$ as in the original BPP (Bloembergen, Purcell and Pound, 1948) and KT (Kubo and Tomita, 1954) analyses. The spectrum, $J(\omega)$, of frequencies present in a fluctuating field having the correlation function $G(\tau)$ is given by its Fourier transform

$$J(\omega) = \int G(\tau) \exp\left(-i\omega\tau\right) d\tau$$

For the exponential form of $G(\tau)$ described in eq 2.24, this leads to

$$J(\omega) = \langle b^2 \rangle_{eq} 2\tau_c/(1 + \omega^2 \tau_c^2) \qquad (2.25)$$

The behaviour of this function is illustrated in fig. 2.14, where it can

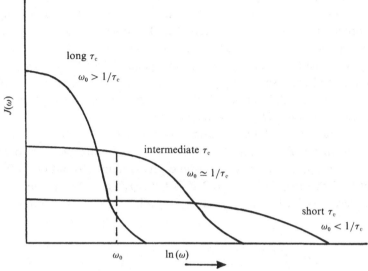

Figure 2.14. $J(\omega)$ as a function of $\ln(\omega)$.

be seen that the power available is roughly constant up to a frequency $\omega \simeq \tau_c^{-1}$, after which it falls off, eventually as $(\omega^2\tau_c)^{-1}$.

Spin–lattice relaxation rates are generally expressed as a sum of contributions of the form $a_i J(\omega_i)$, that is

$$R_j = (T_j)^{-1} = \sum_{i=1}^{n} a_i J(\omega_i) \tag{2.26}$$

where a_i is the coefficient of the spectral density at ω_i and where $J(\omega_i)$ is determined by the mechanism of relaxation involved and the particular relaxation time under consideration, as indicated by the subscript j ($j = 1$ for normal spin–lattice relaxation; $j = 1\rho$ for spin–lattice relaxation in the rotating frame and so on).

As examples we look in a little more detail at the expressions for spin–lattice relaxation arising from dipolar couplings along with nuclear electric quadrupole interactions, since these are the mechanisms most frequently encountered in the study of polymers in the solid state. For dipole–dipole coupling between like spins, which is typical of proton spin–lattice relaxation in polymers, the general expressions for both spin–lattice and transverse relaxation rates are

$$R_1 = T_1^{-1} = \tfrac{3}{2}K_I[J^{(1)}(\omega_0) + J^{(2)}(2\omega_0)] \tag{2.27}$$

$$R_2 = T_2^{-1} = \tfrac{3}{2}K_I[\tfrac{1}{4}J^{(0)}(0) + \tfrac{5}{2}J^{(1)}(\omega_0) + \tfrac{1}{4}J^{(2)}(2\omega_0)] \tag{2.28}$$

The sum over all interacting nuclei is inferred and $K_I = (\mu_0/4\pi)^2 \gamma^4 \hbar^2 I(I+1)$. The superscripts label the various spectral densities $J(\omega)$ according to the order of the terms in the second-rank dipolar interaction tensor from which they arise (cf Chapter 3). Only the frequencies 0, ω_0 and $2\omega_0$ appear in these equations, because the dipolar interaction is a second-rank tensor and because the dipolar couplings can only connect Zeeman states with energies differing by 0, $\hbar\omega_0$ and $2\hbar\omega_0$ in the high-field limit characteristic of most NMR experiments.

For a situation in which the same exponential correlation function determines all the $J^{(n)}(\omega)$ and the motion is one of isotropic rotational diffusion with $|r_k|$, the internuclear distance, remaining constant (as in the original BPP theory), $\langle b^2 \rangle_{eq}$ in eq 2.25 is $\frac{4}{5}r_k^{-6}$, $\frac{2}{15}r_k^{-6}$, and $\frac{8}{15}r_k^{-6}$ for $nq = 0$, 1 and 2, respectively. For an isolated pair of spins the relaxation rates are then given by

$$R_1 = T_1^{-1} = \frac{2}{5r^6} K_I \left(\frac{\tau_c}{1 + \omega_0^2 \tau_c^2} + \frac{4\tau_c}{1 + 4\omega_0^2 \tau_c^2} \right) \tag{2.29}$$

$$R_2 = T_2^{-1} = \frac{3}{5r^6} K_I \left[\tau_c + \frac{5}{3} \left(\frac{\tau_c}{1 + \omega_0^2 \tau_c^2} \right) + \frac{2}{3} \left(\frac{\tau_c}{1 + 4\omega_0^2 \tau_c^2} \right) \right] \tag{2.30}$$

For a more general assembly of coupled spins, the relaxation rates may be written as

$$R_1 = T_1^{-1} = \frac{2}{3} M_2 \left[\left(\frac{\tau_c}{1 + \omega_0^2 \tau_c^2} \right) + \left(\frac{4\tau_c}{1 + 4\omega_0^2 \tau_c^2} \right) \right] \tag{2.31}$$

$$R_2 = T_2^{-1} = M_2 \left[\tau_c + \frac{5}{3} \left(\frac{\tau_c}{1 + \omega_0^2 \tau_c^2} \right) + \frac{2}{3} \left(\frac{\tau_c}{1 + 4\omega_0^2 \tau_c^2} \right) \right] \tag{2.32}$$

where M_2 is that fraction of the second moment of the dipolar couplings which is fluctuating and characterised by a correlation time τ_c.

In a similar way, the spin–lattice relaxation in the on-resonance rotating frame can be described by (Jones, 1966)

$$R_{1\rho} = T_{1\rho}^{-1} = \frac{3}{2} K_I [\frac{1}{4} J^{(0)}(2\omega_1) + \frac{5}{2} J^{(1)}(\omega_0) + \frac{1}{4} J^{(2)}(2\omega_0)]$$

$$= \frac{2}{5r^6} K_I \left[\frac{3}{2} \left(\frac{\tau_c}{1 + 4\omega_1^2 \tau_c^2} \right) + \frac{5}{2} \left(\frac{\tau_c}{1 + \omega_0^2 \tau_c^2} \right) + \left(\frac{\tau_c}{1 + 4\omega_0^2 \tau_c^2} \right) \right] \tag{2.33}$$

Note that some authors (Doddrell *et al.*, 1972; Heatley, 1979, 1989) use different conventions in specifying nuclear position functions, which give rise to a different set of coefficients on the spectral density functions in the foregoing expressions.

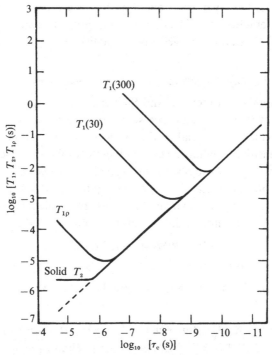

Figure 2.15. The ideal response of T_1, $T_{1\rho}$ and T_2 as a function of τ (or temperature) for a pair of protons 0.1 nm apart. $\omega_1 = 100$ kHz. Equation 2.30 is inappropriate in the low τ_c regime (dashed line) and is replaced by eq 3.36. The numbers in parenthesis denote ω_0 in megahertz.

It can be seen that this equation resembles that for R_2 except that the spectral density at zero frequency is replaced by one at the frequency $2\omega_1 (= 2\gamma B_1)$. This is a crucial distinction since this low-frequency term has the characteristic behaviour of a spin–lattice process. When $\omega_1 \tau_c = 0.5$ the contribution of this term to $R_{1\rho}$ goes through a maximum, compared with $\omega_0 \tau_c = 0.62$ for R_1. Typical behaviour of T_1, T_2 and $T_{1\rho}$ for a simple model system is illustrated in fig. 2.15. Under extreme narrowing conditions (very rapid motion), $\omega_0^2 \tau_c^2 \ll 1$, all three relaxation rates equal $2K_I \tau_c / r_k^6$. Relaxation rates are independent of ω_0 in the extreme narrowing regime. T_1 and $T_{1\rho}$ minima, however, are raised and shifted to shorter τ_c (higher temperatures) with increase in ω_0. At longer correlation times, where $\omega_0^2 \tau_c^2 \gg 1$, $T_1 = R_1^{-1}$ is proportional to ω_0^2 whereas $T_2 \ll T_1$ and is insensitive to ω_0. Near the R_1 maximum or T_1 minimum, T_1 is also relatively insensitive to τ_c and there may, in fact, be an appreciable range of molecular mobilities associated with a T_1 that is not changing very much. Note too, that T_1 generally has two solutions for τ_c, albeit of very different magnitudes. In practice, this is not a serious problem since a parallel measurement of T_2 usually removes any ambiguity in assignments.

2.7.2 Dipolar relaxation between unlike spins

The dipolar couplings between ^{13}C and ^1H spins in a polymer play a vital role in the spin–lattice relaxation processes for the ^{13}C spins as well as in the spin dynamics which are crucial in the experimental techniques used to achieve both sensitivity enhancement and high-resolution described in Chapter 4. For two spins I and S with corresponding resonance frequencies ω_I and ω_S, their mutual fluctuating dipolar couplings produce spin–lattice relaxation and their longitudinal magnetisations are coupled. Relaxation is governed by the pair of differential equations (Solomon, 1955)

$$dM_z^I/dt = R_1^{II} \Delta M_z^I + R_1^{IS} \Delta M_z^S$$
$$dM_z^S/dt = R_1^{SI} \Delta M_z^I + R_1^{SS} \Delta M_z^S \tag{2.34}$$

where $\Delta M_z^{I/S} = M_0^{I/S} - M_z^{I/S}(t)$ and the spin–lattice relaxation rates, R_1^{II} and R_1^{IS} are given by

$$R_1^{II} = (T_1^{II})^{-1} = K_S'[\tfrac{1}{12}J^{(0)}(\omega_I - \omega_S) + \tfrac{3}{2}J^{(1)}(\omega_I) + \tfrac{3}{4}J^{(2)}(\omega_I + \omega_S)] \tag{2.35}$$
$$R_1^{IS} = (T_1^{IS})^{-1} = K_I'[-\tfrac{1}{12}J^{(0)}(\omega_I - \omega_S) + \tfrac{3}{4}J^{(2)}(\omega_I + \omega_S)] \tag{2.36}$$

The prime on K denotes the replacement of γ_I^4 by $\gamma_I^2 \gamma_S^2$ in the expression for K_I, the subscript on K denotes the spin type and I/S denotes either I or S. The expressions for R_1^{SS} and R_1^{SI} are obtained by exchanging the I and S labels in these last equations.

The homo-relaxation rates $R_1^{II/SS}$ contain spectral terms at the sum and difference frequencies as well as the individual resonance frequencies ω_I and ω_S. The cross-relaxation rates, on the other hand, contain only the former. These sum and difference frequencies arise from the $I^+S^- + I^-S^+$ and I^-S^-/I^+S^+ terms in the heteronuclear dipolar coupling interaction since these are the energy differences required to balance the flip–flop, flop–flop and flip–flip processes they represent. It is clear that cross relaxation and hence the coupling of the M_z magnetisations can only occur through such terms. Transverse relaxation under these circumstances is not a coupled process and the relevant expression is

$$R_2^I = (T_2^I)^{-1} = K_I'[\tfrac{1}{6}J^{(0)}(0) + \tfrac{1}{24}J^{(0)}(\omega_I - \omega_S)$$
$$+ \tfrac{3}{4}J^{(1)}(\omega_I) + \tfrac{3}{2}J^{(1)}(\omega_S) + \tfrac{3}{8}J^{(2)}(\omega_I + \omega_S)] \tag{2.37}$$

Relaxation of ^{13}C nuclei *via* dipolar interactions with N attached equivalent protons a fixed distance r_{CH} away is a case of special interest.

Relaxation by other means such as chemical shift anisotropy, scalar coupling or spin rotation is limited to a few special cases (Heatley, 1979). When the protons are decoupled through saturation (cf Chapter 4), ^{13}C relaxation is described by

$$R_1(^{13}\text{C}) = [T_1(^{13}\text{C})]^{-1}$$

$$= \frac{N}{12} K'_{\frac{1}{2}}[J^{(0)}(\omega_\text{H} - \omega_\text{C}) + 18J^{(1)}(\omega_\text{C}) + 9J^{(2)}(\omega_\text{H} + \omega_\text{C})]$$

$$= \frac{2N}{15r_\text{CH}^6} K'_{\frac{1}{2}}\left(\frac{\tau_\text{c}}{1 + (\omega_\text{H} - \omega_\text{C})^2 \tau_\text{c}^2} + \frac{3\tau_\text{c}}{1 + \omega_\text{C}^2 \tau_\text{c}^2} + \frac{6\tau_\text{c}}{1 + (\omega_\text{H} + \omega_\text{C})^2 \tau_\text{c}^2}\right) \quad (2.38)$$

$$R_2(^{13}\text{C}) = [T_2(^{13}\text{C})]^{-1} = \frac{N}{6} K'_{\frac{1}{2}}[J^{(0)}(0) + \tfrac{1}{4}J^{(0)}(\omega_\text{H} - \omega_\text{C}) + \tfrac{3}{2}J^{(1)}(\omega_\text{C})$$

$$+ 9J^{(1)}(\omega_\text{H}) + \tfrac{9}{4}J^{(2)}(\omega_\text{H} + \omega_\text{C})]$$

$$= \frac{N}{15r_\text{CH}^6} K'_{\frac{1}{2}}\left(4\tau_\text{c} + \frac{\tau_\text{c}}{1 + (\omega_\text{H} - \omega_\text{C})^2 \tau_\text{c}^2} + \frac{3\tau_\text{c}}{1 + \omega_\text{C}^2 \tau_\text{c}^2}\right.$$

$$\left. + \frac{6\tau_\text{c}}{1 + \omega_\text{H}^2 \tau_\text{c}^2} + \frac{6\tau_\text{c}}{1 + (\omega_\text{H} + \omega_\text{C})^2 \tau_\text{c}^2}\right) \quad (2.39)$$

The nuclear Overhauser enhancement factor (NOEF), η_c, is a useful additional parameter which measures the enhancement of the integrated ^{13}C signal intensity when the protons are fully saturated (Kuhlmann and Grant, 1968; Doddrell *et al.*, 1972; Schaefer, 1974; Lyerla and Levy, 1974). The NOEF is

$$\eta_\text{c} = \frac{I^\text{d} - I^0}{I^0} = \frac{\gamma_\text{H}}{\gamma_\text{C}}\left(\frac{9J^{(2)}(\omega_\text{H} + \omega_\text{C}) - J^{(0)}(\omega_\text{H} - \omega_\text{C})}{J^{(0)}(\omega_\text{H} - \omega_\text{C}) + 18J^{(1)}(\omega_\text{C}) + 9J^{(2)}(\omega_\text{H} + \omega_\text{C})}\right) \quad (2.40)$$

where I^d and I^0 are the ^{13}C intensities with and without proton saturation. Results are also reported in terms of nuclear Overhauser enhancements (NOE) where

$$\text{NOE} = I^\text{d}/I^0 = 1 + \eta_\text{c} \quad (2.41)$$

T_1, T_2 and η_c for ^{13}C relaxation are portrayed as a function of τ_c in fig. 2.16. When $\omega_0 \tau_\text{c} \ll 1$ (rapid motion), $\eta_\text{c} = 2$ and

$$R_1(^{13}\text{C}) = R_2(^{13}\text{C}) = \frac{4N}{3r_\text{CH}^6} K'_{\frac{1}{2}} \cdot \tau_\text{c} \quad (2.42)$$

For aliphatic protons $r_\text{CH} = 0.109$ nm, in which case

$$R_1(^{13}\text{C}) = R_2(^{13}\text{C}) = 2.13 \times 10^{10} N\tau_\text{c} \quad (2.43)$$

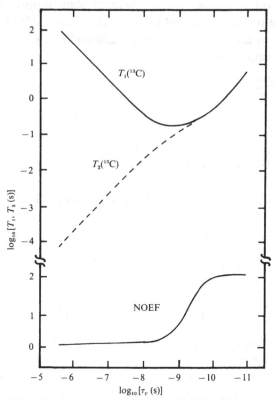

Figure 2.16. The ideal response of T_1, T_2 and NOEF for ^{13}C relaxation as a function of τ_c for a ^{13}C nucleus with one directly-bonded proton. $\omega_H = 300$ MHz and $\omega_C = 75.45$ MHz.

It must be emphasised that these plots describe interactions involving a single pair of unlike nuclei with motions characterised by a single correlation time. In solid polymers there may well be other homonuclear and impurity spin–lattice relaxation mechanisms that do not contribute to cross relaxation. While such interactions are included in the Provotorov (1962) and Solomon (1955) analysis, they are ignored in eq 2.40.

The interpretation of ^{13}C rotating frame relaxation data in solids is complicated by the competing spin–spin contributions involving mutual spin flips between spin-locked carbons and protons and spin–lattice contributions which effectively probe molecular motion (Schaefer *et al.*, 1980). VanderHart and Garroway (1979) examined the conditions under which the observed $T_{1\rho}(^{13}C)$ reflects molecular motion as opposed to equilibration of the spin-locked carbon and proton dipolar systems. Their results are best illustrated in the subsequent discussion of experimental data.

The dynamics of cross polarisation between ^{13}C and ^{1}H nuclei (cf Chapter 4) can provide useful added information as illustrated in Chapters 5 and 6. The ^{13}C magnetisation, $S(\tau)$, develops with contact time, τ, under dipolar interactions according to

$$S(\tau) = S_0 \lambda^{-1}[1 - \exp(-\lambda\tau/T_{CH}^{(SL)})]\exp[-\tau/T_{1\rho}(^{1}H)] \qquad (2.44)$$

where

$$\lambda = [1 + T_{CH}^{(SL)}/T_{1\rho}(^{13}C) - T_{CH}^{(SL)}/T_{1\rho}(^{1}H)]$$

The ^{1}H–^{13}C cross relaxation time constant under Hartman–Hahn matched, spin-lock conditions, $T_{CH}^{(SL)}$, characterises the initial growth of ^{13}C magnetisation whereas $T_{1\rho}(^{1}H)$ describes the decay of the same ^{13}C magnetisation at long CP contact times. In short, the ^{1}H and ^{13}C spin systems initially exchange energy to achieve a common spin temperature (equilibrium) followed by interaction with the lattice.

Another experiment, initially devised by Solomon (1955) to investigate cross relaxation in anhydrous hydrofluoric acid, has been applied to solid polymers (McBrierty and Douglass, 1977; Douglass and McBrierty, 1978). In this transient Overhauser experiment a 180° pulse is applied at the resonant frequency of the $S(^{1}H)$ spins and the magnetisation of the $I(^{19}F)$ spins is monitored by the application of a 90° pulse at subsequent times. Under these conditions

$$I_z - I_0 = \frac{2\sigma S_0}{\Delta}(e^{-t/D_1} - e^{-t/T_1}) \qquad (2.45)$$

where in this analysis $\rho = R_1^{II}$, $\rho' = R_1^{SS}$, $\sigma = R_1^{IS} = R^{SI}$ and

$$\Delta = [(\rho - \rho')^2 + 4(N_I/N_S)\sigma^2]^{\frac{1}{2}}$$
$$D_1 = 2(\rho + \rho' + \Delta)^{-1}; \qquad T_1 = 2(\rho + \rho' - \Delta)^{-1} \qquad (2.46)$$

These expressions predict non-exponential decay and yield values for the mean cross relaxation σ and direct ρ, ρ' relaxation rates, respectively. In this experiment, the Overhauser ratio is

$$\frac{\sigma}{\rho} = \frac{9J^{(2)}(\omega_I + \omega_S) - J^{(0)}(\omega_I - \omega_S)}{9J^{(2)}(\omega_I + \omega_S) + 18J^{(1)}(\omega_I) + J^{(0)}(\omega_I - \omega_S)}$$
$$= \frac{5 + 0.77(\omega_I\tau_c)^2 - 4.23(\omega_I\tau_c)^4}{10 + 24.06(\omega_I\tau_c)^2 + 4.33(\omega_I\tau_c)^4} \qquad (2.47)$$

The original Solomon expressions are retrieved when $\gamma_I \approx \gamma_S$, $\rho = \rho'$ and $N_I \approx N_S$. It is assumed in deriving eq 2.45 that the spin temperature is uniform within each spin system.

This experiment is interesting for two reasons. First, it provides enhanced sensitivity in determining T_1 components, which sensitivity arises from the observation of the difference of two exponentials (eq 2.45) rather than their sum as is usually the case. T_1 values within a factor of two of each other can be routinely resolved in this way. Second, σ becomes large when the spectral density function $J(\omega_I - \omega_S)$, corresponding to mutual spin flips of I and S spins, in the numerator of eq 2.47 becomes large (McBrierty and Douglass, 1977) (cf Section 6.6.2). This is analogous to the development of a $T_{1\rho}$ minimum at the difference frequency which, in this case, is $(\gamma_H - \gamma_F) B_0 / 2\pi = 1.889\ \text{MHz}$ when $\omega_0(^1\text{H}) = 30\ \text{MHz}$. Correlation frequency/temperature points determined in this fashion locate conveniently on the transition map between the T_1 and $T_{1\rho}$ sensitive regions (cf Section 1.5).

2.7.3 Relaxation due to electric quadrupole interactions

Relaxation arises when the interaction between the nuclear quadrupole moment and electric field gradients in the polymer is modulated by molecular motions such as molecular tumbling. Spin–lattice relaxation produced by fluctuations in nuclear electric quadrupole coupling is also potentially multi-exponential in character. However, for $I = 1$, a single exponential process is predicted where the rate is given by (Abragam, 1961)

$$R_{1Q} = T_{1Q}^{-1} = \frac{3\omega_Q^2}{80}\left(1 + \frac{\eta_Q^2}{3}\right)[J(\omega_0) + 4J(2\omega_0)] \qquad (2.48)$$

where, in this case, J is the Fourier transform of the reduced correlation function $J^{(n)}(\omega)/\langle b^2 \rangle_{\text{eq}} = 2\tau_c/(1 + \omega^2 \tau_c^2)$ which is independent of n. Spin–spin relaxation is similarly described by

$$R_{2Q} = T_{2Q}^{-1} = \frac{3\omega_Q^2}{160}\left(1 + \frac{\eta_Q^2}{3}\right)[3J(0) + 5J(\omega_0) + 2J(2\omega_0)] \qquad (2.49)$$

ω_Q is the quadrupole coupling constant defined in Section 2.6.1.3 and η_Q is the asymmetry parameter which approximates to zero for ^2H nuclei. In the extreme narrowing condition, eqs 2.48 and 2.49 reduce to the simple expression

$$R_{1Q} = R_{2Q} = \frac{3\omega_Q^2}{8} \cdot \tau_c \qquad (2.50)$$

Table 2.3. *Correlation frequency relationships and the ranges of ν_C and τ_C appropriate to each measurement. The values are typical of protons*

NMR feature	Expression for ν_c	Range of ν_c (Hz)	Range of τ_c (s)
T_1 minimum	$\nu_c = (2)^{\frac{1}{2}}\nu_0$	$4 \times 10^6 - 4 \times 10^8$	$4 \times 10^{-8} - 4 \times 10^{-10}$
$T_{1\rho}$ minimum	$\nu_c = \gamma B_1/2\pi$	$10^4 - 5 \times 10^5$	$2 \times 10^{-5} - 3 \times 10^{-7}$
S–A region of $T_{1\rho}$ [a]	$\nu_c = 1/2\pi T_{1\rho}$	$10 - 10^3$	$2 \times 10^{-2} - 2 \times 10^{-4}$
T_2 transition	$\nu = 1/2\pi T_{2LT}$	$10^4 - 10^5$	$2 \times 10^{-5} - 2 \times 10^{-6}$
Linewidth transition	$\nu = \gamma \delta B_{LT}/2\pi$ [b]	$10^4 - 10^5$	$2 \times 10^{-5} - 2 \times 10^{-6}$

[a] Slichter–Ailion (1964) region.
[b] δB is the linewidth between points of inflection.

McConnell (1983) has derived expressions to deal with the more complicated case of asymmetric molecules undergoing rotational Brownian motion.

2.8 Summary

Regarding the more conventional expressions in the foregoing discussion, it often suffices for the general pracitioner to use the following approximate formulae.

$$T_{1\text{min}} = \sqrt{2}\,\pi\nu_0\,T_{2LT}^2 \tag{2.51}$$

$$T_{1\rho\text{min}} = 4\gamma B_1 T_{2LT}^2 \tag{2.52}$$

where

$$T_{2LT}^{-2} = T_{2lt}^{-2} - T_{2ht}^{-2} \tag{2.53}$$

T_{2lt} and T_{2ht} denote the relaxation times below and above the T_2 transition; $T_{2LT} \sim T_{2lt}$ if the change in T_2 across the transition is large. Table 2.3 lists the expressions by which ^{1}H NMR data are translated into correlation frequencies and times. Figure 2.17 gives an overview of the range probed by the various NMR parameters covering about nine decades, bounded on the low side by quadrupole spin alignment measurements, certain two-dimensional exchange NMR experiments (Kentgens *et al.*, 1985) or $T_{1\rho}$ measured under Slichter–Ailion conditions and, at the upper end of the scale, by T_1 at the highest available laboratory field strengths.

The foregoing presentation is designed to give a broad perspective of the fundamentals of one-dimensional NMR: multidimensional NMR is not considered in this rudimentary treatment. While the approach undoubtedly provides a reasonable interpretive framework for NMR relaxation, it is not rigorously applicable to heterogeneous systems such as polymers in the

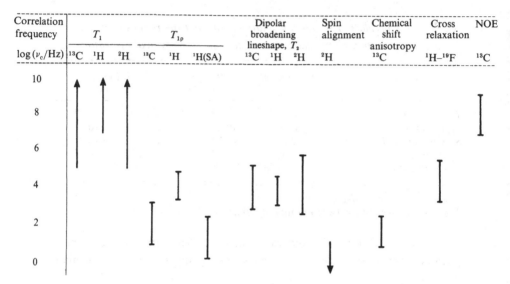

Figure 2.17. Typical motional frequency ranges spanned by different NMR measurements. (Reprinted with permission from McBrierty (1989). Copyright (1989), Pergamon Press Ltd.)

solid state, where spin relaxation processes are not amenable to the simple, single correlation time, exponential correlation function description. Further complications arise from distributions of correlation frequencies and superposition of slow and fast motions, often highly anisotropic in nature, which make the detailed interpretation of NMR relaxation complex and occasionally disputed. Refinements to the basic theory which address the more important of these deficiencies are treated in the next chapter.

3

Nuclear spin interactions

3.1 General considerations

The general conceptual framework of NMR presented in the foregoing chapter is now consolidated through further elaboration of the theory which describes the way in which nuclear spins interact with each other and with their surroundings. This is achieved through formal examination of contributions to the nuclear spin Hamiltonian. In the spirit of our deliberations on polymers, the treatment is largely restricted to diamagnetic non-conducting systems.

The Hamiltonian which describes the spin system has contributions

$$\mathcal{H} = \mathcal{H}_Z + \mathcal{H}_{RF} + \mathcal{H}_D + \mathcal{H}_{CS} + \mathcal{H}_Q + \mathcal{H}_J + \mathcal{H}_{SR}$$

where

$\mathcal{H}_Z$ is the Zeeman interaction which accounts for the coupling of nuclear spins with the external static magnetic field $\mathbf{B}_0$

$\mathcal{H}_{RF}$ represents the coupling of nuclear spins with the applied rf field $\mathbf{B}_1(t)$

$\mathcal{H}_D$ is the dipole–dipole interaction of nuclear spins with each other *via* their magnetic dipole moments

$\mathcal{H}_{CS}$ describes the chemical shift associated with the electronic screening of nuclei, which is generally anisotropic

$\mathcal{H}_Q$ accounts for the coupling between nuclear spins with quadrupole moments ($I > \frac{1}{2}$) and electric field gradients

$\mathcal{H}_J$ describes the indirect electron-coupled nuclear spin interaction

$\mathcal{H}_{SR}$ is the coupling between nuclear spins and the magnetic moment associated with molecular angular momentum, the spin–rotation interaction

51

The component Hamiltonians fall into two general categories. $\mathcal{H}_Z$ and $\mathcal{H}_{RF}$, for a given nucleus, depend only on external parameters such as the strength of $\mathbf{B}_0$ and $\mathbf{B}_1(t)$ and are thus referred to as *external Hamiltonians*: $\mathcal{H}_Z$ establishes the resonance condition (eq 2.6) whereas $\mathcal{H}_{RF}$ describes the coupling of the rf field to the spins. The remaining terms are in a second category of *internal Hamiltonians* since they depend on internal interactions and clearly contain the critical molecular information of interest. Coincidentally, they have the same structural form. Note that $\mathcal{H}_Z$, $\mathcal{H}_{RF}$, $\mathcal{H}_{CS}$, $\mathcal{H}_Q$ and $\mathcal{H}_{SR}$ involve sums of single-spin interactions whereas $\mathcal{H}_D$ and $\mathcal{H}_J$ couple every spin in the system with all others.

An important aspect of internal Hamiltonians is the way in which various tensor quantities transform between coordinate frames. It is useful therefore to specify the various coordinate frames of reference likely to be encountered in the subsequent analysis of polymers.

3.2 Coordinate systems

The most general situation envisaged in the analysis of NMR data requires four coordinate systems:

- Laboratory system (XYZ): invariably, $\mathbf{B}_0$ is along the Z axis
- Sample system $(X_0 Y_0 Z_0)$: typically, in a uniaxially drawn polymer, the Z_0 axis is collinear with the draw axis
- Molecular system (xyz): defines the orientation of a particular molecule or structural unit with respect to $(X_0 Y_0 Z_0)$
- Principal axes system (PAS), (123): the components of the spin coupling parameters such as chemical shift anisotropy or quadrupole coupling constant are defined in this frame.

The four coordinate systems are related to each other by appropriate Euler transformations, shown schematically in fig. 3.1. For example, the PAS (123) is generated from the laboratory frame (XYZ) by successive Euler rotations $(\alpha_0 \beta_0 \gamma_0)$ as in fig. 3.2. The abbreviated notation $\Omega_i \equiv \alpha_i \beta_i \gamma_i$ is used where convenient. It is important at the outset to appreciate the symmetry implications of the three Euler angles α_0, β_0 and γ_o: α_0 and γ_0 denote angular rotations about the Z and 3 axes respectively and β_0 is the polar angle of the 3 axis with respect to the Z axis.

The most elegant device for effecting rotational transformations derives from the representation of rotations by matrices with normalised spherical harmonics as the basis functions (Appendix 2). This gives compact and

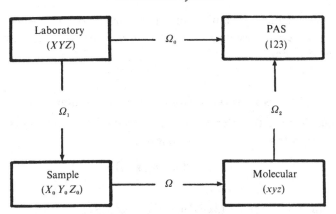

Figure 3.1. Description of the four coordinate frames used to model polymer behaviour.

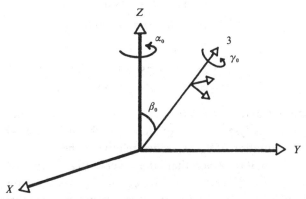

Figure 3.2. Generation of the PAS (123) from the laboratory frame (XYZ) by successive Euler rotations (Ω_0).

well-established expressions for transformation of tensors of any rank (Appendix 3). There are two principal advantages in using this approach: first, it facilitates a treatment of crystallographic and textural symmetry in the polymer; second the transformation of tensors from one coordinate system to another may be effected *via* successive rotations involving intermediate coordinate systems as required. Thus, in the most general case, the rotation $R(\Omega_0)$ which generates the PAS from the laboratory frame (fig. 3.1) may be effected in successive steps

$$R(\Omega_0) \equiv R(\Omega_1)\, R(\Omega)\, R(\Omega_2) \tag{3.1}$$

as described by eq A3.2. Note that the transformation contains important information on the distribution of tensors in a polymer. Appendix 4 develops a general procedure for modelling the distributions commonly encountered in non-isotropic polymers.

3.3 General expression for the internal Hamiltonians

The secular form of the internal Hamiltonian derived in Appendix 5 may be expressed in frequency terms as follows:

$$\omega = \mathfrak{w} + f(\beta, \gamma, \beta_1, \gamma_1, \eta)\,\mathfrak{d} \tag{3.2}$$

where

$$\begin{aligned}
f(\beta, \gamma, \beta_1, \gamma_1, \eta) = \{ & \tfrac{1}{2}(3\cos^2\beta_1 - 1)(3\cos^2\beta - 1) \\
& - \tfrac{3}{2}\sin 2\beta_1 \sin 2\beta \cos(\gamma_1 + \alpha) \\
& + \tfrac{3}{2}\sin^2\beta_1 \sin^2\beta \cos 2(\gamma_1 + \alpha) \\
& - \eta[\tfrac{1}{2}(3\cos^2\beta_1 - 1)\sin^2\beta \cos 2\gamma \\
& + \tfrac{1}{2}\sin 2\beta_1 \sin 2\beta \cos 2\gamma \cos(\gamma_1 + \alpha) \\
& - \sin 2\beta_1 \sin \beta \sin 2\gamma \sin(\gamma_1 + \alpha) \\
& + \tfrac{1}{2}\sin^2\beta_1(1 + \cos^2\beta)\cos 2\gamma \cos 2(\gamma_1 + \alpha) \\
& - \sin^2\beta_1 \cos \beta \sin 2\gamma \sin 2(\gamma_1 + \alpha)]\}
\end{aligned} \tag{3.3}$$

The parameters $\mathfrak{w}$ and $\mathfrak{d}$ for the various interactions are listed in table 3.1.

On occasion, analyses are encountered in the literature where, for example, the orientation of $\mathbf{B}_0$ is specified relative to the sample coordinate frame as in fig. 3.3(a) rather than the inverse (fig. 3.3(b)). Replacing $(\alpha_1 \beta_1 \gamma_1)$ with $(-\gamma_4 - \beta_4 - \alpha_4)$ in eq 3.3 accounts for this situation.

Quite often it is sufficient to refer the tensors under consideration directly from the PAS to the laboratory system using Euler angles $\alpha_0 \beta_0 \gamma_0$ (fig. 3.2). Appropriate expressions are obtained from eq 3.3 with $\Omega_1 \equiv 0$ and $\Omega \equiv \Omega_0$.

$$\omega = \mathfrak{w} + \mathfrak{d}[3\cos^2\beta_0 - 1 - \eta \sin^2\beta_0 \cos 2\gamma_0] \tag{3.4}$$

Now consider each of the interactions in turn.

3.3.1 The chemical shift

The chemical shift relative to the resonance frequency $\omega_0 = \gamma B_0$ is obtained by inserting the appropriate expressions for $\mathfrak{w}$ and $\mathfrak{d}$ from table 3.1 into eqs 3.2 and 3.3.

$$\omega = \omega_0[\tilde{\sigma} + \tfrac{1}{3}\Delta\sigma f(\beta, \gamma, \beta_1, \gamma_1, \eta)] \tag{3.5}$$

Table 3.1. *Specification of parameters used in eq 3.2 (Abragam, 1961;
Hentschel et al., 1978)*

Interaction	Assumption	ϖ	δ
Anisotropic chemical shift	$I = \frac{1}{2}$	$\omega_0 + \tilde{\sigma}\gamma B_0$	$\frac{1}{3}\Delta\sigma\gamma B_0$
Dipolar	$I_j = I_k = \frac{1}{2}$	ω_0	$\pm\frac{3}{4}\gamma^2\hbar r_{jk}^{-3}$
Quadrupolar[a]	$I = 1$	ω_0	$\pm\dfrac{3\omega_Q(2m-1)}{8I(2I-1)}$

[a] m is the magnetic quantum number.

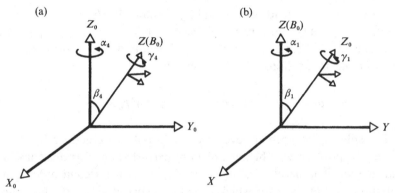

Figure 3.3. (a) Orientation of laboratory frame (XYZ) relative to the sample frame ($X_0 Y_0 Z_0$). (b) Orientation of the sample frame ($X_0 Y_0 Z_0$) relative to the laboratory frame (XYZ) which is the inverse of (a).

The analogue to eq 3.4 is

$$\omega = \omega_0[\tilde{\sigma} + \tfrac{1}{3}\Delta\sigma\,(3\cos^2\beta_0 - 1 - \eta\sin^2\beta_0\cos 2\gamma_0)] \tag{3.6}$$

or, equivalently

$$\sigma_{zz} = \tilde{\sigma} + \tfrac{1}{3}\Delta\sigma\,(3\cos^2\beta_0 - 1 - \eta\sin^2\beta_0\cos 2\gamma_0)$$
$$= \sigma_{11}\sin^2\beta_0\cos^2\gamma_0 + \sigma_{22}\sin^2\beta_0\sin^2\gamma_0 + \sigma_{33}\cos^2\beta_0 \tag{3.7}$$

where σ_{zz} is the chemical shift component along $\mathbf{B}_0$ in the laboratory frame.

When the chemical shift tensor is axially symmetric in the PAS, $\sigma_{11} = \sigma_{22}$ and $\eta = 0$ and eqs 3.5–3.7 simplify accordingly.

3.3.2 Nuclear quadrupole interactions

The secular quadrupolar Hamiltonian of which eq 2.22 is a special case (axial symmetry, $\eta_Q = 0$) may be obtained by substituting the appropriate

parameters from table A5.1 into eq A5.6. Recall that non-secular terms
may be neglected in the strong-field limit where quadrupolar couplings are
much smaller than the Zeeman splittings. Inserting appropriate data from
table 3.1 into eq 3.2 yields

$$\omega = \omega_0 \pm \frac{3\omega_Q}{8} \frac{2|m|-1}{I(2I-1)} f(\beta, \gamma, \beta_1, \gamma_1, \eta) \tag{3.8}$$

The analogue of eq 3.4 is

$$\omega = \omega_0 \pm \frac{3\omega_Q}{8} \frac{2|m|-1}{I(2I-1)} (3\cos^2\beta_0 - 1 - \eta_Q \sin^2\beta_0 \cos 2\gamma_0) \tag{3.9}$$

ω_Q, as before, is the quadrupole coupling constant. The first-order splitting
is independent of $\mathbf{B}_0$ and vanishes for $|m| = \frac{1}{2}$; that is, for $\frac{1}{2}$-integer spins, the
frequency of the central transition $-\frac{1}{2} \leftrightarrow \frac{1}{2}$ is not affected in first order by
quadrupolar interactions. For $I = 1$, eq 3.9 is

$$\omega = \omega_0 \pm \frac{3\omega_Q}{8} (3\cos^2\beta_0 - 1 - \eta_Q \sin^2\beta_0 \cos 2\gamma_0) \tag{3.10}$$

The splitting of Zeeman levels by quadrupolar interactions has been
treated to progressively higher orders of perturbation. For quadrupolar
interactions it is necessary, on occasion, to invoke second-order contributions, expressions for which have been worked out in the literature
(Bersohn, 1952; Volkoff, 1953; Abragam, 1961; Taylor *et al.*, 1975).

As pointed out in Section 2.6.1.3, polymer applications have largely
centred on deuterium NMR ($I = 1$) where spectra in the solid state are
dominated by deuterium quadrupole interactions. The asymmetry parameter η_Q is usually zero and, for a given orientation of a C–^{2}H bond in the
sample, the observed symmetric splitting is described by

$$\omega = \omega_0 \pm \frac{3\omega_Q}{8} [3\cos^2\beta_0 - 1] \tag{3.11}$$

The $+$ and $-$ signs account for the two allowed NMR transitions (fig.
2.11).

3.3.3 Dipolar interactions

Since the dipolar interaction between two magnetic moments is axially
symmetric about the internuclear vector, it follows that $\eta = 0$ and therefore
$m' = 0$ in eq A5.3. For convenience the internuclear vector is presumed to
lie along the polar axis of the PAS in fig. 3.2 in which case eq A5.3 reduces

to the more familiar form in eq 2.19 in terms of contributions A to F (table 2.2). Recall that C to F are off-diagonal and therefore make only second-order contributions to the energy. Aside from those cases where thermal motion generates or absorbs the requisite energy for transitions where m changes by ± 1 or ± 2, the contributions C to F can be neglected. The resulting truncated diagonal Hamiltonian forms the basis of the discussion on lineshapes which follows.

3.4 Spectral lineshapes

NMR powder patterns represent superpositions of the resonance condition for individual transitions over all possible orientations. By and large, spectra are smooth, structureless profiles, particularly when dipolar interactions predominate. Information on structure and motion is accessible by means of a moments analysis of the spectra (Van Vleck, 1948).

Bloembergen and Rowland (1953) first calculated the chemical shift powder pattern portrayed in fig. 2.12 by randomising the angular dependence of σ over a sphere. The components of the chemical shift tensor can, in principle, be extracted from the frequency positions of the discontinuities, ω_1, ω_2 and ω_3. Recall that for axial symmetry ($\eta = 0$ and $\sigma_{11} = \sigma_{22}$) the pattern assumes the shape of half a Pake doublet (fig. 2.11).

Powder lineshapes can be portrayed as contours of constant chemical shift (or constant ω) traced onto a sphere (Haeberlen, 1976). Each orientation of $\mathbf{B_0}$ in the PAS corresponds to a definite spectral line position. For an axially symmetric tensor ($\sigma_{11} = \sigma_{22}$), the orientational dependence of ω is given by eq 3.4 with $\eta = 0$ where β_0 in this case is the polar angle of $\mathbf{B_0}$ in the PAS. The lines of constant chemical shift are parallel circles on a sphere; their projections onto the 23 or 13 plane are shown in fig. 3.4. Successive lines represent fixed increments in chemical shift. Note that the spectral intensity $I(\omega)$ integrated over any interval is proportional to the number of nuclei whose NMR lines are within that interval. The incremental areas bounded by successive contour lines are greatest for $\beta_0 = \pi/2$ ($\omega = \omega_1 = \omega_2$) and least for $\beta_0 = 0°$ ($\omega = \omega_3$), mimicking closely the axially symmetric powder spectrum of fig. 2.12(b).

The non-axially symmetric case is more complex but may be treated in similar fashion. Contour patterns are no longer transversely isotropic about the 3 axis as typically reflected in contour profiles computed by Duncan and Dybowski (1981) for acetaldehyde. Recall that more complicated expressions are required in the unlikely event that differences between ω_1, ω_2 and ω_3 are not small (Bloembergen and Rowland, 1953).

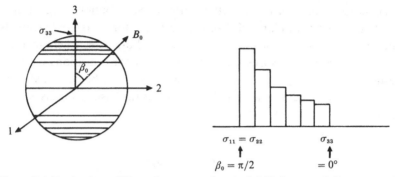

Figure 3.4. Projection of lines of constant chemical shift for an axially symmetric tensor onto the 23 or 13 plane. Successive lines represent fixed increments of chemical shift. The incremental areas bounded by successive contours are presented as a histogram which generates the shape of the axially symmetric powder lineshape.

The spectral lineshape analysis for non-isotropic polymers developed by Hentschel and coworkers (1978) and outlined in Appendix 6 is directly compatible with the general theoretical approach adopted here. The analysis considers (i) a polymer with fibre symmetry and comprising structural units which are also transversely isotropic as described in Chapter 7; (ii) axially symmetric coupling tensors; and (iii) ω_0 arbitrarily set equal to zero. The lineshape $S(\omega)$ is described by eq A6.6 as follows

$$S(\omega) = 8\pi^2 \sum_l P_{l00} \, P_l(\cos\beta_1) \, P_l(\cos\beta_2) \, S_l(\omega) \qquad (3.12)$$

P_{l00} are moments of the distribution defined in eq A4.5. β_2 defines the orientation of the polar 3 axis of the PAS relative to the polar z axis of the molecular coordinate frame (fig. 3.1); β_1 specifies the orientation of the sample Z_0 axis in the laboratory frame (XYZ); and $S_l(\omega)$ are 'subspectra' defined in eq A6.7. For an isotropic spatial distribution of structural units in the sample ($l = 0$), eq 3.12 reduces to

$$S(\omega) = S_0(\omega) = [2\sqrt{3}\,\mathfrak{d}]^{-1} [\omega/\mathfrak{d} + 1]^{-\frac{1}{2}} \qquad (3.13)$$

which describes the expected spectral shape portrayed in fig. 2.12.

Recall that these spectra correspond to the $0 \leftrightarrow 1$ transition for quadrupolar splitting with $I = 1$. The $-1 \leftrightarrow 0$ transition yields the mirror image of these patterns about $\omega = 0$. The lineshape is, in fact, centred at ω_0, presumed for convenience to be zero. The total powder spectrum is the sum of both transitions, which generates the doublet spectrum (fig. 2.11) where the separation between the two singularities in this case is $\Delta v_Q = 2a = 3\omega_Q/4$. For $l > 0$, the subspectra add and subtract equally to the total spectrum intensity. Note that a given contribution defined by l can be set to zero by choosing β_1 such that $P_l(\cos\beta_1) = 0$.

In practice, the sharp features of the lineshape are smoothed by a variety of factors which include residual dipolar interactions, vibrational motions, sample inhomogeneities and spectrometer noise as portrayed in fig. 2.11. It is normal to convolute $S(\omega)$ with a Gaussian or Lorentzian broadening function $g(\omega)$

$$S(\omega)_{\text{broadened}} = \int_{-\infty}^{\infty} S(\omega')\,g(\omega-\omega')\,\mathrm{d}\omega' \qquad (3.14)$$

to facilitate a more realistic comparison between theory and experiment.

3.5 Moments of rigid spectral lineshapes

The nth moment, M_n, of the lineshape $S(\omega)$ under high field conditions is

$$M_n = \int \omega^n S(\omega)\,\mathrm{d}\omega \qquad (3.15)$$

$S(\omega)$ is defined in Appendix 6. Expressions for the first and second moments for samples with axial (fibre) symmetry are

$$M_1(\omega) = \frac{8\pi^2}{5}\,\eth\,P_{200}\,P_2(\cos\beta_1)(3\cos^2\beta_2 - 1 - \eta\sin^2\beta_2\cos 2\gamma_2) \quad (3.16)$$

$$
\begin{aligned}
M_2(\omega) = \frac{32\pi^2}{5}\eth^2\Big\{ & P_{000}[1+\eta^2/3] \\
& +\tfrac{2}{7}P_{200}\,P_2(\cos\beta_1)\Big[(1-\eta^2/3)\,P_2(\cos\beta_2) \\
& \qquad +\Big(\frac{8\pi}{15}\Big)^{\frac{1}{2}}\eta(Y_{22}(\beta_2\,\gamma_2)+Y_{2-2}(\beta_2\,\gamma_2))\Big] \\
& +\tfrac{2}{7}P_{400}\,P_4(\cos\beta_1)\Big[(1+\eta^2/18)\,P_4(\cos\beta_2) \\
& \qquad -\Big(\frac{10\pi}{81}\Big)^{\frac{1}{2}}\eta(Y_{42}(\beta_2\,\gamma_2)+Y_{4-2}(\beta_2\,\gamma_2)) \\
& \qquad +\frac{(70\pi)^{\frac{1}{2}}}{54}\eta^2(Y_{44}(\beta_2\,\gamma_2)+Y_{4-4}(\beta_2\,\gamma_2))\Big]\Big\}
\end{aligned}
\qquad (3.17)
$$

It is evident from the functional form of eq 3.16 that average chemical shift information is contained in the first moment of the lineshape.

Table 3.2. *Coefficients a_{nl} in eq 3.18*

		l			
n	0	2	4	6	8
1	0	1	—	—	—
2	$\frac{1}{5}$	$\frac{2}{7}$	$\frac{18}{35}$	—	—
4	$\frac{3}{35}$	$\frac{20}{77}$	$\frac{1836}{5005}$	$\frac{72}{385}$	$\frac{72}{715}$

For axial symmetry, which also describes dipolar interactions, $\eta = 0$, in which case

$$M_n(\omega) = 8\pi^2(2\mathfrak{d})^n \sum_{l=0}^{2n} \frac{a_{nl}}{(2l+1)} P_{l00} P_l(\cos\beta_1) P_l(\cos\beta_2) \qquad (3.18)$$

Table 3.2 list the coefficients a_{nl}. The second moment, M_2, of the lineshape or, equivalently, T_2 of the free induction decay (FID), when dipolar interactions dominate, is a case of general interest (Van Vleck, 1948; McBrierty and Douglass, 1970). For homonuclear interactions between nuclei $I = \frac{1}{2}$ (table 3.1), the second moment in units of gauss^2 is

$$M_2 = \frac{9\gamma^4\hbar^2}{4} \sum_k \left[\frac{P_2(\cos\beta_{0k})}{r_k^3}\right]^2$$

$$= \frac{9\gamma^4\hbar^2}{4} \sum_{l=0,2,4} a_l \langle P_l(\cos\beta)\rangle P_l(\cos\beta_1) S_l \qquad (3.19)$$

In the summation, k is shorthand for $j > k$ and the internuclear vector $r_k (\equiv r_{jk})$ is deemed to lie along the 3 axis of the PAS as in fig. 3.2. $\langle P_l(\cos\beta)\rangle = (8\pi^2/2l+1) P_{l00}$ are moments of the distribution of structural units or molecules in the oriented polymer relative to Z_0 (eq A4.5); β_1 specifies the orientation of the sample Z_0 axis relative to $\mathbf{B}_0$; and S_l are lattice sums in the molecular coordinate frame (xyz)

$$S_l = \sum_k \left[\frac{P_l(\cos\beta_{2k})}{r_k^6}\right] \qquad (3.20)$$

For an isotropic distribution of internuclear vectors, M_2 reduces to the familiar form

$$M_2 = \frac{9\gamma^4\hbar^2}{20} \sum_k \frac{1}{r_k^6} \qquad (3.21)$$

Other magnetic nuclei ($I' = \frac{1}{2}$) in the sample which are not simultaneously at resonance with the first type also make a contribution to the second

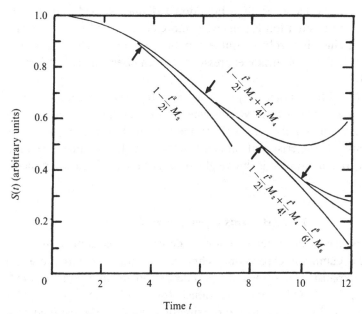

Figure 3.5. Goodness of fit of successive approximations to a Gaussian decay curve. The arrows indicate the points at which the error in each of the approximations reaches 1 % (Reproduced with permission from Powles and Strange (1963). Copyright (1963), IOP Publishing Ltd.)

moment but reduced by a factor of $\frac{4}{9}$ (Van Vleck, 1948). For this contribution eq 3.19 becomes

$$M'_2 = \gamma^2 \gamma'^2 \hbar^2 \sum_{k'} \left[\frac{P_2(\cos \beta_{0k'})}{r^3_{k'}} \right]^2 \tag{3.22}$$

where the primed notation is used to label the second type of nucleus; $r_{k'}$ joins the resonant reference nucleus j to the non-resonant k'th neighbouring nucleus. The total second moment is the sum of 3.19 and 3.22. Equation 3.21 is similarly modified. VanderHart and coworkers (1967) analysed the contribution to the second moment of coupling to nearby quadrupolar spins and showed that as the strength of the interaction increases relative to B_0, the contribution to M_2 increases. The fourth moment, M_4 has also been used to characterise non-isotropic polymers (McBrierty et al., 1971b; McBrierty and McDonald, 1973). The corresponding Van Vleck expression contains three types of lattice sum, B^4_{ij}, $B^2_{ij} B^2_{ik}$ and $B^2_{ij} B_{ik} B_{jk}$. Only the first two lattice sums make significant contributions. A formal treatment of the third is onerous and in view of its small contribution to the overall fourth moment, typically a few percent, either an approximation is used or it is neglected altogether. Equation 3.18 may be used to transform the first, but the situation is somewhat more complicated for the second lattice sum since it involves double summations over l, $l' = 0, 2, 4$. The derivation of the appropriate expression is given elsewhere (McBrierty *et al.*, 1971b).

Note that these moment expressions describe only one component or phase in a heterogeneous system such as the crystalline regions of a partially crystalline polymer. In those polymers with resolvable lineshape components characteristic of different regions in the polymer, the total second or fourth moment is the weighted sum of the resolved components (cf Chapter 7).

3.5.1 *Moments expansion of the FID*

In the absence of an exact function to describe a free induction decay (FID), the cumulant expansion technique has been used to obtain a moments expansion of a free induction decay (Powles and Strange, 1963; Mansfield, 1972). Odd moments vanish for dipolar and quadrupolar interactions and the successive approximations to the FID, denoted $S(t)$, are

$$S(t) = 1/2\pi \sum_{n=0}^{\infty} (-1)^n M_{2n} \, t^{2n}/(2n)!$$

$$= 1/2\pi \left\{ 1 - \frac{t^2}{2!} M_2 + \frac{t^4}{4!} M_4 - \frac{t^6}{6!} M_6 + \ldots \right\} \qquad (3.23)$$

M_n are the usual moments of the corresponding absorption spectrum. Figure 3.5 shows the goodness of fit of successive approximations to the decay curve. Obviously it is important to determine the FID accurately near $t = 0$, which requires special techniques to overcome problems of receiver recovery time following a pulse of rf power (cf Chapter 4). This approach allows the determination of moments from FIDs, which is of particular importance, for example, in monitoring orientation distributions in mechanically deformed polymers (Chapter 7): M_2 is more sensitive to sample orientation in B_0 than T_2 since $T_2^{-1} \propto M_2^{\frac{1}{2}}$.

In an alternative approach, the FID is fitted to the expression

$$S(t) = e^{-(a^2 t^2/2)} \cdot \frac{\sin bt}{bt} \qquad (3.24)$$

where $M_2 = a^2 + \frac{1}{3} b^2$ and $M_4 = 3a^4 + 2a^3 b^2 + \frac{1}{5} b^4$ (Abragam, 1961). The fit determines the parameters a and b from which M_2 and M_4 can be calculated.

3.6 Effects of motion on spectra

Confusion can arise in distinguishing between averages that relate to molecular *motion* and *structural* averages which account for textural symmetries in the spatial distribution of molecules (for example, isotropic distributions, fibre symmetry and partially ordered systems). The effects of textural symmetry are examined in Chapter 7. Here we focus upon motional considerations and recall from Chapters 1 and 2 that, in general, lineshapes are sensitive to motions with correlation times τ_c comparable to or less than the inverse linewidth of the spectrum. In such cases, η and $\mathfrak{d}$ in eq 3.2 assume their averages over the motion. In solids the range of molecular motions is exceedingly wide, bounded at very low temperatures by immobility, or near immobility, and at the other extreme by motions approaching liquid-like character. Both the nature and the timescale of the motion are important.

In calculating motionally averaged spectral moments or T_2 values, $P_2(\cos \beta_{0k})/r_k^3$ in the lattice sum (eq 3.19) must be averaged over the motion prior to squaring (the square of the average is not the average of the square). A few simple cases have been treated analytically while others have been simulated on a computer. A rather novel analytical procedure for dealing with complicated relative motions of nuclei (as encountered, for example, in the evaluation of *inter*molecular contributions to the lattice sum) has been developed (McBrierty and Douglass, 1980). The method is based on the two-centre multipole expansion in electrostatics (Carlson and Rushbrooke, 1950).

McCall and Anderson (1963) used a rather simple, though graphic, model to illustrate the effects of molecular motion on the local field and therefore on linewidth. Assuming that motion only affects the orientation, β_{2k}, of the internuclear vector $\mathbf{r}_k$ and not its length, the time-averaged local field

$$B_{\text{loc}} \propto T_2^{-1} \int_0^{T_2} [3 \cos^2 \beta_{2k}(t) - 1] \, dt \qquad (3.25)$$

For rapid isotropic motion, the time average approximates to the space average.

$$B_{\text{loc}} \propto \int_0^{\pi} [3 \cos^2 \beta_{2k} - 1] \sin \beta_{2k} \, d\beta_{2k} \qquad (3.26)$$

which equates to zero as expected. More interesting is the result for partially narrowed lines arising from incomplete time averaging where the motions are not fast enough or there are spatial restrictions on β_{2k}.

Consider the case of $\beta_{2k} > \beta$, that is, all orientations are accessible outside the cone defined by β. The ratio of the narrow spectral width to the rigid lattice width is approximately $\cos\beta\sin^2\beta$. For polymers above and below their glass transition temperature, this ratio is typically 0.02, in which case $\beta < 10°$, demonstrating that virtually all orientations are accessible above T_g, assuming arbitrarily rapid motions (cf Section 6.3).

By virtue of the range of linewidths encountered in dipolar, quadrupolar and chemical shift spectra, it is clear that information on a wide range of motional frequencies is available. As an initial working hypothesis, it is usually assumed that the motions are ideal in the sense that they can be characterised by a single correlation time.

We now examine the way in which a number of ideal molecular motions, both internal and external to the sample, influence the lineshape and first consider the case of sample rotation about an axis inclined at an angle β_1 to the magnetic field $\mathbf{B_0}$ (Carr, 1953; Andrew and Newing, 1958; Lowe, 1959; Andrew, 1975). This serves a twofold purpose: first, the analysis underpins a treatment of magic-angle spinning (MAS) (Chapter 4) and, second, the results give some important indications as to the way in which motions generally affect lineshapes.

3.6.1 Rotation about an axis and magic-angle spinning

Rotation of the sample at an angular frequency ω_r about the axis Z_0 inclined at β_1 to $\mathbf{B_0}$ (fig. 3.6) confers a periodic time dependence on the Hamiltonian. To describe this situation, consider $\mathscr{H}$ as

$$\mathscr{H} = \overline{\mathscr{H}} + \mathscr{H}'(t) \tag{3.27}$$

where $\overline{\mathscr{H}}$ represents the time-averaged Hamiltonian over the motion and $\mathscr{H}'(t)$ is the time-dependent part with zero mean. The situation is formally equivalent to that described in fig. 3.3(b) with $\Omega_1 \equiv (\alpha_1,\beta_1,\omega_r t)$ and eqs A5.6, 3.2 and 3.3 apply. The Zeeman contribution is unaffected by the imposed rotation and the time-dependent terms are easily recognized functions in $\gamma_1 = \omega_r t$.

Inspection of eq 3.3 reveals that chemical shift, dipolar and quadrupolar contributions have terms periodic in ω_r and $2\omega_r$. These contributions in the spectrum generate satellite spectra on either side of the central line at intervals of $n\omega_r/2\pi$. The second moment of the total spectrum is invariant with respect to rotation and the intensity of the satellites falls off as ω_r^{-n} provided that the spatial distribution of nuclei does not change (Anderson, 1954; Andrew and Newing, 1958; Andrew and Jenks, 1962). For

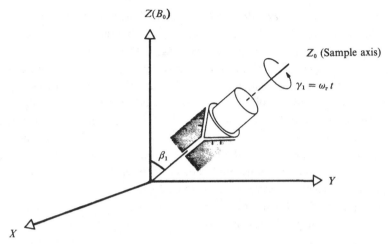

Figure 3.6. Rotation of sample about its polar axis $\mathbf{R} = \mathbf{Z}_0$, inclined at an angle β_1 to the magnetic field $\mathbf{B}_0(Z)$.

sufficiently rapid rotation ($\omega_r \gg \gamma B_{loc}$), satellites are unobservably weak, leaving the narrowed central spectrum described by $\mathscr{H}$ which, as we shall see, is identical to the Hamiltonian appropriate to an isotropic fluid. The isotropic chemical shift $\bar{\sigma}$ is the only term to survive in $\mathscr{H}$ irrespective of the textural character of the sample or the symmetry of σ. Note, too, that weaker chemical shift anisotropies can be removed at much smaller spinning speeds than can dipolar interactions.

Recall that the coefficient $(3\cos^2\beta_1 - 1)$ in eq 3.3 is zero when β_1 is $\cos^{-1} 1/\sqrt{3} = 54.7°$, the magic angle; this leads to zero averages for $\mathscr{H}_D$ and $\mathscr{H}_Q$. The suppression of $\mathscr{H}_Q$, however, requires that the quadrupolar interaction is sufficiently weak to be treated as a first-order perturbation. Second-order quadrupole effects remain in modified form under MAS because of their different angular dependence. The price paid, or course, is a loss of information on the anisotropy of the nuclear interactions involved. This information may, however, be retrieved in the rather more sophisticated approach involving slower spinning speeds, as described in the section which follows.

3.6.2 Rotational spin echoes in solids

Rotational spin echoes and their Fourier transforms, spinning sidebands, observed in the slow spinning regime ($\omega_r < \gamma B_{loc}$) contain information on anisotropic contributions to the Hamiltonian, which is otherwise forfeited in achieving the enhanced liquid-like resolution that results from rapid spinning (Lippmaa *et al.*, 1976; Maricq and Waugh, 1979; Olejniczak *et al.*, 1984). A moments analysis of spinning sidebands was developed by Maricq and Waugh to obtain the principal values of the chemical shift tensor for a single chemical species in a powder. M_2 and M_3 (the latter by virtue of rotation at the magic angle) are invariant with respect to sample rotation and chemical shift parameters can be extracted from M_2 and M_3; but all sidebands are involved and problems can arise when the sidebands from one chemical species overlap another. These problems can be circumvented in an alternative approach (Herzfeld and Berger, 1980) which analyses individual sideband intensities as outlined in Appendix 7, where an expression for the sideband intensities I_N has been derived (eq A7.6). Results are available as a suite of plots in Herzfeld and Berger's original paper with $N, \mu = \gamma B_0(\sigma_{33} - \sigma_{11})/\omega_n$ and $\rho = (\sigma_{11} + \sigma_{33} - 2\sigma_{22})/(\sigma_{33} - \sigma_{11})$ as disposable parameters. Information is extracted by comparing experimental line intensities with theoretical estimates. Alternatively, one may use the method of Tycko and Dabbagh (1991) to construct the full asymmetric chemical shift pattern from the spinning spectra.

Expressions for sideband intensities I_{MN} of two-dimensional MAS spectra $S(\Omega_1, \Omega_2)$ (cf Section 4.5.1) are also given in Appendix 7. Various symmetry considerations apply to I_{MN} which provide an internal check on experimental procedures and, as expected, the sum of the intensities for any slice parallel to the Ω_2 axis excluding the centre slice is zero (Harbison *et al.*, 1987). This implies a zero net intensity contribution to the spectrum from off-centre components, leaving only the centre band slice as observed in random unsynchronised signal acquisition. Thus, the two-dimensional MAS experiment extracts a part of the signal which averages to zero in a powder average or rotor cycle. It is further observed that the *n*th moment of the orientation distribution function must be taken into account if sidebands with $M = n$ are significant in the recorded two-dimensional spectrum (Blümich *et al.*, 1987).

Pulse sequences have been devised to suppress rotational sidebands if desired (Dixon, 1982; Raleigh *et al.*, 1987; Hagemeyer *et al.*, 1991).

3.6.3 Rapid rotation about an arbitrary molecular axis

A situation commonly encountered in chemical shift analysis involves rapid, thermally driven, rotation about an arbitrary axis R in the molecule whereby the chemical shift tensor assumes axial symmetry which is described in terms of two components $\sigma_{\parallel}^{(R)}$ and $\sigma_{\perp}^{(R)}$ representing, respectively, chemical shielding parallel and perpendicular to the axis of rotation. To determine the relationship between the averaged and unaveraged components of σ, consider R to be the Z axis in fig. 3.2. Rapid rotation of the 3 axis about R implies that α_0 is *motionally* averaged over a circle. This leads to an expression for $\sigma_\parallel^{(R)}$ which is identical to eq 3.7.

$$\sigma_\parallel^{(R)} = \sigma_{11}\sin^2\beta_0\cos^2\gamma_0 + \sigma_{22}\sin^2\beta_0\sin^2\gamma_0 + \sigma_{33}\cos^2\beta_0 \qquad (3.28)$$

where β_0 is the angle between the 3 axis of the PAS and the axis of rotation. The situation is somewhat more complicated for $\sigma_\perp^{(R)}$ since it involves the σ_{XX} and σ_{YY} analogues of eq 3.15. The result is

$$\sigma_\perp^{(R)} = \tfrac{1}{2}\sigma_{11}(1-\sin^2\beta_0\cos^2\gamma_0) + \tfrac{1}{2}\sigma_{22}(1-\sin^2\beta_0\sin^2\gamma_0) + \tfrac{1}{2}\sigma_{33}\sin^2\beta_0$$
$$(3.29)$$

and

$$\Delta\sigma^{(R)} = \sigma_\parallel^{(R)} - \sigma_\perp^{(R)}$$
$$= \tfrac{1}{2}(3\cos^2\beta_0 - 1 - \eta\sin^2\beta_0\cos 2\gamma_0)\,\Delta\sigma \qquad (3.30)$$

When $\beta_0 = 0°$, rotation is about the polar axis of the PAS itself and, as expected, $\sigma_\parallel^{(R)} = \sigma_{33}$ and $\sigma_\perp^{(R)} = (\sigma_{11}+\sigma_{22})/2$.

3.6.4 Some idealised motions encountered in polymers

Rapid isotropic reorientation as in liquids averages the orientational dependence of the tensor under examination to zero. Reference to fig. 3.1 shows that the required motional average is taken over functions in Ω_0 which specify the orientation of the PAS relative to XYZ. Noting that the isotropic motional average of $\mathscr{D}^{(2)}(\Omega_0)$ is zero in eq A5.4

$$\mathscr{H}_{\text{secular}} = CT_{0,0}P_{0,0} \qquad (3.31)$$

and only the isotropic chemical shift term $\bar{\sigma}$ survives.

When motion is not isotropic there is, in general, more structure in the line. Lineshape calculations are correspondingly more involved. There is a plethora of reported lineshape simulations to describe specific motions in polymers (Spiess, 1985b). Figure 3.7 illustrates one case of motion wherein the chemical shift (CS) or electric field gradient (EFG) principal axis frame is jumping about an axis with approximate threefold symmetry: the single coherence spectrum and the symmetrised spectrum, as would be observed respectively in a chemical shift and deuterium powder spectrum, are shown. Specifically, an axially symmetric EFG tensor reorients about an axis fixed in a local frame with the principal axis

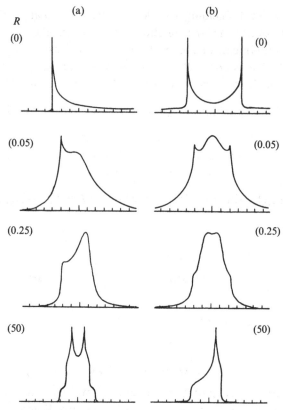

Figure 3.7. (a) Single coherence (for example, chemical shift powder) and (b) symmetrised (for example, deuterium powder) spectra for the case where the CS or EFG principal axis frame is jumping about an axis which only approximates to three fold symmetry (see text). The coupling constant is 125 kHz and the jump rates R (MHz) are as indicated. All simulations use the same inhomogeneous line broadening (D. C. Douglass, unpublished results).

of the EFG at 109.47° to the fixed axis for all sites. Sites are located in the local frame at azimuthal angles of 0 and $\pm 116°$. As the jump rate increases, the broad axially symmetric spectrum converts to a narrowed spectrum characterised by a non-axial EFG for the averaged Hamiltonian. At each orientation the spectrum consists of a sum of Lorentzian lines, each characterised by an intensity and a width. When motion is either extremely slow or extremely fast, the Lorentzian lines are very narrow. All simulations have the same amount of inhomogeneous line broadening. These spectra are equivalent to those for which the polar angle in the local frame is about 42° or 70°, or for a six-site model where jumps take place between sites with symmetric combinations of any of these angles.

As pointed out by Spiess and Sillescu (1981), the Fourier transform of the solid echo, as opposed to the FID, leads to lineshapes that differ from the true absorption spectrum when motion is present. Lineshape distortion arising from the analysis of solid echoes (cf Section 5.4) must be taken into account when seeking information on the type and timescale of the motion.

3.7 NMR relaxation in solids

The theory of NMR relaxation as discussed in Section 2.7 has been developed in seminal papers by Bloembergen, Purcell and Pound (BPP) (1948), Kubo and Tomita (1954), Solomon (1955) and Provotorov (1962). Typically, the BPP analysis makes three assumptions: (i) pairwise dipolar interactions, (ii) the assignment of a single correlation time τ_c to the molecular motion responsible for relaxation, and (iii) isotropic motions described by a Hamiltonian which has zero average over the motion, as in ideal liquids. These assumptions are inappropriate for the solid or near-solid state, but the theory can be modified to overcome the more serious limitations.

In the first modification, a perturbing Hamiltonian is constructed for the solid whose average value is zero. It is fortunate that this procedure does not alter the form of the expressions for the relaxation times but merely requires a change in the definition of the spectral densities (McBrierty and Douglass, 1970). The fact that no account is taken of the mixing of the Zeeman levels by the invariant part of the dipolar Hamiltonian is of negligible importance except possibly at the lowest fields used in the $T_{1\rho}$ experiment. The second modification introduces a distribution of correlation times $I(\tau_c)$ into the spectral density function $J(\omega)$ (Connor, 1963; Beckmann, 1988).

3.7.1 Dipolar relaxation in solids

The construction of a Hamiltonian with zero average value involves modification of the time correlation function $G(\tau)$ (eq 2.24)

$$G^{(n)}(\tau) = \langle F_n(t)^* F_n(t+\tau) \rangle$$
$$= \langle |F_n(t)|^2 \rangle \exp(-\tau/\tau_c) \tag{3.32}$$

$F_n(t)$ are time-dependent functions of the internuclear vector orientation in the laboratory coordinate frame.

In the original BPP theory in which, it is recalled, the same exponential correlation function determines all the $J^{(n)}(\omega)$ and the motion is one of isotropic rotational diffusion of internuclear vectors r_k, the functions of position coordinates are

$$\langle|F_n(t)|^2\rangle = C_n\{\langle|Y_{2n}(\beta_{0k}\,\alpha_{0k})/r_k^3|^2\rangle\}$$

$$= \tfrac{4}{5}r_k^{-6} \qquad\qquad \text{for } n = 0$$

$$= \tfrac{2}{15}r_k^{-6} \qquad\qquad \text{for } n = 1$$

$$= \tfrac{8}{15}r_k^{-6} \qquad\qquad \text{for } n = 2 \qquad\qquad (3.33)$$

Recall that $\beta_{0k}\,\alpha_{0k}$ are the angles which specify the orientation of the internuclear vector relative to $\mathbf{B}_0$. These results are used in the derivation of eqs 2.29–2.33.

In solid polymers, where isotropic rotational diffusion is seldom if ever encountered, the above analysis no longer applies and the following modified expressions are required (McBrierty and Douglass, 1970).

$$R_1 = T_1^{-1} = 3K_I\left\{\left(\frac{\tau_c}{1+\omega_0^2\tau_c^2}\right)\sum_k\langle|f_{k1}|^2\rangle + \left(\frac{\tau_c}{1+4\omega_0^2\tau_c^2}\right)\sum_k\langle|f_{k2}|^2\rangle\right\} \quad (3.34)$$

$$R_{1\rho} = T_{1\rho}^{-1} = \tfrac{3}{4}K_I\left(\frac{\tau_c}{1+4\omega_e^2\tau_c^2}\right)\sum_k\langle|f_{k0}|^2\rangle \qquad\qquad (3.35)$$

$$R_2 = T_2^{-2} = \frac{3K_I}{8}\left\{\sum_k\langle|f'_{k0}|^2\rangle\right\} = \tfrac{1}{2}M_2(\omega) \qquad\qquad (3.36)$$

The effective frequency ω_e (in radians per second) in the rotating frame approximates to $\gamma[B_1^2 + B_{loc}^2]^{\frac{1}{2}}$. Noting that B_{loc}^2 is one-third the powder second moment (Douglass and Jones, 1966) and typical values of B_1 are 5–100 G, $B_1^2 \gg B_{loc}^2$, in which case $\omega_e \sim \gamma B_1$. The lattice sums $\langle|f_{kn}|^2\rangle$ and $\langle|f'_{k0}|^2\rangle$ are

$$\langle|f_{kn}|^2\rangle = C_n\{\langle[Y_{2n}(\beta_{0k}\,\alpha_{0k})/r_k^3]_{lt}^2\rangle - \langle[Y_{2n}(\beta_{0k}\,\alpha_{0k})/r_k^3]_{ht}^2\rangle\} \quad (3.37)$$

$$\langle|f'_{k0}|^2\rangle = C_0\langle[Y_{20}(\beta_{0k}\,\alpha_{0k})/r_k^3]^2\rangle \qquad\qquad (3.38)$$

where $C_0 = 16\pi/5$, $C_2 = 8\pi/5$ and $C_4 = 32\pi/5$. The brackets [ ] denote the motional average over rapid molecular motions at temperatures below (subscripted lt) and above (subscripted ht) the transition, respectively. Recall that T_2 derives from the static part of the local field at the reference nucleus ($T_2 \ll T_1$ for solid polymers) whereas T_1 and $T_{1\rho}$ reflect the time-dependent part of the local field. The derivation of the T_2 expression follows Bloembergen's (1949, 1961) treatment for $\tau_c > T_2$ since, under these conditions, the approximations inherent in the BPP analysis are poor. In deriving eq 3.36, which is equivalent to the expression derived by

Van Vleck (1948) for the second moment M_2 of the resonance absorption spectrum, the exponential correlation function used in the BPP analysis is replaced by a Gaussian one which is more appropriate to the solid state. Note too that eq 3.36 is a special case of eq 3.19 and it applies to plateau regions of the T_2 versus τ_c (or temperature) curve between transitions. The situation is more complex in the vicinity of the transition itself. Andrew and Lipofsky (1972) have treated the case of partial narrowing by restricted motions in the transition region.

In the $T_{1\rho}$ expression (eq 3.35), the contributions corresponding to $\Delta m = \pm 1$ and ± 2 are neglected since $T_{1\rho}$ is dominated by that part of the spectral distribution around $2\omega_1$ when $\omega_0^2 \tau_c^2 \gg 1$. Jones (1966) has extended the analysis to include all terms for homonuclear interactions in describing off-resonance relaxation in the rotating frame (eq 2.33).

In NMR parlance, the analysis leading to the expressions listed above is based upon a weak-collision theory which is valid even for low fields provided that $\tau_c \ll T_{2RL}$ where T_{2RL} is the rigid lattice T_2 and $B_1 > B_{loc}$. When $\tau_c > T_2$ and $B_1 \approx B_{loc}$ the above analysis no longer applies (Wolf, 1979). This is the strong collision, Slichter–Ailion region (Slichter and Ailion, 1964; Ailion and Slichter, 1965) for which

$$R_{1\rho} = T_{1\rho}^{-1} = \frac{2(1-p)}{\tau_c}\left\{\frac{B_{loc}^2}{B_1^2 + B_{loc}^2}\right\} \approx \frac{1}{\tau_c} \qquad (3.39)$$

The parameter p depends on the details of the motion and, to a reasonable approximation, $(1-p)$ is the fraction of the second moment made time-dependent by the molecular motion under consideration (Douglass and Jones, 1966). The Slichter–Ailion conditions are infrequently encountered in solid polymers, largely due to the effects of overlapping motions. The α-relaxation in PE is a notable exception (McCall and Douglass, 1965).

A useful range of motional frequencies can be explored because of the experimental facility to choose B_1 within about 5–100 G. However, caution must be exercised when working at low $\mathbf{B}_1$ fields since multiple relaxation rates can arise because of Zeeman and dipolar contributions. This can be especially important when $T_{1\rho}$ data are used to determine spatial dimensions in blended polymers (cf Section 6.2.1).

3.7.2 *Distribution of correlation times*

The inadequacy of the BPP analysis to deal with the complexities of motion in all or part of a solid polymer has been addressed by incorporating

a distribution of correlation times $I(\tau_c)$ into the spectral density function
(Connor, 1963)

$$J(\omega) = \int_0^\infty \frac{2\tau_c\, I(\tau_c)}{1+\omega^2\tau_c^2}\, d\tau_c \qquad\qquad (3.40)$$

where

$$\int_0^\infty I(\tau_c)\, d\tau_c = 1$$

Note that the distribution of correlation times, each characterising an
exponential correlation function, cannot be distinguished from single non-
exponential correlation functions describing non-random motion of
internuclear vectors (Sillescu, 1971).

Many of the analytical distributions used to analyse NMR data have
been carried over from dielectric studies with the attendant inadequacies
discussed in Section 1.5. The results nevertheless illustrate the consequences
of $I(\tau_c)$ on the position and shape of relaxation minima and the limiting
slopes which define them. These are discussed in detail in the review
literature (McBrierty and Douglass, 1980; Beckmann, 1988).

The $\log\chi^2$ distribution introduced by Schaefer (1973) is especially useful.
The distribution is skewed and has a tail towards longer correlation times
typical of the low-frequency cooperative motions encountered in polymers.
The logarithmic timescale can also accommodate a very broad distribution
of correlation times. Narrow distributions produce little deviation from
BPP behaviour whereas broad distributions lead to shallow, flatter and
asymmetric T_1 versus τ_c (or temperature) dependence. The response of the
NOEF to a change in τ_c is much less pronounced and the observed
maximum NOEF can be less than the theoretical maximum (Schaefer,
1973; Heatley, 1979).

^{2}H NMR (Schnauss *et al.*, 1990) and multidimensional ^{13}C NMR
(Schmidt-Rohr and Spiess, 1991a) have been used to investigate the
correlation time distribution, $I(\tau_c)$, in the vicinity of the glass transition. By
probing the loss of correlation at three or more points in time in poly(vinyl
acetate) at $T_g + 20$ K, two-dimensional ^{13}C NMR revealed non-exponential
loss of correlation. More sophisticated multidimensional analysis further
revealed that $I(\tau_c)$ was heterogeneous on the time scale of the average
correlation time but became homogeneous at later times as a result of
fluctuations of a much slower nature. Clearly, the ability to control mixing
times in multidimensional NMR affords an excellent probe of correlation
time distributions.

As pointed out in Section 1.5, one of the aims of NMR is to determine
the activation energy, ΔE, which characterises the kinetics of the motion.

Distributions in τ_c introduce considerable uncertainty into the determination of ΔE from limiting slopes either side of individual T_1 or $T_{1\rho}$ versus temperature minima (McCall, 1969; McBrierty and Douglass, 1980). In low-resolution NMR it is better to use combined T_1, $T_{1\rho}$ and T_2 data since the temperatures of T_1 and $T_{1\rho}$ minima are insensitive to many of the distributions encountered and a value of τ_c can be assigned to the minimum which describes a particular relaxation process. At the midpoint of a linewidth transition, $\tau_c \simeq 10^{-4}–10^{-5}$ s. These results are then correlated with relaxation data from other techniques on transition maps as described in Section 1.5.

3.8 Spin diffusion

The discussion on the dipolar Hamiltonian in Section 2.6.1.1 introduced spin diffusion which involves the spatial transfer of nuclear magnetisation *via* energy conserving flip–flop transitions in a homonuclear spin system to an extent defined by the order of the dipolar couplings (the B term in $\mathscr{H}_D$). When $T_1 \gg T_2$, as is usually the case in solids, excess energy can remain in the spin system for a time which is long relative to T_2 before being transferred to the lattice. Thus, spin diffusion has time to even out the spatial gradients in the longitudinal magnetisation which are typically observed in heterogeneous polymers. This is analogous to the presence of a temperature gradient in the spin system whereby energy is transferred from hot to cold regions, or to the flow of water between a network of reservoirs (fig. 3.8). The time required to equilibrate two regions in a spin system is of practical importance since it provides a measure of the spatial dimension involved in the transport process (cf Chapter 6).

A rigorous treatment of magnetisation transfer, requiring the analysis of a many-body problem, has not been developed thus far for NMR. An approximate model which is adequate for the interpretation of data from dense spin systems (for example, protons) has been constructed in terms of a Gaussian random walk wherein the fundamental step is visualised in terms of the exchange of magnetisation between two spins. For the case where the mean difference in resonant frequencies at the two sites is due, for example, to differences in chemical shift, ξ, Douglass (1991, private communication) has shown that the probability, P, for a spin, $I = \frac{1}{2}$ initially in state $m = -\frac{1}{2}$ to exchange into state $m = \frac{1}{2}$ is

$$P = \frac{1}{(1+S^2)} \cdot [\sin^2 at] \qquad (3.41)$$

where $S^2 = 4\xi^2/d^2$, $a = d(1+S^2)^{\frac{1}{2}}/\hbar$ and $d = \gamma^2\hbar^2(1-3\cos^2\theta)/r^3$ is the

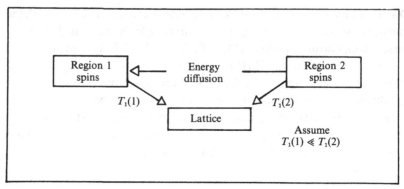

Figure 3.8. Schematic illustration of the way in which efficiently relaxing spins in region 1 of a polymer can relax spins in region 2 which are only weakly coupled to the lattice.

dipolar coupling between the two nuclei. The universal random walk result for diffusion is

$$D_S = \langle r^2 \rangle / n\tau \qquad (3.42)$$

where $\langle r^2 \rangle$ is the mean square diffusive path length, τ is the time for a fundamental step in the random walk and $n = 2, 4, 6$ for diffusion in 1, 2 and 3 dimensions, respectively. Different interfacial geometries involving spherical and lamellar configurations, used to provide better models of polymer morphology (Cheung *et al.*, 1980; Packer *et al.*, 1984; Havens and VanderHart, 1985; Henrichs *et al.*, 1988), assign different values to n which are generally within the range of 1 to 6. Equation 3.41 affords a rough estimate of both the time of the fundamental step and the amount of magnetisation transferred. τ may be identified as half the period for exchange, and $1/(1 + S^2)$ as the fraction of magnetisation transfer in that time, in which case

$$D_S \approx 2(\gamma^2 \hbar^2 / n\pi r) \langle |(1 - 3\cos^2\theta)| / (1 + S^2)^{\frac{1}{2}} \rangle \qquad (3.43)$$

This approach parallels more sophisticated treatments in the literature (Lowe and Gade, 1967; Henrichs *et al.*, 1984, 1986; Suter and Ernst, 1985; Connor *et al.*, 1985). In the absence of chemical shifts, for example, eq 3.43 is structurally similar to that of Lowe and Gade (1967) for dipolar interactions only (eq 84 of their paper). Note also that diffusion is inefficient if S^2 is large, that is, if d is small (weak dipolar coupling) and/or ξ is large (large differences in resonance frequencies at the two sites). Thus, in the absence of phonons or photons to compensate for energy mismatch, there is no exchange between unlike nuclei such as ^{1}H and ^{19}F for which ξ is large in the context of this calculation (table 2.1); but, as described earlier, cross relaxation between these two spin systems can take place in the presence of lattice motions at the difference frequency. This is known as phonon-assisted spin diffusion (McBrierty and Douglass, 1977; Douglass and McBrierty, 1978). In this special case where the lattice is involved, energy is not conserved in the strict sense of the word.

Spin diffusion between nuclei whose resonances do not overlap may be facilitated or enhanced under a number of conditions.

- In phonon-assisted spin diffusion when the interaction is modulated by molecular motions at the difference frequency of the two nuclei, as described in Section 2.7.2. The Hartman–Hahn condition (cf Chapter 4) is, in a sense, another form of assisted spin diffusion where photons supplied by the rf field rather than phonons ensure the necessary energy balance.
- Sample spinning which induces overlap of resonance lines and accordingly 'switches on' spin diffusion for at least part of the rotation period (Alla *et al.*, 1980).
- In dilute spin systems such as ^{13}C when spin diffusion is 'rotor driven', that is, when the rotor frequency is adjusted to an integer fraction of ξ, the chemical shift difference between the two participating spins (Colombo *et al.*, 1988).

The general claim that deuterium spectra are ideal site-specific probes, unaffected by spin diffusion, is not strictly true. In their study on a selectively deuterated single crystal, Schajor and coworkers (1980) concluded that spin diffusion can take place between ^{2}H nuclei even when their lines do not overlap. Of course, some mechanism to deal with the energy mismatch must be invoked. This effect has not been explored thus far in polymers.

While it is tempting to think that inhibition of spin diffusion simplifies the interpretation of relaxation behaviour in all circumstances, this is not always so. For relaxation by spin diffusion to rotating methyl groups which act as sinks, removal of spin diffusion leaves isolated methyl groups with the full anisotropy of their relaxation behaviour and non-exponential decay of magnetisation. Corresponding complications may also be anticipated for other relaxation modes. Consider now a number of special circumstances.

3.8.1 Off-resonance experiments

For off-resonance experiments, the secular interaction and therefore eq 3.43 is altered. The relevant effective Hamiltonian is

$$\mathscr{H}_{D(\text{effective})} = \tfrac{1}{2}(3\cos^2\Theta - 1)\mathscr{H}_D \qquad (3.44)$$

when the irradiating field strength B_1 is greater than the resonance linewidth. $\mathscr{H}_D$ is the normal secular term describing the unsaturated lineshape in the absence of the rf field and the angle Θ is defined in fig 2.4. Since $D_S \propto \mathscr{H}_{D(\text{effective})}$, it is clear that D_S is also Θ-dependent. This is reflected in spin–lattice relaxation data recorded in the laboratory frame (T_1), the rotating frame $(T_{1\rho})$ or in the off-resonance rotating frame $(T_{1\rho}^\Theta)$. Note two cases: (i) when $\Theta = 90°$, as in the resonant rotating frame experiment, D_S is one-half of its magnitude in the laboratory frame and (ii) when $\Theta = 54.7°$, the magic angle, dipolar coupling and therefore spin diffusion is suppressed, in principle, to zero (Tse and Hartmann, 1968).

3.8.2 Dilute spin systems

While spin diffusion effects in solid protonic polymers tend to be strong, it is usually difficult to identify the relaxation pathways directly. ^{13}C NMR on the other hand does yield site specific information but homonuclear dipolar coupling is weak and therefore spin diffusion is correspondingly slower. There is inevitably a broad distribution of diffusion rates and large internuclear distances between ^{13}C carbons in natural abundance incur excessively long diffusion times ($\tau \approx 10$–100 s) for the detection of observable effects (Caravatti *et al.*, 1982). There are ways to improve the situation. In ^{13}C-enriched polymers $\tau \approx 1$ s, which opens up new possibilities for ^{13}C spin diffusion to probe extremely short dimensional scales between labelled sites (Henrichs and Linder, 1984; Linder *et al.*, 1985). Spin diffusion can be significantly enhanced between naturally abundant ^{13}C spins when the rotor frequency in the MAS experiment is adjusted to an integer fraction of the chemical shift difference of the participating spins (Colombo *et al.*, 1988; Kubo and McDowell, 1988). This couples the spin-flip processes to the mechanical rotation energy which provides the necessary energy balance. Spins which do not satisfy this condition do not participate in rotor-driven spin diffusion, which points to interesting possibilities for spin-pair selection.

The finer points of the meaning of diffusion (Anderson, 1958) – that is, true transport as opposed to local confinement of magnetisation – as a function of spin concentration constitute an intriguing area of research but one that is beyond the level of approximation considered here.

3.8.3 Diffusion-limited relaxation

The effects of spin diffusion in circumstances where the source of relaxation is incapable of fully relaxing all the surrounding nuclei to give a single observed relaxation time can give rise to subtle interpretive difficulties. Consider the flow of energy associated with the magnetisation M_i between the two regions of a two-phase system and to the lattice for which (Douglass and McBrierty, 1971)

$$\left. \begin{array}{l} dM_1/dt = -R_1 M_1 - K_1 M_1 + K_2 M_2 \\ dM_2/dt = -R_2 M_2 - K_2 M_2 + K_1 M_1 \end{array} \right\} \qquad (3.45)$$

The subscripts denote regions 1 and 2, K_1 and K_2 are parameters which control the strength of the diffusion coupling, and R_1 and R_2 are, respectively, the relaxation rates in the two regions. It is assumed that the rate of energy flow between the two regions, each characterised by a uniform spin temperature, is proportional to the difference in their reciprocal spin temperatures, that is

$$K_1 M_1^0 = K_2 M_2^0 = D_S \qquad (3.46)$$

where $M_{1,2}^0$ are the initial total magnetisation intensities in the respective regions. When $D_S = 0$, the two regions relax independently. The solution to eqs 3.45 is

where

$$M_1(t) = \{-K_2 M_2^0 + (\alpha - R_2 - K_2) M_1^0\} e^{-\alpha t/\Delta}$$
$$+ \{K_2 M_2^0 + (R_2 + K_2 - \beta) M_1^0\} e^{-\beta t/\Delta} \qquad (3.47)$$
$$\Delta = \{(R_1 - R_2)^2 + (K_1 + K_2)^2 + 2(R_1 - R_2)(K_1 - K_2)\}^{1/2}$$
$$\alpha = (R_1 + R_2 + K_1 + K_2 + \Delta)/2$$
$$\beta = (R_1 + R_2 + K_1 + K_2 - \Delta)/2$$

A corresponding solution can be written down for $M_2(t)$ by interchanging the subscripts 1 and 2.

In modelling the temperature dependence of relaxation data, R_1 and R_2 can be approximated by

$$R_1 = C_1[\chi_1/(1+\chi_1^2)]; \qquad R_2 = C_2[\chi_2/(1+\chi_2^2)] \qquad (3.48)$$

where $\chi_{1,2}$ determine the correlation times associated with each region and $C_{1,2}$ are parameters defining the magnitude of the fluctuating magnetic fields causing relaxation. The theoretical predictions of the model are compared with typical experimental data in Section 6.8.1.

It is important to note that when diffusion coupling between the regions is comparable to the intrinsic relaxation times, the intensities associated with the two relaxation modes are not simply given by the relative intensities of the two regions as in the uncoupled case. The short relaxation time now reflects the internal establishment of an approximately steady state for energy flow between the two spin systems whereas the longer relaxation time is associated with energy flow to the lattice through the region with the fastest intrinsic relaxation time.

3.8.4 Geometrical considerations

Of course, the assumption of a uniform spin temperature and disregard for interfacial geometry is crude when the diffusion rate is comparable to the relaxation rates: the spatial dependence of the magnetisation cannot be neglected since material near the interface will relax more rapidly than material which is more remote. To illustrate this point, consider the reasonably tractable model of a sphere of crystalline material coated with a layer of amorphous polymer (fig. 3.9). This has the desirable feature of placing a large fraction of the crystalline material near the boundary interface. For simplicity, assume that the crystalline regions are not relaxing at all while the amorphous layer is behaving as a sink with an extremely short relaxation time. Solution of the diffusion equation describes this situation as follows

$$M(t) = 2a^3/\pi^2 \sum_{n=1}^{\infty} \frac{1}{n^2} \cdot \exp\{-[D_s(n\pi/a)^2 + R_1]\,t\} \qquad (3.49)$$

The first term of the series represents the long T_1 (or $T_{1\rho}$) component and the remaining terms collectively reflect the short component which is relaxing at least four times faster. Note that the 60% intensity of the long component (the intensity of the first term in the series) bears little or no resemblance to the actual amount of material in the crystalline sphere, which is important, for example, in the interpretation of component T_1 and $T_{1\rho}$ intensities when spin diffusion is operative.

3.8.5 Relaxation to paramagnetic centres

The contribution of paramagnetic ions to relaxation in solid polymers is complex. As explained by Blumberg (1960), nuclei in the immediate vicinity of *fixed* ions do not contribute to the resonance line because of the

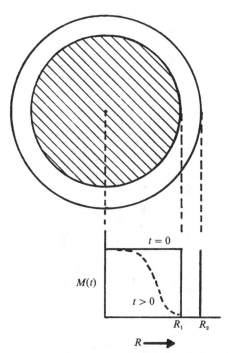

Figure 3.9. Sphere of crystalline material of radius $R_1 = a$ (shaded area) coated with a uniform layer of amorphous material which acts as a sink. The magnetisation profile, as described by eq 3.49, is shown for $t = 0$ and for $t > 0$. It is assumed that $T_1 \gg D_s(n\pi/a)^2$.

large shift in Larmor frequency; nor do they participate in spin–spin transitions with those remaining nuclei which do contribute to the resonance line. In this region, the magnetic field due to the ion is comparable to the local dipolar field in the polymer. A sphere of radius b defines the effective radius of the relaxation sink where b is referred to as a 'diffusion barrier radius'. Blumberg examined the following situations:

- Rapid spin diffusion where nuclear spin energy can diffuse more rapidly to the paramagnetic ion than the ion can transmit energy to the lattice. In this situation $M(t)$ decays exponentially.
- Diffusion-limited relaxation where $M(t)$ exhibits an initial $t^{\frac{1}{2}}$ dependence at short times before developing asymptotically into exponential decay.

The two cases are distinguished by the initial time dependence and by a different dependence on $\mathbf{B}_0$ for long times. In keeping with the earlier

Table 3.3. *Summary of nuclear magnetic interactions in solids*

Interaction	Field dependence	Isotropic average	Orientational dependence
Homonuclear dipolar	None	0	$(1 - 3\cos^2\beta_0)$
Heteronuclear dipolar	None	0	$(1 - 3\cos^2\beta_0)$
Chemical shift	Linear	$\tilde{\sigma}$	$[(3\cos^2\beta_0 - 1) - \eta\sin^2\beta_0\cos 2\gamma_0]$
Quadrupole (first order)	None	0	$[(3\cos^2\beta_0 - 1) - \eta_Q\sin^2\beta_0\cos 2\gamma_0]$

conclusions of Khutsishvili (1956) and de Gennes (1958), the final rate of exponential decay, R_1, can be described as

$$R_{1\,\text{observed}} = R_{1\,\text{intrinsic}} + R_{1\,\text{paramagnetic}} \qquad (3.50)$$

as long as $R_{1\,\text{intrinsic}}$ is not too large.

In a related study where there is no diffusion barrier, Douglass and Jones (1966) investigated $T_{1\rho}$ relaxation in n-alkanes where methyl groups on each end of the chain act as relaxation sinks. For the spin-diffusion-limited case, $T_{1\rho} = 4L^2/\pi D_s^2$ where $2L$ is the length of the chain. When the sinks are rate determining, $T_{1\rho} = \frac{1}{3}(N+1)\,T_{1\rho A}$, as expected, where $T_{1\rho A}$ is the relaxation time of a methyl proton and N is the number of carbon atoms. The analysis predicts a slope of 1 for $T_{1\rho\,\text{min}}$ versus N when methyl group relaxation is rate-determining and 2 when spin diffusion is rate-determining. In practice, the slope is 1.7 indicating that both mechanisms contribute in n-alkanes.

The situation is different when the paramagnetic species are mobile (Bloembergen, 1949). In this case, all nuclei see equivalent fluctuating fields from the electrons and all relaxation times are shortened. This is typically observed in polymers with adsorbed oxygen, particularly at low temperatures (cf Section 5.4.2).

3.9 Overview and summary

Table 3.3 summarises the various nuclear magnetic interactions in solids and highlights their more important characteristics. The expressions which derive from these interactions provide the essential framework within which NMR relaxation data from solid polymers are interpreted. It clearly does not represent a comprehensive exposition since some of the more sophisticated experimental approaches involving non-routine pulse sequences require modifications of the general theory (cf Chapter 4). Subsequent chapters will illustrate the way in which the theory of NMR is used to extract detailed information from experimental data on solid polymers.

4

Experimental methods

The study of solid polymers by NMR requires spectrometers with rather special characteristics. In this chapter we outline their principal features and describe a number of experiments that explore different aspects of the way in which resonant nuclei behave: measurements on ^{1}H, ^{2}H and ^{13}C predominate. The low natural abundance of ^{13}C and the unusually broad deuterium linewidths pose special technical problems.

Although the emphasis throughout is on pulse methods of excitation, it is important to recall that, for the determination of spectra, the only method widely used for many years was the field or frequency sweep technique, often referred to as continuous wave (CW) excitation. In this method an rf field of small amplitude is applied continuously and either the $\mathbf{B_0}$ field or the rf frequency is swept across the resonance absorption. Because of the very different requirements of high-resolution measurements in liquids and the broad, low-resolution spectra characteristic of abundant spins in solids (for example ^{1}H and ^{19}F), quite different spectrometer designs evolved for these two areas. The so-called 'broad-line' CW spectrometer used field modulation techniques which generate the derivative of the absorption lineshape. Corrections were often necessary to account for the fact that component lines of the spectrum with substantially different spin–lattice relaxation characteristics and linewidths responded differently to the imposed modulation.

CW spectrometers, generally, were not well suited to spin–lattice relaxation measurements; low-resolution, high-power pulse spectrometers were preferred. It was only the availability of relatively accessible on-line computing power that made possible the combination of both spectrum measurement and relaxation studies *via* pulse excitation methods and Fourier transformation. Very little NMR work is done today using CW methods, either for liquids or solids. The following sections accordingly focus on spectrometers employing pulse excitation methods, but it should be borne in mind that high-power, short-pulse excitation is really only the other extreme from CW methods and that various intermediate forms of time dependence of exciting fields which exploit the characteristics of both may be appropriate. As an example, the process of adiabatic demagnetisation in the rotating frame (ADRF) alluded to in Section 2.5 may be accomplished by spin-locking followed by a slow reduction of the rf field amplitude towards zero.

4.1 Spectrometer characteristics

The essential features of a spectrometer which can be used to investigate solids are shown in fig. 4.1. The various functions illustrated are similar to those of any routinely available high-resolution Fourier transform NMR spectrometer. Increasingly, commercially produced spectrometers tend to have the capability to observe both liquid and solid samples because so many of the functions are common. However, the particular specifications to be met in a solids spectrometer will be highlighted as each section is discussed in turn.

Experiments in solids, as in liquids, may involve a number of different frequencies generated by a suitable frequency synthesiser. Single- or double-resonance irradiation is commonly used, extending to triple frequencies in experiments such as those involving double cross polarisation (Schaefer *et al.*, 1979, 1984a–d). In fig. 4.1 only a single frequency channel is illustrated; the addition of other frequencies simply involves duplicating most, if not all, the features shown as part of the single channel.

The spectrometer's central computer controls a range of functions such as frequency synthesiser output, receiver gain, receiver and filter bandwidths and so on. The pulse programming function, which is one of the most important features of any spectrometer, is incorporated into the overall computer-controlled system which ideally allows for the production of rf irradiation with chosen time-dependent amplitude and phase. In the majority of experiments, rectangular pulses of defined phases which occur at the relevant times dictated by the pulse programmer are generated.

In solid-state applications, the operational characteristics of the pulse programmer are particularly stringent: pulses of microsecond duration or less, with width control and stability of better than 1 %, are routinely required. Of course, since the pulse programmer only produces the logic signals to inject the rf radiation into the amplifiers, probe and so on, it is the combined operation of the whole system that is important, not just

Experimental methods

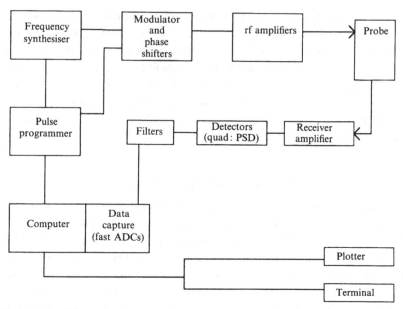

Figure 4.1. Schematic diagram showing the essential features of an NMR spectrometer.

individual components. The pulse programmer should be sufficiently versatile to accommodate the diverse range of pulse programmes required for solid-state studies. Commercial liquid-state spectrometers may not have the necessary range of capabilities but, increasingly, these differences are being removed in the newer research grade spectrometers.

The outputs from the pulse programmer are used to 'gate' the signal from the rf source. This process is carried out in the box labelled 'modulators and phase-shifters'. In the majority of solid-state spectrometers, this unit will provide for four separate quadrature phases of rf with respect to the source, namely 0°, 90°, 180° and 270° which are sufficient for the majority of experiments, but not for a particular class of experiment involving excitation and detection of multiple-quantum coherences (Baum *et al.*, 1985) which requires phase control on a finer interval scale than 90°. Again, this capability is being routinely introduced into the newer spectrometers.

The final stage in rf pulse generation is amplification. Signal strengths coming from the modulators are likely to be less than 1 W whereas, for solid state NMR, power of about 1 kW is needed to generate the required B_1 field strengths to give 90° pulses of the order of microseconds in samples within rf coils which are typically about 1 cm in diameter. Such rf field amplitudes are needed to excite the broad spectra typical of many solids and for such techniques as dipolar decoupling as described in Section 2.7.2. This level of amplification can be routinely achieved in several ways, the details of which are not of any great concern here. However, selection and control of the rf power level is important in many applications and the degree of linearity of the amplification process will have a bearing on whether any sensible control is achievable. The choice is between linear, broadband amplification and non-linear tuned amplification. The latter is generally cheaper and easier to use but linear amplifiers are increasingly preferred.

The amplified rf pulses are coupled to the sample in the probe which is positioned in the B_0 field. Probe design will be considered separately. The means by which the rf pulses and the induced NMR signals are coupled into and out of the probe is a technical matter which, again, will not be discussed in any great detail except to indicate that the way in which this is carried out can have a significant effect on the performance of the system. It can, for example, have important bearing on the recovery or dead time t_R, that is, the period following the switching off of a high-power pulse during which the signal-receiving channel is overloaded and unable to amplify faithfully the weak NMR signal.

The low-level NMR signals excited by the rf irradiation are amplified with low-noise, high-gain amplifiers, collectively called 'the receiver'. At this stage they are themselves rf signals and before recording they are 'detected' or 'demodulated'. This process is tantamount to observing the NMR signal in a rotating frame of reference, as discussed in Section 2.2, where the rotating frame frequency is the reference frequency delivered to the phase-sensitive detectors which carry out this function. Whereas a single phase detector is often used for simple NMR investigations on solids such as low-resolution relaxation measurements, spectrometers nowadays usually have two separate detectors which are provided with reference signals having the same frequency but with a phase difference of 90°. This method, known as quadrature detection, has a number of advantages particularly in regard to the efficient use of rf power in covering a given spectral width and in distinguishing NMR signals which are at both higher and lower frequencies than the reference signal. Following detection, the NMR signals are now audiofrequency in character and, prior to recording, they are usually passed through audiofrequency filters which may encompass a variety of designs. Butterworth and Bessel filters are the most popular. The bandwidth of these filters is chosen to match the digitisation rate with which the signals are captured by the data-recording system.

It is in the data-coupling system that a solid-state spectrometer has particular requirements not usually met by systems designed for recording liquid-state spectra. Simply stated, since the resonance lines in solids are often broad, corresponding to short-lived FIDs, a fast analogue-to-digital converter (ADC) is required to allow the capture of sufficient points to define the FID or, equivalently, the spectral width. Modern fast ADCs can operate at dwell times (the time between each successive data point) of 100 ns or less, corresponding to sweep widths of many megahertz. Since the bandwidths of typical NMR probe circuits are usually less than 1 MHz these ADCs are quite acceptable. If quantitative measurements are to be made, the resolution of the ADC is important. Typically, the faster devices are limited to 8-bit resolution, which can limit the precision with which good-quality signals may be measured.

Since each spectrometer facility has its own special operational philosophy within the wide range of options available, it is inappropriate to go into more detailed discussion. An in-depth treatment of many experimental aspects of NMR is available in the relevant literature (Fukushima and Roeder, 1981).

4.2 NMR probes for solids

The probe is the heart of any NMR spectrometer and the performance of the system depends critically on its characteristics, the important aspects of which are now considered.

The role of the probe is to couple rf energy to the sample and to sense the NMR rf signal induced in the sample by the radiation. The sample is positioned in a coil which is generally made up of several turns of copper wire (sometimes also silver or gold). At the highest frequencies, other forms of resonator may be appropriate. The number of turns in a typical coil will depend on the detailed design of the circuit: for 50–100 MHz there could be between 10 and 4 turns. The coil acts as the transmitting/receiving 'aerial' for the rf energy and is tuned to the relevant (resonance) frequency as in broadcast radio technology. The coil circuit is matched in impedance (usually 50 Ω) to the transmitter and receiver output/input circuits. For the simple probe circuit illustrated in fig. 4.2, the tuning and impedance matching of the coil of inductance L (usually of order 1 μH) is achieved using capacitive elements. There are various ways in which these functions may be implemented and the interested reader is referred elsewhere (Fukushima and Roeder, 1981).

The important property of such a probe circuit is its efficiency in converting rf energy into $\mathbf{B_1}$ field strength on transmission and vice-versa on

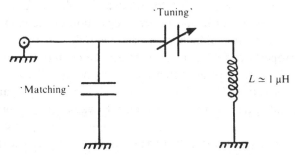

Figure 4.2. Typical tuned circuit used in an NMR probe.

receiving. Efficiency is highest when the circuit has a high quality factor, Q. This is a measure of the ability of the circuit to store energy relative to the dissipative losses in the components. A compromise is usually necessary since a high-Q circuit takes a longer time to return to equilibrium following a pulse but is more efficient at converting rf power into $\mathbf{B}_1$ field strength. A comprehensive discussion of circuit design for NMR is not appropriate but one must recognise that this compromise exists and that it affects the quality of the NMR information obtained unless due regard is paid to its consequences. If, for example, it is required to excite uniformly a broad range of frequencies, as in the case for ^{1}H spectra of many solid polymers, the pulse width must be kept much shorter than the inverse of the spectrum width. Pulse widths less than 1 μs are commonly used in these circumstances. To obtain maximum signal strength, this width is usually chosen to correspond to a 90° pulse which determines the required $\mathbf{B}_1$ field. Typically, for protons ($\gamma = 2.67 \times 10^8$ rad s^{-1} T^{-1}) a 1 μs 90° pulse requires a $\mathbf{B}_1$ field of 6 mT (60 gauss). The rf power needed to produce this $\mathbf{B}_1$ field depends on the Q of the circuit which, in turn, determines the speed with which the circuit 'rings down' following the pulse. Thus, in order to detect the NMR signal excited by the pulse, the rf energy stored in the probe circuit at the time the rf is switched off must be dissipated to the extent that the voltage it presents at the receiver input is much less than the corresponding NMR signal. Typically, the voltage across the sample coil when a pulse of $\mathbf{B}_1$ field of order 6 mT is applied will be several kilovolts whereas the NMR signal may be 0.1 mV or less. In order to capture faithfully the FID corresponding to typical broad spectra such as ^{1}H spectra of solid polymers, the recovery time of the probe should be very much less than the inverse of the spectrum width. This may require recovery times of the order of a few microseconds at most.

In reality, for the broadest spectra, it is usually not possible to obtain

sufficiently short recovery times to achieve completely faithful repro-
duction of the spectrum *via* Fourier transformation of the FID. For any
probe there is some unavoidable recovery time following pulse excitation
which is a property of the circuit. To restate the compromise mentioned
above, the achievement of high efficiency in rf energy utilisation requires
high Q but this leads to longer recovery times. The problem is particularly
acute for broad spectra which require high $\mathbf{B}_1$ values to give broadband
excitation (that is, short pulses of significant flip angle) and fast recovery
because of the short transverse relaxation time. In practice also, the
recovery time following the pulse is a function of the strength of the NMR
signal, as already noted. It is important, therefore, to consider the effects of
these constraints if any attempt is to be made to undertake quantitative
investigations of lineshapes, relaxation behaviour and so on. A number of
pulse sequences, to be discussed later, have been developed which help to
reduce or obviate the effective recovery time of the spectrometer.

In many experiments there is a requirement to irradiate the sample with
two or even three frequencies, sometimes simultaneously. In such cases, the
probe circuit must present appropriate characteristics to both transmitters
and receivers at the relevant frequencies. One of the principal challenges to
the development of cross polarisation techniques and the direct observation of signals from ^{13}C and similar spins in the presence of high-
power ^{1}H decoupling, was the design of probe circuits that (i) were efficient
in their use of rf energy, (ii) had the required recovery properties, (iii)
allowed the measurement of microvolt NMR signals at one frequency
while kilovolts at another frequency were present on the same coil and (iv)
provided for sample spinning at kilohertz frequencies at a carefully set and
stable angle with respect to the external $\mathbf{B}_0$ field and preferably with the
facility to change the temperature. Again, it is not appropriate to go into
the detailed design of such circuits here, but it is important to understand
the characteristics and limitations of any probe of this type if optimum
performance is to be achieved.

4.3 Sample-spinning devices

Many NMR experiments applicable to solid polymers are carried out using
sample spinning as one of the techniques to achieve partial or complete
averaging of anisotropic spin interactions, particularly the shielding
anisotropy and relatively weak heteronuclear dipolar couplings. The most
frequently used method is magic-angle spinning (MAS) discussed earlier in
Chapters 2 and 3. Recall that the requirement is to effect sample rotation

around an axis which is adjusted to 54.7° relative to the direction of $\mathbf{B}_0$. Spinning speeds ideally should range between some arbitrary small value, say 100 Hz, and values sufficient to produce complete averaging of the relevant anisotropic interactions. Complete averaging in this context is achieved when the spectrum shows no further change on increasing the spinning speed. Typically, for interactions such as shielding anisotropy, this occurs when spinning sidebands are no longer distinguishable which, of course, is dependent on both sample and experimental conditions. Currently, spinning rates up to 10–15 kHz are readily achievable with some commercially available probe systems. Realise, however, that spinning at these rates subjects sample and container to considerable forces and the routine use of these techniques therefore requires robust design and the necessary experience and skill in sample preparation and manipulation.

There have been a number of sample-spinner designs and all, to a greater or lesser degree, utilise low-friction gas bearings. Such a bearing is created when two solid surfaces are held apart by the flow or pressure of a thin layer of gas. The construction materials must have the necessary mechanical stability under conditions of use, they must preferably not contribute to the spectra under examination, and they must not interfere with the performance of the NMR experiment in terms of rf probe characteristics, $\mathbf{B}_0$ homogeneity, and so on. Modern designs use high-performance polymer resins such as Torlon®, Vespel®, Derlin® and ceramics such as alumina, zirconia and machinable glass (Macor®). ^{13}C NMR cannot use proton-containing organic polymers for the sample container, or at least that part of it in the active region of the rf coil, since these materials give rise to strong signals in the ^{13}C spectrum. Ability to vary and control sample temperature under MAS conditions is also important and this is usually achieved by conditioning the temperature of either drive and/or bearing gases.

As already indicated, MAS is most often used for averaging chemical shift anisotropy contributions to lineshapes. These contributions scale with the value of $\mathbf{B}_0$ so that higher fields require proportionally higher spinning speeds to achieve the same degree of averaging. In ^{13}C NMR, for example, the shielding anisotropies of carbonyl and aromatic carbons are amongst the largest, being typically of the order of 100–200 ppm (table A1.2). This means that, at a ^{13}C resonance frequency of 50 MHz for which $B_0 = 4.7$ T, typical bandwidths for the resonances from these carbons can be of the order of 10 kHz. Spinning at 3–5 kHz, therefore, still gives rise to spinning sidebands of significant intensity, which can complicate interpretation.

In summary, the sample probe which is at the heart of NMR must have the required characteristics appropriate to the experiment: it must be capable of accommodating the sample in its available form; temperature variation and control are desirable; and it must facilitate sample spinning at the selected speed with a stable and reproducible setting of the magic angle in MAS experiments. These are demanding specifications and ones which are often only partially met in practice. The intending polymer NMR spectroscopist wishing to make the most of investigations of the solid state should become familiar with many of the important features and compromises mentioned above.

4.4 Pulse sequences

This section examines some of the more important and frequently used pulse sequences for NMR of solid polymers. It is not possible or desirable to attempt to be comprehensive and, since this book is aimed principally at the polymer scientist, it will avoid the more sophisticated experiments and merely note their existence and their potential.

4.4.1 Single-resonance experiments

4.4.1.1 Single-pulse determination of spectra

The simplest single-resonance experiment records the NMR spectrum of a sample by applying a single pulse to excite an FID which is then Fourier-transformed. Figure 4.3(a) illustrates this pulse sequence in a generalised form. A $\theta°$ (usually 90°) pulse creates some transverse magnetisation which produces an FID. This is shown as a simple monotonic decay in the figure but could be more complicated or be set off-resonance from the detector reference frequency, which would impose oscillatory behaviour on the FID, reflecting the differences between the detector reference and the spectral frequencies. The recovery time, t_R, following the pulse is indicated and the FID is shown as a broken line during that period to emphasise that it may not be measured accurately. The undistorted FID is portrayed at a digitising rate set by the dwell time, t_D, the time between successive measurements of the signal. This sequence is repeated after a time T_R, the recycle delay, and the signals from the second and subsequent repetitions of the sequence are co-added in the spectrometer computer memory until sufficient improvement in signal-to-noise ratio is achieved. The details of this experiment prompt a few general comments, many of which apply to subsequent situations.

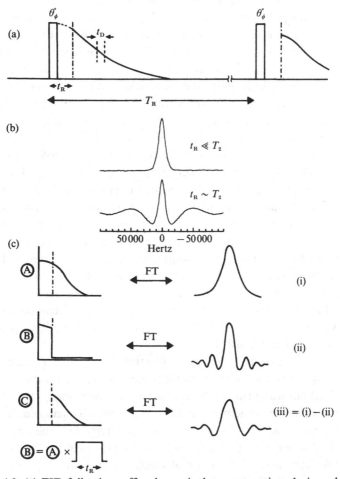

Figure 4.3. (a) FID following a θ_ϕ° pulse. t_R is the recovery time during which no signal is detected and t_D is the dwell time (see text). (b) Effect of recovery time on the ^{1}H spectrum of solid hexamethylbenzene recorded with the sequence shown in (a). (c) Signal distortion introduced when the recovery time is excessively long. Compare (c) with the corresponding (lower) experimental spectrum in (b).

The recycle delay is chosen to ensure that the longitudinal nuclear magnetisation has recovered to equilibrium following the preceding pulse. If a 90° pulse is used then, in general, $T_R > 5T_1$ for a system in which a single T_1 is applicable. In some circumstances, particularly if T_1 is very long, it can be advantageous to use a pulse-flip angle which is less than 90° in conjunction with a faster repetition rate. The dwell time t_D determines the spectral or sweep width in the transform domain. For quadrature detection, the spectral width is $(t_D)^{-1}$. Data acquisition may be started at any convenient point provided that its location is known relative to the exciting pulse. It is usual to pass the detected signal through audio-frequency filters whose bandwidth is set to match the spectral width and, if such is the case, then data recording normally begins at an integral multiple of dwell times from the excitation pulse. This simplifies the application of the so-called first-order (linear frequency-dependent) phase-shift correction to the spectrum. Each pulse in successive sequences in fig. 4.3(a) is portrayed as having a phase associated with it. Though not essential for a general discussion of experimental methods, note that almost all NMR measurements using pulse excitation are carried out with some form of phase cycling and associated data routing. The latter term refers to the fact that the outputs from the two quadrature phase detectors are switched between the computer memory locations assigned to the two parts of the complex FID and may be negated prior to their addition into the memory for successive pulse sequences with different phases. In the case illustrated here, the purpose of such phase cycling is to reduce the build up of artefact signals which are not relevant to the NMR experiment, to correct for pulse imperfections and so forth. In other circumstances, the phase cycling may be used to select signals arising from various quantum orders (Wokaun and Ernst, 1977).

The effect of the recovery time, t_R, is illustrated in fig. 4.3(b) which shows ^{1}H NMR spectra of solid hexamethylbenzene recorded with the sequence in fig. 4.3(a) but using two artificially chosen recovery times. One of these is much less than the effective T_2 whereas the other is a significant fraction of it. The spectra were recorded with matched filters, the same dwell time, and a first-order phase correction appropriate to the start of data acquisition. It is clear that the effect of a recovery time which is comparable to T_2 causes inevitable distortion of the spectrum. This takes the form of oscillatory sidelobes and an apparent narrowing of the central feature for this single-line spectrum. The origin of this effect is readily appreciated by noting that the experimental spectrum may be thought of as the ideal spectrum minus the spectrum obtained by Fourier transformation of that part of the FID which is missing because of the recovery time. In other words, this will be the ideal spectrum convoluted with the Fourier transform of a rectangular time function of width t_R as illustrated schematically in fig. 4.3(c). Clearly, attention must be paid to all of these complicating factors if any form of quantitative investigation is contemplated.

4.4.1.2 Spectral moments

As mentioned in Chapter 3, the moments expansion of an NMR lineshape can be useful in relating such quantities to the structure of a solid. Second and fourth moments have found considerable application in the study of polymers, notably in the study of orientation effects (cf Chapter 7). Traditionally, these moments were determined from the spectrum obtained by CW techniques. In the time domain, they can be determined directly from the FID expressed in terms of the moments (eqs 3.23 and 3.24) as described in Section 3.5.1.

To determine moments accurately, it is important to record faithfully the FID close to zero time. This is precisely where it is most difficult to obtain accurate measurements because of the recovery time. It was a consequence of both this requirement in the determination of moments and the need to be able to generate undistorted spectra that led to the development of certain echo techniques for use in solids, as described in the following section.

4.4.1.3 Spin-echo experiments

There have been a number of experimental approaches to circumvent or, at least, to minimise the effects of the inevitable recovery time following pulse excitation. In one of the more commonly used procedures, the time evolution of the FID is reversed to form a spin echo. In solids the time course of the FID may be caused by a number of spin interactions, the main ones being dipole–dipole couplings, electric quadrupole couplings and shielding anisotropy. Of these, the last is the simplest to use as an example.

In a polycrystalline sample the resonance of an axially symmetric site gives rise to the powder pattern shown in fig. 4.4(a) which comprises a superposition of resonance lines of a much narrower width, each arising from separate, differently oriented, crystallites. Following a 90° pulse, the transverse magnetisation decays in a time of the order of Δ^{-1}, that is, the inverse of the anisotropy. A 180° pulse applied at a time $\tau < T_2$, where T_2 is the transverse relaxation time of an individual component of the lineshape, causes a refocussing of the relative precession of the different components arising from the separate crystallites and a maximum in the signal, the spin echo, which occurs around the time 2τ (fig. 4.4(b)). Again, the interested reader is referred elsewhere for a more detailed discussion of spin-echo formation (Farrar and Becker, 1971; Slichter, 1990). It is sufficient to note here that if the condition $t_R < \tau \ll T_2$ can be satisfied then

(a)

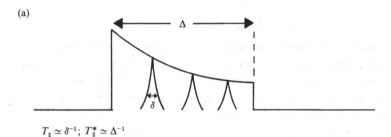

$$T_2 \simeq \delta^{-1}; \; T_2^* \simeq \Delta^{-1}$$

(b)

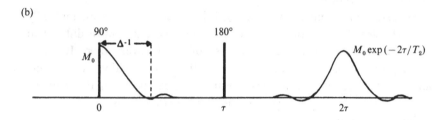

(c)

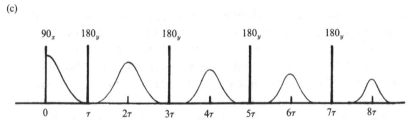

Figure 4.4. (a) Powder pattern generated from a superposition of narrow resonance lines from separate, differently oriented, crystallites. (b) Formation of an echo at 2τ in the $90°-\tau-180°$ pulse sequence. (c) Carr–Purcell/Meiboom–Gill pulse sequence.

a close approximation to an undistorted spectrum may be obtained by Fourier transformation of the echo signal for times greater than 2τ. Echo formation may be repeated by application of successive 180° pulses to give an echo train. Thus so-called Carr–Purcell/Meiboom–Gill sequence (Meiboom and Gill, 1958) is illustrated in fig. 4.4(c).

The case of second-rank tensor interactions, such as the homonuclear dipole–dipole and quadrupole interactions, is more difficult to appreciate. It must suffice to note that the use of a $90_x-\tau-90_y$ sequence, referred to as a solid or quadrupole-echo sequence plays a similar role for these interactions: the subscripts x and y denote the orthogonal axes along which the pulses are applied. Indeed, the majority of ^{2}H NMR spectra and relaxation studies in polymers make extensive use of this experiment. Even in this case, there is a distinction between the response of the typical quadrupolar spin-1 deuterium powder pattern in solids and that of the many-spin dipolar couplings between spin-$\frac{1}{2}$ nuclei such as ^1H in a solid. The quadrupole case has certain similarities with that of shielding anisotropy discussed above. As pointed out in Section 3.4, the powder spectrum of a polycrystalline sample containing ^2H spins in axially symmetric environments – for example, C–^2H bonds in polymers – is a superposition of a continuous range of doublet spectra in which the splittings depend on crystallite orientation in the $\mathbf{B}_0$ field. To the extent that each doublet splitting is time-independent, the response to the quadrupolar echo sequence is very similar to that described for the shielding anisotropy. This is because each doublet frequency is independent of the others and the spins have essentially a time-independent distribution of such frequencies. Indeed, in situations where molecular motion may interchange such frequencies, for example, by conformational jumps on a timescale of the order of τ, this sequence forms a powerful addition to the range of NMR experiments which can be used to quantify such processes, as is discussed in more detail in Section 3.6.4 and in Chapter 5.

For a complex assembly of ^1H spins, typical of a solid polymer, the homonuclear dipole–dipole couplings are more akin to the spin-1 quadrupole case with motional modulation. This is because the flip–flop transitions couple together the range of frequencies characteristic of the local dipolar fields (cf Chapter 2). Thus, for the solid-echo sequence to be effective, the second pulse must be applied as close as is practical to the top of the FID following the first pulse. In fact, this sequence only reproduces the FID as an echo in the limit $\tau \rightarrow 0$. Again, the reader is referred to other sources for further, more detailed, discussion of these questions (Mehring, 1976; Slichter, 1990). It must be emphasised that in order to obtain satisfactory and quantitative results using these sequences, some degree of understanding of the factors affecting their performance is necessary.

4.4.1.4 Spin–lattice relaxation measurements

In order to determine the spin–lattice relaxation properties of a spin system it must first be placed in a defined non-equilibrium state. There are a variety of ways to achieve this which depend on the particular spin–lattice process under consideration. For spin–lattice relaxation in the laboratory frame, that is, relaxation of magnetisation along the field $\mathbf{B}_0$, there are two main methods. The first, often referred to as inversion-recovery, starts by inverting the magnetisation. For a system initially at equilibrium, ($M_z =$

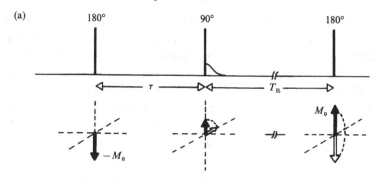

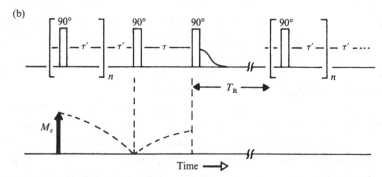

Figure 4.5. (a) $180°$–τ–$90°$ pulse sequence used to determine T_1. (b) Saturation recovery sequence used to measure long T_1 relaxation times.

M_0), a $180°$ pulse rotates the magnetisation such that $M_z(0) = -M_0$. This inversion of the population distribution of the spin energy levels produces a very 'hot' spin system in spin-temperature terms (although formally a negative temperature) and at a time τ later, a $90°$ pulse creates an FID which is representative of the value of $M_z(\tau)$ (fig. 4.5(a)). This process may be characterised either by measuring the total amplitude of the FID as a function of τ or by Fourier transforming it to give a series of partially T_1-relaxed spectra at different τ values. This latter representation can be particularly informative for complex and heterogeneous systems such as partially crystalline polymers or block copolymers and polymer blends which contain regions of differing composition, structure and dynamics.

A second starting point for measuring laboratory-frame spin–lattice relaxation is to set the magnetisation reproducibly to zero at the beginning of each sequence. This method is called saturation-recovery and is illustrated in fig. 4.5(b). The sequence is applicable when the timescale of intrinsic transverse relaxation is much less than that for the spin–lattice relaxation process, that is, $T_2 \ll T_1$, a condition which often applies in solids. The initial burst of closely spaced 90° pulses reduces M_z to zero as long as $T_2 < \tau' \ll T_1$. After allowing a time τ to elapse, a further 90° pulse produces an FID which monitors the recovery of M_z towards M_0 as a function of τ. This sequence has particular advantage in situations where the spin–lattice relaxation process is inefficient, that is, for large values of T_1. This is because there is no requirement for a recycle time of any particular magnitude between successive sequences as the saturation preparation puts the magnetisation to zero whatever its value. This is not so for the inversion-recovery sequence which requires a time of the order $5T_1$ between repetitions. When this recycle time is not prohibitively long, the inversion-recovery sequence is preferred as it gives a wider range of magnetisation change when investigating the relaxation process. In complex systems relaxation may not be described by a single value of T_1 and it may be important to have as accurate a representation of the process as possible in order to model its behaviour in terms of component relaxation times.

Rotating frame relaxation is a further type of spin–lattice relaxation which is important, both in its own right and for double-resonance experiments involving cross polarisation. The pulse sequence which is most commonly used for its determination is illustrated in fig. 4.6. As discussed in Chapter 2, this process denotes relaxation towards zero of magnetisation which has been aligned along the rf magnetic field $\mathbf{B}_1$. In this aligned state it is 'spin-locked'. In the simplest version of the pulse sequence, the rf field is applied exactly at resonance. A $90_{x'}$ pulse rotates $M_z (= M_0)$ into the y' direction following which an rf field $\mathbf{B}_{1y'}$ is applied. To ensure that spin-locking is established, the amplitude of $\mathbf{B}_{1y'}$ must be much larger than the frequency spread in the spectrum of the spin system. This ensures that any tendency for the magnetisation to dephase under the influence of the spectral width will be refocussed by the rotation of the magnetisation components about y' at frequency $\omega_1 = \gamma B_1$. The relaxation process is mapped out by measuring the FID – which occurs when the $\mathbf{B}_{1y'}$ pulse is switched off – as a function of the spin-locking pulse length, t. Again, either the initial amplitude of the FIDs or the spectra obtained by Fourier transformation may be used to characterise the relaxation behaviour.

In its more general form, the rotating frame relaxation may be examined with the rf field applied somewhat off-resonance. In this situation the sequence begins with a pulse that rotates the magnetisation through an

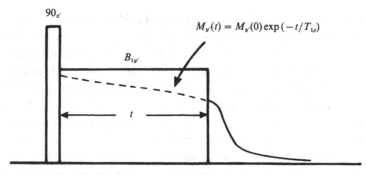

Figure 4.6. Pulse sequence for measuring spin–lattice relaxation in the rotating frame.

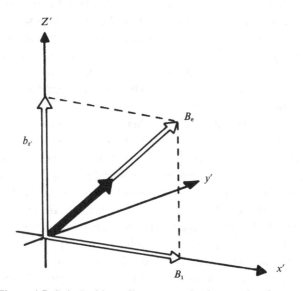

Figure 4.7. Spin locking off-resonance in the rotating frame.

angle less than 90° about the x' axis. The combination of the off-resonance effective z field and the applied $\mathbf{B}_{1y'}$ field is arranged so that the effective field in the rotating frame is aligned with the partially rotated magnetisation, thereby producing a spin-locked state at an angle to the z axis. This is illustrated in fig. 4.7. The purpose of such an experiment can be twofold. Firstly, the relaxation behaviour varies with the angle of spin-locking (Jones, 1966). Secondly, provided the $\mathbf{B}_{1y'}$ rf field is large enough, the secular dipolar couplings between like spins, it is recalled, are scaled under spin-locking by a factor $(3\cos^2\Theta-1)/2$. Thus, close to $\Theta = 54.7°$,

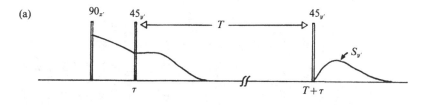

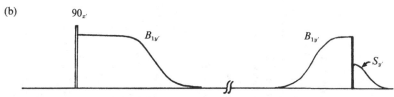

Figure 4.8. (a) Jeener–Broekaert sequence used to achieve dipolar or quadrupolar order in the spin system. (b) Sequence which achieves adiabatic demagnetisation in the rotating frame (ADRF).

there is a substantial reduction in the effectiveness of these dipolar couplings, particularly in promoting spin-diffusion (Tse and Hartmann, 1968).

Finally, there is a form of spin–lattice relaxation which is usually referred to as dipolar–lattice relaxation. In this process, some of the spin order represented by the equilibrium Zeeman magnetisation, M_0, is transformed into a form of order which is internal to the spin system. For dipolar interactions this is referred to as dipolar order whereas for quadrupolar spins such as ^2H, it is called spin-alignment or quadrupole order; the details of these spin-ordered states are described elsewhere (Goldman, 1970). At thermal equilibrium the extent of such order in spin systems is negligible, but, when ordered, these states correspond to spin temperatures which are very different from the lattice. Once created they then relax towards equilibrium with rates that are a form of spin–lattice relaxation in a way that extends the detail with which the system under study may be characterised. For ^2H, spin-alignment has found a particular use in investigating infrequent jump processes in solids (Spiess, 1985a) (cf Section 5.4.2).

The preparation of the spin system for this type of relaxation experiment involves one of two pulse sequences (fig. 4.8). The first is the so-called Jeener–Broekaert (1967) sequence which uses a $90_{x'}–\tau–45_{y'}$ pulse pair, assuming on-resonance conditions. Adjustment of τ varies the amount of internal spin-order produced and it is normally chosen to be of the order of the inverse of the spectral width arising from the interaction into which

Table 4.1. *Some single resonance pulse sequences commonly encountered in solid-state NMR*

Name	Pulse sequence	Application	Reference
Free induction decay (FID)	90_x	All interactions contained in spectrum provides T_2 for solids in the low temperature regime: Problems with t_R [a]	Lowe and Norberg (1957)
Solid echo	$90_x - \tau - 90_y$	Provides complete FID and T_2	Powles and Mansfield (1962), Powles and Strange (1963)
Carr–Purcell spin echo	$90_x - \tau - 180_y$	Removes chemical shift, inhomogeneity and other off-resonance effects	Carr and Purcell (1954)
Carr–Purcell–Meiboom–Gill (modified)	$90_{-x} - (\tau - 180_y - 2\tau - 180_{-y} - \tau)_n$	As Carr–Purcell but with compensation for phase errors	Carr and Purcell (1954), Meiboom and Gill (1958)
Saturation recovery	$(90_{-x})_n - 90_x - \tau' - 90_x$	Yields T_1 and permits discrimination against long T_1 values	Weiss *et al.* (1980), Becker *et al.* (1980)
Inversion recovery	$180_x - \tau - 90_y$	Yields single and component T_1 values	Weiss *et al.* (1980), Becker *et al.* (1980)

Spin locking	$90_x B_{1y}(\tau)$	Yields single and component $T_{1\rho}$ values	Hartmann and Hahn (1962), Jones (1966)
Goldman–Shen	$90_x - \tau_1 - 90_{-x} - \tau_2 - 90_x$	Provides information on coupling between spin systems with different T_2 or $T_{1\rho}$ values	Goldman and Shen (1966)
Jeener–Brockaert	$90_x - \tau_1 - 45_y - \tau_2 - 45_y$	Provides information on direct coupling between dipolar and Zeeman energy baths	Jeener and Brockaert (1967), Packer (1980)
WAHUHA cycle	$90_x(\tau - 90_{-x} - \tau - 90_y - 2\tau - 90_{-y} - \tau - 90_x - \tau)_n$	Removes homonuclear coupling to first order	Waugh et al. (1968)
Mansfield six-pulse cycle	$90_{-y} - (\tau - 90_{-x} \, 90_{-y} - \tau - 90_y - 2\tau - 90_{-y} - \tau - 90_y \, 90_{-x} - \tau)_n$	Removes homonuclear coupling	Mansfield and Ware (1966), Mansfield (1970)
MREV-8 cycle	$90_x(\tau - 90_{-x} - \tau - 90_{-y} - 2\tau - 90_y - \tau - 90_{-x} - 2\tau - 90_{-x} - \tau - 90_{-y} - 2\tau - 90_y - \tau - 90_x - \tau)_n$	Homonuclear decoupling: compensation for pulse defects	Rhim et al. (1973), Rhim et al. (1974)

(a) t_R = recovery time

order is being transferred. The second method starts by spin-locking the magnetisation after which the amplitude of $\mathbf{B}_{1y'}$ is then reduced adiabatically to a value less than the relevant spectral width (fig. 4.8(b)). This process transfers order into this line broadening interaction and is known as adiabatic demagnetisation in the rotating frame (ADRF).

Following the creation of these internally ordered states, a time t is allowed to elapse before an NMR signal is generated whose amplitude is related to the amount of spin order remaining unrelaxed at t. This signal monitors the relaxation and may be effected in both cases by use of a 45° read pulse. While the phase of this read pulse is not important in itself, the required signal is that which is in-phase with it. An alternative method is to reverse the ADRF process by first adiabatically switching on an rf field and subsequently suddenly switching it off to generate an FID which can be used to monitor relaxation. In addition to their intrinsic interest and their use in the study of slow motions, dipolar ordered states also find application in cross-polarisation experiments (Pines *et al.*, 1972). In practice all of the pulse sequences for the various spin-lattice relaxation measurements are used with some form of phase cycling to eliminate instrumental responses, spurious signals and so on. Details of some single resonance pulse sequences are given in table 4.1.

4.4.2 *Double-resonance experiments*

In the context of polymers in the solid state, double-resonance experiments are used in the majority of cases for obtaining ^{13}C NMR spectra and relaxation properties. The two main phenomena involved are dipolar decoupling (DD) and cross polarisation (CP). In this section some of the more important pulse sequences are outlined. The experiments described are generally carried out under MAS conditions.

4.4.2.1 *Single-pulse excitation (SPE)*

The pulse sequence illustrated in fig. 4.9 describes the simplest experiment used for obtaining the NMR spectra of spins such as ^{13}C in solids. The name and acronym are not an accurate description of the sequence except in so far as the ^{13}C signal is excited with a single, often 90°, pulse. It can be seen that the sequence comprises a single pulse applied at the ^{13}C frequency where the subsequent FID is acquired in the presence of the requisite level of irradiation at the ^{1}H resonance frequency to effect dipolar decoupling of the proton and carbon spins. This experiment is entirely equivalent to the one used to obtain similar spectra from solution, the difference being only

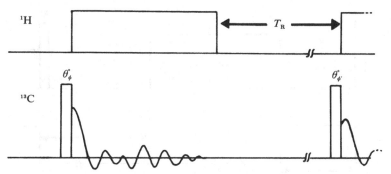

Figure 4.9. Sequence used to determine ^{13}C spectra under dipolar decoupling conditions.

in the amplitude of the decoupling field. Typically, this should be at least 40 kHz ($= \gamma B_1$) and preferably as high as 100 kHz or more if the residual dipolar coupling contribution to the linewidth is to be rendered negligible. This is particularly necessary for systems with larger secular dipolar couplings as found, for example, in the more rigid regions of polyolefins.

As with all pulsed NMR experiments, the final spectrum is obtained by Fourier transforming the FIDs accumulated from a large number of repetitions of the sequence with appropriate phase cycling to reduce artefacts. In this case, the recycle time, T_R, will be dictated by the spin–lattice relaxation times of the ^{13}C nuclei in the sample. If T_R is made much less than $3T_1$ then substantial loss of signal will occur. In some solid polymer systems, T_1 values can become very long, which, in turn, requires very large values of T_R to guarantee complete and fully representative spectra. For example, in the crystalline regions of polyethylenes, T_1 values can be several thousand seconds, which makes this approach to determining the ^{13}C spectrum unrealistic, except in very particular circumstances. However, because of the very different and shorter relaxation times of ^{13}C spins in the more mobile, disordered regions, the SPE experiment is extremely useful for selectively examining these regions.

4.4.2.2 Cross polarisation (CP)

The solution to the problem of long T_1 values which render the SPE experiment so inefficient, is to rely on the abundant ^{1}H spin system to provide the source of the ^{13}C signal. This process, usually called cross polarisation (CP), also has the advantage of giving rise to an increased sensitivity directly through an increased signal amplitude relative to the SPE sequence. A double benefit accrues: the CP experiment is repeatable on a timescale dictated by the ^{1}H spin–lattice relaxation properties which

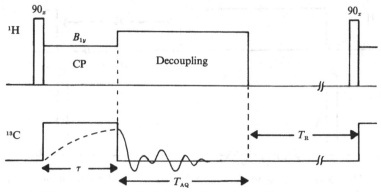

Figure 4.10. Basic cross-polarisation experiment under Hartmann–Hahn matching conditions $\gamma_H B_{1H} = \gamma_C B_{1C}$. The double irradiation makes the ^{1}H and ^{13}C spins behave as 'like' spins.

are usually much shorter than those of ^{13}C in the same system, and a larger signal (in the ratio of the resonance frequencies ^{1}H/^{13}C ≈ 4) is obtained. Cross polarisation requires two steps. Firstly, the ^{1}H spin system must be put into a state suitable to transfer magnetisation and, secondly, the two spin systems must be placed into efficient communication or contact to allow the polarisation exchange to occur.

Figure 4.10 illustrates the basic CP experiment. The ^{1}H spins are initially spin-locked which, as discussed in Chapter 2, effectively assigns them a very low spin temperature in the rotating frame: this suitably drives the polarisation transfer. Following this preparation step, the cross polarisation is brought about by switching on a resonant ^{13}C rf field with an amplitude satisfying the so-called Hartmann–Hahn (1962) matching condition, $\gamma_H B_{1H} = \gamma_C B_{1C}$. This simultaneous double-frequency irradiation of the ^{1}H and ^{13}C spins gives them the same effective precession frequency around their respective $\mathbf{B}_1$ fields. Under this condition, the two species of spin may exchange energy *via* their dipolar interactions. The double irradiation makes the ^{1}H and ^{13}C spins behave as 'like' spins in the sense that part of their dipolar coupling can, under these conditions, induce energy-conserving (in the doubly rotating frame) flip–flop transitions. Further details can be found in both original papers (Hartmann and Hahn, 1962; Pines *et al.*, 1972) and books (Mehring, 1976; Slichter, 1990).

The efficiency of the cross-polarisation process depends on both the ^{13}C/^{1}H and ^{1}H/^{1}H secular dipolar couplings (Mehring, 1976; Demco *et al.*, 1975). Typically, cross-relaxation times vary from a few hundred microseconds to a few milliseconds. Carbons in rigid systems which have

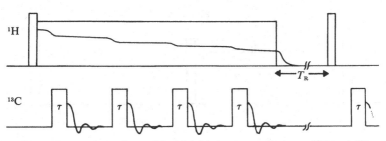

Figure 4.11. Sequence used to achieve multiple contact between ^{13}C and ^{1}H spins to maximise the transfer of ^{1}H polarisation.

protons directly bonded to them cross polarise faster than those which rely on dipolar couplings to non-bonded protons or those in systems where the dipolar contributions to the lineshapes are partially motionally averaged. When CP is carried out in the presence of MAS, as it most frequently is, it is important to note that the rate of CP can be affected by magic-angle spinning when its frequency becomes comparable to the ^{1}H/^{1}H secular dipolar linewidth (Stejskal *et al.*, 1977). Whilst this is not often a problem, the development of MAS spinners which can achieve spinning speeds of 10 kHz or greater makes it necessary to bear this effect in mind, particularly where the ^{1}H linewidth is reduced by anisotropic motions.

There are a number of modifications and variations to the basic CP sequence. First, one of the strong points of CP is that a single contact of ^{13}C with ^{1}H only transfers a small fraction of the available ^{1}H polarisation. It was thus recognised that a multiple contact experiment, in which several contacts were carried out for a single spin-locking, afforded the possibility of substantially increased sensitivity. This modification is illustrated in fig. 4.11. It should be appreciated that the ability to carry out this experiment successfully depends on $T_{1\rho}(^{1}$H) being sufficiently long. For the single-contact experiment, this requires only that $T_{1\rho}(^{1}$H) be longer than $5T_{CR}$, where T_{CR} denotes the cross-relaxation time. For the multiple-contact experiment, $T_{1\rho}$ must allow for both cross relaxation and data acquisition for the several cycles. Since, for high resolution, the acquisition time will usually need to be of the order of 100 ms or longer, this places rather stringent requirements on $T_{1\rho}(^{1}$H). For this reason, multiple-contact experiments have not found general application in polymer NMR but would be of value for the study of static samples of very rigid materials since these might be expected to have long $T_{1\rho}(^{1}$H) and would not require such long data acquisition times.

In the same spirit of improving the efficiency of the cross-polarisation process, the 'flip-back' procedure illustrated in fig. 4.12 was devised

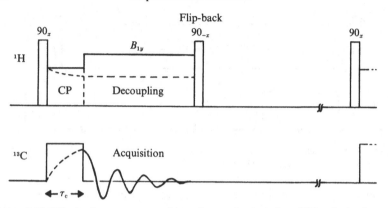

Figure 4.12. Flip-back sequence used to enhance the transfer of ^{1}H polarisation to the ^{13}C spins.

(Tegenfeldt and Haeberlen, 1979). It introduces a $90°_{-x}$ pulse to the ^{1}H sequence immediately following the end of the decoupling pulse which is still maintaining a spin-locked ^{1}H magnetisation. This $90°_{-x}$ flip-back pulse returns the remaining proton magnetisation to the $+Z(B_0)$ axis where it relaxes towards $\mathbf{M}_0$ for the duration of the recycle period, T_R. This sequence is particularly effective for systems with long $T_{1\rho}(^1\text{H})$ since the ^{1}H magnetisation at the time of the flip-back pulse is still close to M_0. What the flip-back sequence achieves is equivalent to a multiple-contact experiment but without the requirement for continuous ^{1}H irradiation. In the case of single-contact experiments, it allows a shorter time T_R to be used since the magnetisation is placed closer to its equilibrium value at the start of the recycle period. It is clear that when $T_{1\rho}(^1\text{H})$ is comparable to the time required for CP and data acquisition, there is no advantage in using the flip-back procedure. Since there is no disadvantage either, it is often used as a standard method.

4.4.2.3 Dipolar dephasing/non-quaternary suppression: NQS

It is always useful in high-resolution NMR to have methods which facilitate the assignment of lines in spectra in particular chemical sites in the molecular structure. One such experiment is the solid-state equivalent of the attached proton test and other similar experiments in liquid-state NMR (Harris, 1983a, b). The experiment for solids makes use of the strong distance dependence of the ^{13}C/^{1}H dipolar coupling and the fact that the ^{13}C NMR linewidth, in the absence of the proton dipolar decoupling, is larger for those ^{13}C spins with ^{1}H spins near to them, usually as directly bonded protons. The only difference from the standard CP sequence is

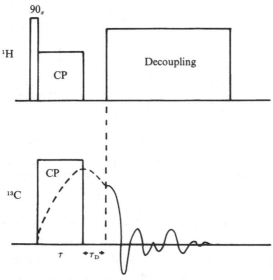

Figure 4.13. NQS sequence. During τ_D the ^{13}C magnetisation evolves in the absence of dipolar decoupling. The ^{13}C signal is recorded under $^{13}C-^1H$ dipolar decoupling conditions.

that, following the CP step, the dipolar decoupling is switched off for a time τ_D (fig. 4.13). The ^{13}C signal is then recorded with dipolar coupling in the usual way. The effect of this extra evolution time in the absence of proton decoupling is to allow the signal from those carbons close to protons with concomitant broad lines to dephase completely. The remaining signal then arises only from quaternary carbons having no nearby protons. The experiment is remarkably successful in distinguishing between the two sets of carbons. For $^{13}C/^1H$ systems a value of $\tau_D \simeq 40\,\mu s$ is usually sufficient to suppress protonated carbon signals completely. The rather clean distinction which this experiment is able to draw between these classes of protons may well depend on the $^1H/^1H$ flip–flop processes as these probably produce a degree of decoupling for ^{13}C spins with only relatively weak dipolar couplings to protons, thus affording them a narrower linewidth than expected on purely geometric grounds.

As the NQS experiment relies on differences in undecoupled linewidths, any circumstance which affects such linewidths such as methyl group motion may complicate the results. In most solids, methyl groups undergo rather facile rotation about their local C_3 axis and, at ambient temperatures, the rates of this motion normally are sufficient to average the $^{13}C/^1H$ and even the $^1H/^1H$ dipolar couplings. For the particular geometry

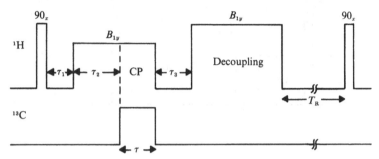

Figure 4.14. Sequence which incorporates extra delays, τ_1, τ_2, and τ_3 into the normal CP experiment. These delays allow discrimination or association of ^{13}C signals with particular ^{1}H relaxation processes.

of the methyl group, the internuclear C–H vector makes an angle close to the magic angle to the rotation axis with the result that the averaged ^{13}C/^{1}H dipolar coupling is small. This leads to methyl carbons behaving rather like quaternaries in the NQS experiment. Any motion which similarly reduces the ^{13}C linewidth arising from proton dipolar coupling will cause the relevant ^{13}C lines to require a longer time to dephase under the NQS experiment. This should be borne in mind when using the method, particularly in polymer systems where various degrees of molecular mobility may be present in a sample at a particular temperature (cf Section 1.4).

4.4.2.4 Delayed contact and related experiments

The NQS experiment described in the previous section is an example of a general class of experiment which discriminates between signals in the ^{13}C high-resolution NMR spectrum by exploiting distinguishing NMR properties. For NQS it is the ^{13}C linewidth in the absence of ^{1}H decoupling which provides the means of discrimination. Figure 4.14 illustrates a general pulse sequence incorporating other possibilities. It is a normal CP experiment except that some extra delays or evolution periods have been introduced. The first of these is a delay, τ_1, between the ^{1}H 90° pulse and the application of the spin-locking pulse. The τ_2 is a delay between the establishment of spin-locking of the protons and the application of the ^{13}C CP contact irradiation. The third is the gating-off of the ^{1}H decoupler for the NQS experiment, τ_3 ($= \tau_D$). The extra delays allow discrimination or association of ^{13}C signals with particular ^{1}H relaxation processes. In that sense they, like the NQS experiment, are equivalent to two-dimensional NMR sequences, although they would not, in general, be used in that

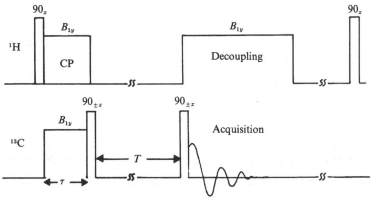

Figure 4.15. Sequence devised by Torchia (1978) to determine $T_1(^{13}C)$ from spin–lattice relaxation only during CP.

fashion. The first delay results in the spin-locking of ^1H magnetisation with a T_2 comparable to or longer than the chosen delay, τ_1. If there were regions of the material which had significantly different proton concentrations or local mobilities, then CP from this delayed spin-locking would give signals from carbons associated predominantly with the more mobile or dilute regions. The second delay, τ_2, allows discrimination based on differences in proton rotating frame spin–lattice relaxation. Again, CP from the resulting, partially relaxed, ^1H magnetisation may associate particular signals in the ^{13}C spectrum with regions having these different relaxation properties. It should be noted that ^1H spin diffusion may complicate the interpretation of such experiments (cf Section 3.6).

4.4.2.5 Dilute-spin spin–lattice relaxation

The principles of spin–lattice relaxation for magnetically dilute spins such as ^{13}C are the same as for any other spins (cf Section 2.7.2). However, there are some practical features which are of consequence. To measure laboratory frame spin–lattice relaxation requires the production of a non-equilibrium value of M_z. In the absence of CP, the only difference from the general descriptions given in Section 4.4.1.4 would be the addition of either or both dipolar decoupling and MAS. Indeed, such measurements of $T_1(^{13}C)$ values would be the natural choice for signals best studied by the SPE experiment (Section 4.4.2.1). However, the use of CP in the measurement of T_1, both for sensitivity enhancement and discrimination purposes, requires a rather different pulse sequence, first described by Torchia (1978) (fig. 4.15). There are two features to note. Firstly, a $90°_\pm$ ^{13}C pulse rotates the transverse ^{13}C magnetisation produced by CP

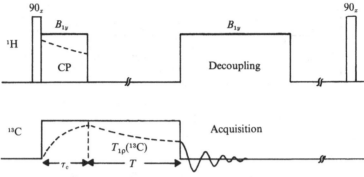

Figure 4.16. Sequence used to determine $T_{1\rho}(^{13}C)$.

into either the $+Z$ or $-Z$ direction. This alternation of orientation is accompanied by alternative addition and subtraction of the ^{13}C signals following the appropriate period of spin–lattice relaxation during which the ^1H dipolar decoupling field is off. This procedure results in the accumulation of signals arising solely from the spin–lattice relaxation of magnetisation from the CP process.

The measurement of rotating frame spin–lattice relaxation of dilute spins such as ^{13}C is more straightforward. As with laboratory frame relaxation, the normal spin-locking sequence may be used for the rotating frame measurements with ^1H dipolar decoupling and/or MAS as required. If CP is used to generate the ^{13}C magnetisation, the rotating frame relaxation is observed by switching off the ^1H rf field, leaving the ^{13}C rf field on for the relaxation period and then measuring the ^{13}C signals with the decoupler switched on again (fig. 4.16). Although the measurement is simple, the interpretation of dilute-spin $T_{1\rho}$ behaviour in the presence of abundant spins such as protons is not straightforward and this is discussed more fully in Chapter 5.

4.4.2.6 Factors affecting resolution

High-resolution ^{13}C spectra of solid polymers can vary considerably in the resolution that appears to be achieved. To the polymer scientist or spectroscopist, poor signal-to-noise ratios arising from broad lines can obscure important information because of the consequent low resolution and can lead to disappointment. However, it is important to realise that, in most cirumstances, such characteristics reflect properties of the polymer rather than intrinsic limitations of the NMR techniques. This section briefly examines the main factors that affect linewidths, and hence resolution, in ^{13}C high-resolution NMR of solid samples. The reader who is interested in a more detailed discussion should consult the paper by Garroway and coworkers (1981). For convenience we consider these factors under two headings.

Instrumental Factors. The two most important considerations, which are largely within the control of the spectroscopist, are the accuracy in setting the magic angle for MAS and the amplitude and centre frequency of the ^{1}H decoupling field.

In simple terms, the spectral distribution arising in a powder sample from anisotropic spin interactions such as the magnetic shielding or dipole–dipole couplings, is scaled under fast rotation about an axis making an angle θ with respect to the field $\mathbf{B}_0$, by the factor $\frac{1}{2}(3\cos^2\theta - 1)$ which of course is the basis of MAS. If θ deviates from $\theta_m = 54.7°$, then fast sample spinning will result in a residual linewidth. If δ is the error, that is, $\theta = (\theta_m \pm \delta)$, then for small δ the residual width is approximately $\delta\Delta$, where Δ is the extent of the full powder pattern. Thus, for a chemical shift anisotropy of 100 ppm, δ must be $\leqslant 0.01$ rad ($< 0.5°$) to ensure that the residual linewidth arising from this interaction is less than 1 ppm. The larger the anisotropic interaction being averaged, the smaller δ must be to maintain a specified resolution. Commercial MAS probes normally allow for adjustment of θ around θ_m. In general, it must be possible to make this adjustment easily and then to know that subsequent manipulations of probe, sample and so on, do not change this setting. Double-bearing MAS systems are mechanically quite stable and require little attention once adjusted to θ_m. Earlier designs of spinner, based on a single gas bearing, are more susceptible to error and usually need to be reset for each sample. To this end, the use of the ^{79}Br resonance in powdered KBr was proposed as a useful and convenient standard and this is often added to samples when regular checking of θ_m is important (Frye and Maciel, 1982).

The effectiveness of the irradiation at frequencies close to the ^{1}H spectrum for decoupling dipolar ^{1}H–^{13}C interactions depends on its amplitude, $B_1(^1\mathrm{H})$ ($= \omega_1(^1\mathrm{H})/\gamma(^1\mathrm{H})$) and its frequency, $\omega_0(^1\mathrm{H})$, with respect to the proton spectral frequencies. This problem is discussed in detail elsewhere (Mehring, 1976; Takegoshi and McDowell, 1986). In short, when $\omega_1(^1\mathrm{H}) \gg [M_2(^1\mathrm{H}\text{–}^1\mathrm{H})]^{\frac{1}{2}}$ and $[M_2(^1\mathrm{H}\text{–}^{13}\mathrm{C})]^{\frac{1}{2}}$, the residual linewidth in the ^{13}C spectrum arising from the ^{13}C–^{1}H dipolar coupling is predicted and observed to be dependent on $\Delta\omega^2$, where $\Delta\omega$ is the resonance offset of the particular ^{1}H signal from the irradiation frequency $\omega_0(^1\mathrm{H})$. This result, rather surprising at first sight, indicates that the efficiency of decoupling depends on the precise frequency location of the decoupling field with respect to the resonance frequency of the ^{1}H spin in the absence of the dipolar interactions. Thus, despite a dipolar-dominated ^{1}H linewidth measured in tens of kilohertz, the precise placing of the frequency $\omega_0(^1$H) within the range of chemical shifts of the ^{1}H spins may influence the effectiveness of the decoupling. For example, using an $\omega_1(^1$H)/$2\pi \approx$ 40 kHz, the CPMAS linewidth of the ^{13}C signal from the crystalline region of high-density polyethylene can be increased by a factor of two on shifting $\omega_0(^1$H) from that characteristic of aliphatic protons to that for liquid water. Such effects must be considered if the ultimate in resolution is required. They are all the more significant the larger the ^{1}H–^{1}H and ^{1}H–^{13}C dipolar couplings relative to $\omega_1(^1$H). Highly crystalline polyethylene is probably one of the most testing cases for this effect. Most other polymers will be less demanding.

Sample Related Factors. For a highly crystalline solid containing only spin-$\frac{1}{2}$ nuclei (for example, ^{13}C and ^{1}H) for which all the anisotropic spin couplings are in the rigid-lattice limit – that is, no motions are present of sufficient rate to effect averaging of any part of the interactions – the resolution achievable in the dilute-spin (^{13}C) spectrum can be of the order of 1 Hz or better if a spectrometer having the requisite characteristics is used. This is rarely seen in practice because these requirements are not all met in most solid samples: there may be molecular motions occurring which can influence the achievable resolution, the sample may be disordered, or quadrupolar spins may be present for which the theoretical basis of the high-resolution techniques of dipolar decoupling and MAS may not be valid. All of these effects can be encountered in polymeric solids as discussed briefly below.

The decoupling of ^{1}H and ^{13}C spins in a solid by strong irradiation around $\omega_0(^1$H) and averaging of the ^{13}C shielding anisotropy by MAS both work by the external imposition of a time dependence on the relevant anisotropic interactions (cf Chapter 2). Thermally activated motions of the molecules containing the spins also introduce such a time dependence. Of course, the limit of fast isotropic internal motions achieves precisely the resolution that DD and MAS are designed to produce. The spins cannot discriminate time dependence generated externally from that arising from thermally driven, internal motions. When both are present the effectiveness of the externally applied schemes for achieving line narrowing may be reduced. This reduction is greatest when the internal and external time dependencies have similar timescales, that is, when $\tau_c^{-1} \simeq \omega_1(^1$H) and ω_{MAS}. For $\omega_1(^1$H)$\tau_c \simeq 1$, the dipolar decoupling becomes much less effective and

^{13}C lines broaden as the dipolar linewidth is reintroduced. Likewise, when $\omega_{MAS}\tau_c \simeq 1$, the chemical shift anisotropy will not be properly averaged, again leading to broader lines. Since, typically, $\omega_1(^1\text{H})/2\pi \simeq 50$–$100$ kHz and $\omega_{MAS}/2\pi \simeq 2$–$10$ kHz, molecular motions for which $\tau_c \simeq 10^{-4}$–10^{-6} s may result in reduced resolution and can even lead to an apparent loss of particular signals if the motions are highly site-specific.

The methyl group is such an example, as illustrated in the CPMAS ^{13}C spectrum of isotactic polypropene as a function of temperature (Fyfe *et al.*, 1982; Cudby *et al.*, 1985). At moderate resolution, the spectrum at 300 K shows three lines of equal intensity (area) assigned to methylene, methine and methyl carbons. As the temperature is lowered, methylene and methine lines show little change, whereas the methyl resonance broadens until, at 105 K, it is no longer visible under the prevailing experimental conditions. At still lower temperatures it reappears, sharpening as the temperature is taken even lower. These effects occur because the methyl group undergoes thermally activated rotation about its threefold axis, as is well known from studies of ^{1}H spin–lattice relaxation in such materials. At room temperature the rate of this motion is much faster than the decoupling field expressed as a frequency. When the temperature is lowered, the methyl hopping rate approaches this frequency and the dipolar decoupling is reduced in its effectiveness until the maximum broadening is reintroduced when the rates match. From the ^{1}H spin–lattice relaxation and associated theory, the maximum in the broadening of the ^{13}C methyl resonance was calculated to occur around 109 K, which is very close to the temperature observed.

With a well-annealed sample, the room-temperature spectrum shows 2:1 splittings of each of the three peaks. These splittings originate in the crystal packing symmetry (cf Section 5.2). Figure 4.17 shows the temperature variation of the spectrum under high-resolution conditions (Cudby *et al.*, 1985). Careful examination of the methyl resonances, the only ones to broaden as the temperature is lowered, shows that the resonance of relative intensity 2 broadens at higher temperatures than that with intensity 1. This is consistent with the assignment of these lines in terms of the crystallographic structure discussed in Section 5.2.

Any motions (for example, phenyl ring flips, alkyl chain rotations, and so on) can affect ^{13}C linewidths through these interference effects and hence limit the resolution attainable. It is important to bear these eventualities in mind when interpreting spectra taken at any one set of experimental conditions of temperature, $\omega_1(^1\text{H})$ and ω_{MAS}, particularly if motions in the relevant frequency ranges are a possibility. Theory (Suwelack *et al.*, 1980;

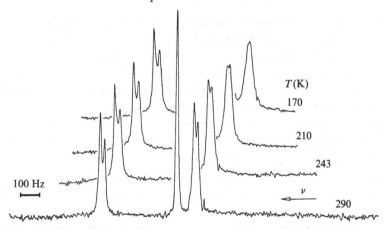

Figure 4.17. Temperature variation of the ^{13}C spectrum of isotactic polypropene (Cudby *et al.*, *Polymer*, **26**, 169 (1985) by permission of the publishers, Butterworth Heinemann Ltd ©.)

Rothwell and Waugh, 1981) predicts characteristic dependencies of the ^{13}C linewidth on $\omega_1(^1H)$ or ω_{MAS} depending on whether $\omega_1(^1H)\tau_c$ or $\omega_{MAS}\tau_c$ is greater or less than unity. Variation of $\omega_1(^1H)$, ω_{MAS} or temperature (τ_c) can assist in identifying these effects.

Another sample-related factor which can affect resolution is the presence in the molecule of quadrupolar spins ($I > \frac{1}{2}$) such as ^{14}N, ^{23}Na and $^{35/37}Cl$. These are present, for example, in polymers such as nylons, polyimides, PVCs and ionomers. Indeed, most of these effects on resolution are not specific to polymers. The behaviour of quadrupolar spins in solids and their impact on high-resolution CPMAS experiments on spin-$\frac{1}{2}$ nuclei were first observed and explained for small molecules (Groombridge *et al.*, 1980; Frey and Opella, 1980; Harris *et al.*, 1985). The effect on the resolution achievable in the ^{13}C spectrum arises from the dipolar coupling between the ^{13}C and quadrupolar spins. Such a heteronuclear dipolar coupling behaves like an additional shift anisotropy for the ^{13}C spin and, if the other nucleus was spin-$\frac{1}{2}$, then MAS would average it as long as $\omega_{MAS} > \omega_{IS}$, where ω_{IS} is the strength of the heteronuclear dipolar coupling. However, the quadrupole coupling (cf Section 2.6.1.3), for a spin $I > \frac{1}{2}$ at a site of less than cubic symmetry, means that its axis of quantisation is not $\mathbf{B}_0(z)$ but is determined by the balance between the Zeeman (ω_0) and quadrupole (ω_Q) interactions. The larger (ω_Q/ω_0) is, the more the quadrupolar interaction, characterised by the direction and symmetry of the local electric field gradient, dominates. For a powder sample there is, thus, a distribution of quantisation directions for the quadrupolar spins.

The spin-$\frac{1}{2}$ ^{13}C nucleus, to which they are dipolar coupled, however, is quantised along $\mathbf{B}_0(z)$ and the ability of MAS to average the contribution of this heteronuclear dipolar coupling to the ^{13}C spectrum is reduced as it requires a common quantisation axis for both spins to be effective. The effect of the presence of quadrupolar nuclei is to leave a residual powder lineshape imposed on the ^{13}C resonances.

The size of the effect depends, *inter alia*, on the magnitude of the dipolar couplings between the quadrupolar and spin-$\frac{1}{2}$ spins, the ratio ω_Q/ω_0 and the orientation of the internuclear vector with respect to the principal axis of the EFG tensor for the quadrupolar spin. For ^{14}N, typically, ω_Q may range up to a few megahertz and for $\omega_0(^1\text{H}) \simeq 200$ MHz, the residual effects in ^{13}C spectra are usually smaller than other linewidth contributions in solid polymers. For $^{35/37}$Cl, on the other hand, linewidths of ^{13}C in close proximity to the Cl spins can be about 10 ppm at $\omega_0(^1\text{H}) = 90$ MHz (Komoroski, 1983). Realise, too, that the ^{13}C-quadrupolar spin dipolar coupling does not require the spins to be directly bonded. The higher the concentration of quadrupolar spins, the more general the effects on resolution become (Harris *et al.*, 1985).

Specifically in the case of polymers, the presence of distributions of conformational states, interchain packing densities and so on, such as would be expected in glassy disordered or cross-linked systems, can lead to significant linewidth contributions. Typically, in cured epoxy resins, linewidths may range up to about 10 ppm with much of this width attributable to inhomogeneous broadening arising, for example, from a distribution of isotropic chemical shifts (Garroway *et al.*, 1979).

4.4.2.7 *Factors affecting quantitative measurements*

NMR in the liquid state is frequently used in a quantitative manner, for which certain experimental conditions have to be met. In the same way, quantitative investigations in the solid state are obviously desirable. In this section we look at some of the factors influencing quantitative measurement in high-resolution NMR of solid polymers.

The simplest experiment to consider is SPE (Section 4.4.2.1). In this, a single pulse is applied to excite the ^{13}C signal which is recorded in the presence of DD and MAS. This is equivalent to the gated decoupling sequence used in liquid-state NMR (Harris, 1983a) where the factors affecting quantitative measurements are essentially the same. The recycle time, T_R, between successive applications of the exciting pulse must be long enough to allow all sites for which quantitative information is sought to return to equilibrium within the required limits. This time is influenced by

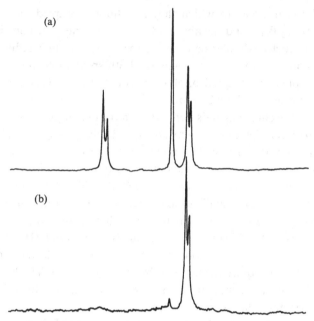

Figure 4.18. Effect of the cycle time T_R on the ^{13}C spectrum of solid isotactic polypropene. (a) a CPMAS spectrum; (b) an SPE spectrum with $T_R = 1$ s.

the $T_1(^{13}\text{C})$ values for the relevant carbon sites, the flip angle of the exciting pulse and, possibly, the effects of transient nuclear Overhauser effects (NOE) produced by the DD irradiation (Harris, 1983b). The latter is not normally significant for, in solids, the NOE is usually small, as is the period of irradiation since acquisition times for FIDs in solids are less than in liquids. Earl and VanderHart (1979) have modified the simple SPE sequence to eliminate possible NOE effects. The difficulty with the SPE and related sequences is that, in many rigid glassy or crystalline polymers in which motions in the megahertz region are severely limited, $T_1(^{13}\text{C})$ values are excessively long for their sensible application to quantitative measurements: $T_1(^{13}\text{C})$ values of the order of 10^2–10^4 s are not unusual. In highly crystalline polyethylene, $T_1(^{13}\text{C})$ for the crystalline region can be several thousand seconds which would imply T_R values of the order of hours, which is clearly impractical. An interesting example in which the local character of $T_1(^{13}\text{C})$ behaviour is apparent is the SPE-determined ^{13}C spectrum of solid isotactic polypropene (fig. 4.18), where it can be seen that with the T_R value used, only the methyl carbon gives a significant signal.

Dipolar decoupling (DD) exerts no specific effects on quantitation but MAS may do. These effects are applicable to both SPE and CP experiments and arise when ω_{MAS} is not large enough to reduce to a negligible amplitude the spinning sidebands arising from ^{13}C sites with the largest anisotropies, such as carbonyl and aromatic carbons. Under these conditions, apart from the possible confusion between spinning sidebands with isotropic peaks, the intensities of centrebands for sites with differing anisotropies will not reflect a quantitative relationship, as spinning maintains a constant second moment for the overall pattern of centre and sidebands. Methods such as total sideband suppression (TOSS) (Dixon, 1982; Hagemeyer *et al.*, 1991), whilst returning intensity from sidebands to centrebands, are not readily rendered quantitative. For truly quantitative applications, fast MAS is advisable, with sideband plus centreband integration a second best, providing sidebands can be unambiguously identified and assigned to their corresponding centrebands.

A major difference between the liquid and solid states is the routine use of cross polarisation in solids to generate the ^{13}C signals. Use of experiments involving CP for quantitative studies requires careful evaluation of a number of factors. The spin dynamics of CP between an abundant ^1H spin system and a dilute ^{13}C system has been dealt with in great detail elsewhere (Mehring, 1976; Garroway *et al.*, 1979). We merely note that during the spin-lock CP process the ^{13}C signal intensity may be taken as given by eq 2.44 in which the equilibrium ^{13}C signal, S_0, is achievable only when $T_{CH}^{(SL)} \ll T_{1\rho}(^{13}C/^1H)$, $\lambda \simeq 1$ and $\tau > 5T_{CH}^{(SL)}$, with a magnitude given by

$$S_0 = [\gamma(^1H)/\gamma(^{13}C)]\, s_0 \tag{4.1}$$

where s_0 is the ^{13}C signal which would be obtained from a quantitative SPE experiment using a ^{13}C 90° pulse, and τ is the contact time under the spin-lock, Hartmann–Hahn matched conditions. In the limit $T_{CH}^{(SL)} \ll T_1(^1H/^{13}C)$ for all carbon sites in a solid, when $\tau > 5T_{CH}^{max}$ (the longest T_{CH} in the sample), the spectral intensities will be quantitative. The CP signal gain, which should be $\gamma(^1H)/\gamma(^{13}C) \simeq 4$, is not always achieved even under conditions which meet the above limit. Values greater than 3.5 can be routinely obtained, however, but the need for calibration is clear.

Typical values of T_{CH} are 50–100 μs for ^{13}C spins directly bonded to ^1H spins, with values up to 500 μs and longer for carbonyl and other non-proton-bonded carbons. The latter rely on non-bonded protons for their polarisation transfer. Based on this range of T_{CH} values, contact times used for typical organic polymer solids will be 1–2 ms. Whether or not these generate quantitative spectra, however, depends crucially on the pertinent magnitudes, particularly of $T_{1\rho}(^1H)$. For proton densities typical of organic solids, $T_{1\rho}(^1H)$ can be as small as 10^{-4} s if there are motions modulating the $^1H-^1H$ dipolar couplings with substantial frequency components at $\omega_1(^1H)$. Under these conditions, the maximum in $S(\tau)$ occurs at small values of τ and is only a fraction of the full enhanced equilibrium value S_0. In any situation for which quantitative information on signal intensities is required and for which the limiting conditions $T_{CH} \ll T_{1\rho}(^{13}C/^1H)$ are not valid, a determination of $S(\tau)$ as a function of τ must be carried out and the $S(\tau)$ for each site fitted to the relevant expression to yield, *inter alia*, values of S_0 for each signal. Many polymer systems require such treatment if quantitative data is the aim. Figures 5.16 and 6.24 portray the typical dependence of $S(\tau)$ on τ for solid polymers.

A particular problem which must be considered for polymers is that they are often spatially heterogeneous solids, the simplest example being the existence of crystalline and disordered regions (cf Section 1.3). These regions usually have different values for the relevant spin–lattice relaxation times and hence exhibit different behaviour under SPE and CP experiments. Thus, for example, the maximum CP signal obtainable at room temperature for a well-annealed sample of isotactic polypropene was found to represent about 80 % of the total carbon in the sample (Aujla *et al.*, 1982). The missing signal was associated largely with the disordered regions for which $T_{1\rho}(^1H) \simeq 0.5$ ms and for which T_{CH} is lengthened relative to that for the crystalline region because of motional averaging of the dipolar interactions.

4.5 Multidimensional NMR in solids

The power of NMR spectroscopy in liquids has been much enhanced by the development of two-dimensional experiments. This same development has been extended to the study of solids including polymers (Ernst *et al.*, 1987; Blümich and Spiess, 1988; Spiess, 1991). In this section we introduce the concepts involved in these experiments which, although highly specialised in nature, can often provide definitive answers to problems of structure and motion in solids which the simpler one-dimensional experiments cannot address.

4.5.1 Two-dimensional NMR in solids

A one-dimensional FTNMR experiment generates a signal which is a function of a single time variable. In the simplest case – excitation with just a single 90° pulse – the FID denoted $S(t)$ is Fourier transformed to give the

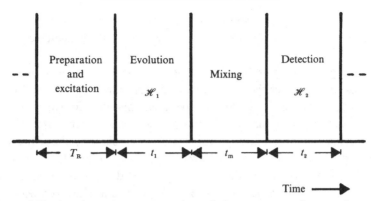

Figure 4.19. A schematic representation of the sequence of events in a two-dimensional FTNMR experiment. The cycle shown is repeated for different values of t_1. The NMR signal is acquired during the time t_2.

unperturbed spectrum $I(\Omega)$. Using the spin physics we have already discussed (for example, MAS, decoupling, multiple-pulse experiments and so on) it is possible to modify the evolution of $S(t)$, and hence the spectrum, to yield selected information about the interactions experienced by the spins. Two-dimensional FTNMR takes this a stage further and introduces two separate time variables within the same NMR experiment, generating a signal $S(t_1, t_2)$ which is a function of both. Fourier transformation with respect to both time variables generates a two-dimensional spectrum $I(\Omega_1, \Omega_2)$. The essence of such a spectrum is that it allows the interactions determining the evolution of the spin system in the time t_1 to be correlated with those operating in time t_2. Again, using the plethora of NMR tricks available, these can be chosen and manipulated to yield the desired information.

Figure 4.19 illustrates the general form of a two-dimensional NMR experiment. The initial preparation period includes the recycle time to ensure that the spin system starts out at thermal equilibrium together with the initial excitation, involving, for example, a single 90° pulse, cross polarisation and so on. The system is then allowed to undergo evolution for time t_1 under the influence of the Hamiltonian $\mathcal{H}_1$. As already indicated, the nature of $\mathcal{H}_1$ can be chosen through application of appropriate pulses: we might, for example, apply a multiple-pulse homonuclear decoupling sequence to the protons in a $^1H/^{13}C$ system in a solid which would allow ^{13}C transverse magnetisation to evolve under the influence of the $^{13}C-^1H$ dipolar coupling, unaffected by $^1H-^1H$ dipolar couplings. We might also ensure that the evolution of this same magnetisation arising from ^{13}C chemical shielding during t_1 is zero through use of echo formation. The mixing period, t_m, is used principally in experiments where dynamics are the goal. The role of t_m will be revisited below but, for the present, consider $t_m = 0$.

Following evolution under $\mathscr{H}_1$ for time t_1, the system is now allowed to evolve for time t_2 under the Hamiltonian $\mathscr{H}_2$. It is during t_2 that a full FID is recorded, as in a one-dimensional experiment, and therefore at $t_2 = 0$ we must have transverse magnetisation for the spins under observation. Again, $\mathscr{H}_2$ may be chosen through application of the chosen pulse sequences (MAS and so on). Following the example above of a solid ^{1}H/^{13}C system, we might choose to record this ^{13}C FID with ^{1}H dipolar decoupling and MAS which will give the high-resolution ^{13}C spectrum. However, because of the evolution in the t_1 time period, each peak in this ^{13}C spectrum will be affected in its phase by the precession caused by the heteronuclear dipolar coupling operating during t_1. In this way, the heteronuclear dipolar coupling is correlated with the chemical shift interaction of the ^{13}C spins from which the dipolar coupling strength between the different ^{13}C spins and their closest protons can be deduced. This can yield direct information on either bond lengths or motions that are sufficiently fast to cause some averaging of these dipolar couplings (Schaefer *et al.*, 1981). To generate the complete two-dimensional spectrum, this sequence must be repeated for an appropriate number and range of values of t_1, which is exceedingly time-consuming. Having generated the matrix $S(t_1, t_2)$, a double Fourier transform yields the two-dimensional spectrum $I(\Omega_1, \Omega_2)$ which can be presented as a contour plot or as a relief map.

To complete this brief introduction to two-dimensional NMR, we return to the role of the mixing time, t_m. If spins can exchange between different frequencies in their spectrum then such processes may be studied in detail using two-dimensional NMR in which a mixing time is employed. This is particularly valuable when the exchange process is not fast enough to affect the spectrum directly but is faster than typical spin–lattice relaxation times. To illustrate this, consider the simple two-line spectrum and pulse sequence shown in fig. 4.20. During the time t_1, each component of the transverse magnetisation develops a phase shift proportional to its relative precession frequency. The second pulse of the sequence stores a component of each magnetisation, with amplitude $\cos \Omega_i t_1$ ($i = $ A, B), along the $Z(B_0)$ axis. Following the mixing period t_m, the third pulse returns these components to the transverse plane where they continue their precession. The appearance of a two-dimensional spectrum from such a system will depend on whether there is a process which can exchange spins between the two frequencies on the timescale t_m, the upper limit of which is set by the

(a)

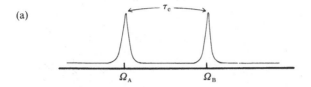

(b)

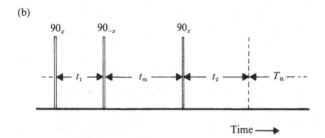

(c)   (d)

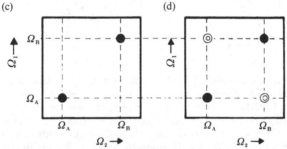

Figure 4.20. A simple two-dimensional FTNMR experiment employing a mixing time, t_m, to investigate spin exchange: (a) a two-line spectrum with exchange time τ_e; (b) three-pulse two-dimensional NMR experiment generating $S(t_1, t_2) t_m$; (c) two-dimensional NMR spectrum for $t_m \ll t_e$; and (d) the creation of cross peaks (open circles) arising from spin exchange when $t_m > \tau_e$.

spin–lattice relaxation timescale. If there is no exchange on this timescale then the two-dimensional spectrum is uninteresting, comprising only two peaks on the diagonal of the two-dimensional plot, that is, only at coordinates (Ω_A, Ω_A) and (Ω_B, Ω_B). Exchange during t_m, however, means that some spins which precessed at Ω_A during t_1 will be precessing at Ω_B during t_2 and vice-versa. This leads to the appearance of cross peaks in the two-dimensional spectrum at coordinates (Ω_A, Ω_B) and (Ω_B, Ω_A), typically as portrayed in fig. 6.16.

This type of experiment affords a particularly graphic demonstration of

Isotropic sample

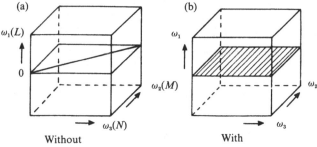

Ordered sample

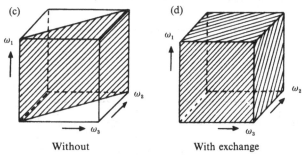

Figure 4.21. ^{13}C three-dimensional MAS-NMR correlating molecular structure, order and dynamics. The schematic representation of the general features of the three-dimensional CORD spectra are as indicated: (a) powder sample without exchange; (b) powder sample with exchange; (c) ordered sample without exchange; and (d) ordered sample with exchange. The orders of the sidebands in ω_1, ω_2, and ω_3 are labelled by integers L, M, and N, respectively. (Reprinted with permission from Spiess (1991). Copyright (1991) American Chemical Society.)

such exchange processes and can be used to follow them in considerable detail, often yielding unambiguous information on the precise mechanism of the motional process. The exchange processes which can be investigated include both direct physical exchange of atoms containing spins and the exchange of magnetisation through effects such as spin diffusion.

4.5.2 Three-dimensional NMR in solids

Addition of a further evolution time and mixing time to the two-dimensional experiment portrayed in fig. 4.20 permits the acquisition of three-dimensional spectra, as described by Spiess (1991). This allows the trajectory of a molecule undergoing complex motion to be sampled at least three times which greatly eases interpretation of complex molecular dynamics. In this way, for example, the motions responsible for α-relaxation in PVF_2 (cf Section 5.4.5) and at the glass transition in amorphous polymers (Schmidt-Rohr and Spiess, 1991a, b) have been clarified.

Three-dimensional NMR can also correlate molecular dynamics or spin diffusion with structure and order using a rotor-synchronised three-dimensional approach (Yang *et al.*, 1990). The experiment which has been termed CORD (correlation of order and dynamics) together with the type of information which it provides, are illustrated in fig. 4.21. For an isotropic sample, the NMR signal remains unmodulated during t_1 and the sidebands are all in the horizontal plane ($\omega_1 = 0$); in the absence of exchange (short t_m) the spectrum becomes diagonal. When the sample is ordered and there is no interchange of NMR frequencies, for whatever reason, during t_m, the signals are in the cross-diagonal plane ($\omega_2 = \omega_3$), referred to as the structure and order plane. This is the analogue of the two-dimensional rotor-synchronised experiment. When exchange occurs during t_m, signals are no longer confined to planes but sample the whole cube. This can only occur when the residual NMR signal is due both to order and dynamics. This correlated information gives essential geo-metrical insight into complex molecular dynamic processes.

5

Structure and motion in solid polymers

5.1 Introduction

Application of NMR principles and experiments to solid polymers described thus far will now be illustrated in this and subsequent chapters. It is not intended to present comprehensive reviews since these appear in the literature on a regular basis and will be referred to as appropriate. Rather, the intention is to cover the main features of the role of NMR in illuminating relevant polymer properties in order to give the reader a clear view of the possibilities and diversity of options. The choices of how to organise the presentation of information are many. Here the interests of the polymer scientist rather than of the NMR specialist are emphasised.

These interests will vary considerably. Those who primarily wish to synthesise new polymers will want to know what polymer has been made, largely in terms of its chemistry. Those more interested in polymer properties and processing will want to use NMR to develop further their understanding of the relationship of mechanical and other properties to chain organisation, orientation, dynamics and so on. NMR of polymer solutions and melts has been of great utility, particularly since the introduction of pulsed Fourier transform methods and the routine availability of high-resolution ^{13}C spectra. Such liquid-state high-resolution spectra have given the synthetic polymer chemist greater insight into the effects of synthesis conditions on microstructure, molecular weight distributions and so forth. However, many polymers are not readily soluble or melted. More importantly, many of the key properties are intrinsic to the as-produced, processed solid and it is in this area that the developments of the last decade and a half in the NMR spectroscopy of solids have had major impact. Of these, access to high-resolution spectra

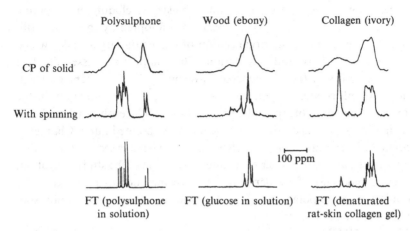

Figure 5.1. The first published CPMAS high-resolution ^{13}C NMR spectra of solid polymers and some relevant solution spectra for comparison. All spectra of solids were obtained using cross polarisation from ^{1}H and dipolar decoupling. The effects on resolution of MAS are clearly demonstrated by comparison of the top and middle rows. Frequency increases to the left, as is usual. (Reprinted with permission from Schaefer and Stejskal (1976). Copyright (1976) American Chemical Society.)

on a routine basis (predominantly ^{13}C but also ^{15}N, ^{29}Si and ^{31}P) and the development of ^{2}H NMR in solids are the most significant. Commercial spectrometers have developed apace in regard to their overall flexibility and sensitivity.

Before moving to the main theme of this chapter it is worth considering what exactly is implied by the term 'solid' in the context of NMR. Mechanically, polymers span a wide range of properties from high-modulus, rigid solids to rubbery semi-liquids. From an NMR viewpoint, the mechanical classification as a solid is only relevant in so far as it is reflected in the nature and time-dependence of the local magnetic and (for quadrupolar nuclei) electric fields experienced by the spins. As we have seen in previous chapters, it is these which determine the NMR spectral and relaxation properties. It is important, therefore, to regard NMR as a means of probing these local fields irrespective of whether we call the material a solid or not. For example, liquid-crystalline materials may appear liquid-like mechanically but may require a solid-state NMR approach to elicit the information sought.

The two topics addressed in this chapter, namely, *structure* and *motion*, require some comment to place them in context. Structure for solid polymers comprises (i) primary chain chemistry (including cross-linking chemistry, if appropriate), (ii) conformations adopted by the chains, (iii) chain packing and (iv) the organisation of chains on larger scales which includes morphology and orientation. This chapter addresses the first three; the topics in (iv) are covered in detail in the two chapters which follow. As information about the first three topics resides mainly in the determination of isotropic chemical shifts, high-resolution methods, principally ^{13}C, will accordingly be the focus. As pointed out in Chapter 1, motion and structure are intimately linked since what is seen as a structure may often be an average over some internal motions. The study of motion, however, makes use of a wider range of NMR methodologies based on the modulation of dipolar, quadrupolar and shielding anisotropic spin interactions, as described later.

The first NMR investigations of polymeric solids, in common with other materials, exploited ^{1}H NMR. However, the lack of chemically resolved information, arising from the usually dominant contributions of the $^{1}H/^{1}H$ dipolar interactions to the linewidths, meant that its use was restricted. This constraint was reinforced by the need for both specialised equipment and NMR knowledge. As indicated above, it was routine access to high-resolution ^{13}C NMR spectra of solids that fuelled the sustained growth of new applications in the field. As many polymer scientists may well encounter this aspect of NMR first, whether in studies of solutions or solids, we shall start with an overview of this area.

5.2 High-resolution ^{13}C NMR spectroscopy of solid polymers

The first demonstration of the combination of magic-angle spinning (MAS) with high-power (dipolar) $^{1}H/^{13}C$ decoupling (DD) to yield liquid-like ^{13}C NMR spectra was performed by Schaefer and Stejskal (1976). Figure 5.1 illustrates their historic spectra from both man-made and natural polymeric solids. Their achievement was based in turn on the crucial pioneering work of Waugh and his coworkers who were the first to show that direct observation of ^{13}C signals in the presence of dipolar decoupling from protons was both feasible and generally the best way to achieve high-resolution (Pines *et al.*, 1973). It also drew on the equally significant work of Andrew and coworkers (1958) and Lowe (1959) who first described and demonstrated MAS, and on the contribution of Hartmann and Hahn (1962) who introduced polarisation transfer in the rotating frame, or cross polarisation (CP).

Many NMR investigations of solid polymer systems have since been carried out using various combinations of techniques, some of which have been described in Chapter 4. Reviews of such applications have appeared regularly, a selection of which is to be found in the listed references for general reading. Here we shall be concerned only with spectra obtained under fast MAS. This, together with DD, generates spectra which retain only the ^{13}C isotropic chemical shifts analogous to, but not necessarily identical with, those in the isotropic liquid state. The choice of CP or single-pulse excitation (SPE) to generate the ^{13}C signal does not in itself influence resolution or the averaging of the anisotropic components of the spin couplings but generates either sensitivity enhancement or varying degrees of discrimination between chemically and/or physically distinct sites or regions in the solid (cf Chapter 4). Such discrimination can arise, for example, because CP tends to favour rigid regions with strong, secular, $^{13}C/^{1}H$ dipolar couplings, whereas SPE can discriminate in favour of morphological regions or individual carbon sites which have short ^{13}C spin–lattice relaxation times, arising usually from greater chain or sidegroup mobility.

5.2.1 ^{13}C chemical shifts in solid polymers

The large number of observations of isotropic chemical shifts in solids, and especially solid polymers (Bovey, 1988; Duncan, 1990), confirm expectations that the factors governing such shifts are essentially the same in solids as in liquids. This is not surprising as the chemical shift reflects the shielding of the nucleus by the immediate electronic environment which would, in general, not be expected to change substantially on solidification except where changes of bonding occur such as the formation of hydrogen bonds or shifts in tautomeric equilibria. This implies that the rank order of shifts and the well-established relationships of these with chemical structure derived from liquid-state studies can be used with some confidence in interpreting and assigning corresponding signals from solids, at least as a first approximation (tables A1.2 and A1.3). There are, however, a number of important differences which are due to the solid state itself, as discussed below.

5.2.1.1 Conformational effects

It has been recognised for many years that the chemical shift may be a function of the conformation of a polymer chain or molecule. In the liquid state, it is usual to find that a single signal is observed even though there may be several magnetically inequivalent conformational sites for a

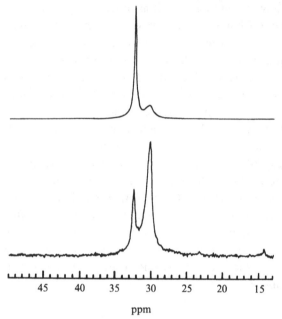

Figure 5.2. ¹³C high-resolution NMR spectra typical of polyethylene. The upper
trace is the CPMAS (1 ms contact time) spectrum; the lower trace is the SPEMAS
spectrum (recycle time, 50 s) of the identical sample (MW = 140000; 68%
crystallinity by X-ray diffraction). The CPMAS spectrum is dominated by the
signal from the all-*trans* conformation, largely crystalline region chains, whereas
the SPE spectrum emphasises the fast-relaxing, disordered regions. The low-
intensity peak at 14 ppm arises from the methyl carbons of a small number of butyl
sidechains in this sample.

particular nucleus. This is normally the result of thermally activated
exchange of spins between their different conformational sites at rates
which are fast relative to the differences in the chemical shifts of the sites.
In the solid state, such exchange processes are, in general, much more
hindered and the molecule is likely to adopt just one of the available
conformations. Such cases may give rise to distinct signals from nuclei
which thus find themselves in different environments. In an early
observation of this effect (Schaefer *et al.*, 1977), it was noted that the ¹³C
CPMAS spectrum of poly(2,6-dimethyl-1,4-phenylene oxide) showed a
1:1 doublet for the 3,5-protonated aromatic carbons which, in solution,
gave a single peak. No other carbons showed any evidence of a splitting. It
was proposed that this splitting was due to restriction in the solid state of
ring rotation about the C–O–C bond, a motion which, in solution, must be
fast compared with the shift difference of 4.3 ppm. A more definitive

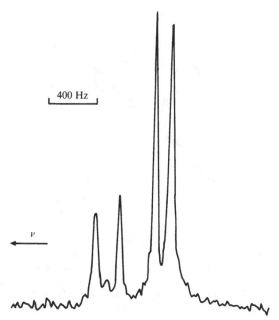

Figure 5.3. The ^{13}C high-resolution CPMAS spectrum of powdered syndiotactic polypropene at 300 K. A single contact CP sequence used a 1 ms contact time and MAS at about 2.3 kHz. About 5×10^4 scans were averaged. The line assignments are (with increasing frequency) methyl, methine and (doublet) methylene carbons. (Reproduced with permission from Bunn *et al.* (1981). Copyright (1981), the Royal Society of Chemistry.)

description of this effect may well involve chain packing in the solid which must both restrict the motion and lead to an asymmetric environment for these carbons. The fact that the other off-axis ring carbons and their attached methyl carbons do not show any resolvable splittings reflects the very local nature of the perturbation producing the change in shielding. Variations in the magnitude of such splittings in situations where symmetry would indicate that more sites should be affected are well known in both solution NMR and in crystalline organic solids (Balimann *et al.*, 1981).

Figure 5.2 shows both CP- and SPE-MAS ^{13}C spectra typical of a linear polyethylene, the interpretation of which was first carried out by Earl and VanderHart (1979). The larger and narrower peak to high frequency is assigned to carbons in the all-*trans* chain conformation and, thus, to the signal from the lamellar orthorhombic crystalline regions. The broader peak is assigned to those carbons in a more disordered, mobile environment normally called the non-crystalline or disordered regions (cf Section 1.3). The lower frequency shifts of these carbons are ascribed to the existence of

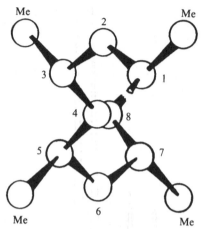

Figure 5.4. The conformation of syndiotactic polypropene. The numbered sequence of the methine/methylene backbone carbons go into the plane of the paper. The two methylene sites, which give rise to the doublet; are 2 and 6 and 4 and 8. These correspond to the higher and lower frequency methylene resonances in fig. 5.3, respectively. (Reproduced with permission from Bunn *et al.* (1981). Copyright (1981), the Royal Society of Chemistry.)

higher concentrations of *gauche* conformations in these regions. Such assignments are based both on information derived from solution-state spectra and on experiments which discriminate between rigid and more mobile chains. The quantitative description of such spectra and their unambiguous interpretation in terms of the conformational, structural and dynamic nature of the chains contributing to the overall spectrum has presented considerable challenge, as will be evident in due course.

In vinyl polymers, helical conformations are common and in certain cases these can give rise to inequivalent sites. Syndiotactic polypropene is such an example as is evident in the ^{13}C CPMAS spectrum portrayed in fig. 5.3 (Bunn *et al.*, 1981). The methyl and methine carbons give rise to single lines but the methylene signal comprises a 1:1 doublet with a splitting of 8.7 ppm. The widths of the doublet components are different and they therefore appear to be of different intensity if judged by their height. Figure 5.4 shows the conformation of the syndiotactic polypropene chain deduced from X-ray diffraction. It is clear that this conformation has two sites for methylene carbons, one lying on the axis of the helix and the other on the outside. On the basis of γ-*gauche* shielding effects (Tonelli, 1989), the shift difference between these two sites was calculated to be about 8 ppm. In addition, the peak to high frequency was assigned on this basis to the carbon site on the outside of the helix. This is the broader of the two methylene peaks, which is probably accounted for by additional splittings arising from *inter*chain interactions, to be discussed in the next section. In solution, the ^{13}C spectrum of this material shows only one peak for the methylene carbons which has a shift close to the average of those for the two peaks in the spectrum of the solid. In this example, the splitting is intrinsic to the *intra*molecular chain structure, with the other methyl and methine moieties having only one site by symmetry.

In the above two examples the presence of fixed conformations is revealed through multiplicities in the spectrum. It is also possible that polymorphs of polymers may exhibit different conformations. If these are present simultaneously in a sample, then, again, multiplicities can be observed. Variation of the relative intensities of these lines with preparation conditions, temperature and so on, or preparation of pure forms of the polymorphs, allow this effect to be identified. Isotactic poly(but-1-ene) exists in three polymorphic forms for which ^{13}C CPMAS spectra have been studied by Belfiore and coworkers (1984). Though X-ray diffraction shows that all three forms have helical conformations, the main chain dihedral angles for the *gauche* and *trans* conformations vary between 60–83° for the *gauche* and 159–180° for the *trans* material. The shifts differ by as much as 4.5 ppm and are interpreted in terms of variations in the γ-*gauche* shielding effect with dihedral angle. Again, the effects on the shielding in these cases are *intra*chain in origin although the adoption of the fixed conformation arises from the *inter*chain constraints in the solid.

Polypivalolactone $(-CH_2C(CH_3)_2COO-)_n$ is a further example which shows the effects of both intrinsic *intra*molecular inequivalence and those arising from polymorphic forms (Veregin *et al.*, 1986). This polymer can also exist in three forms, two of which are presumed to have virtually identical 2(1) helical conformations and the third an extended planar zig-zag chain structure. The 2(1) helix structure should render the two methyl carbons inequivalent in the solid state where interconversion might be expected to be slow. This was observed as a 1:1 doublet in the methyl region of the spectrum with a shift difference of only 1.1 ppm.

5.2.1.2 Crystallographic effects

Even in situations where the chain conformation indicates no inequivalent sites for chemically equivalent carbons, the packing of the chains in the structure may generate inequivalence through crystallographic symmetry. In well-ordered systems, these effects can again appear as multiplicities of spectral lines characterised by simple integer intensity ratios. Isotactic polypropene is a particular example of such effects (Bunn *et al.*, 1982). The

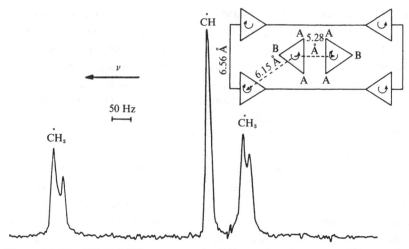

Figure 5.5. The CPMAS ^{13}C spectrum of solid isotactic polypropene. This sample, which was annealed, shows the splittings arising from the unit cell symmetry. The assignments are, as for fig. 5.3, methyl, methine and methylene, with increasing frequency. The arrows represent the handedness of each helix. The labels A and B identify the inequivalent sites discussed in the text. (Bunn *et al.*, *Polymer*, **23**, 694 (1982) by permission of the publishers, Butterworth and Heinemann Ltd ©.)

3(1) helical conformation of the isotactic polypropene chain does not predict other than single resonances for each of the three chemically distinct carbons. However, in well-annealed, highly crystalline samples, all three chemical sites give rise to two lines with a 2:1 intensity ratio. The shifts are small, of the order of 1 ppm or less, the methine carbon doublet being the smallest and barely resolvable but nonetheless detectable. Figure 5.5 shows both the ^{13}C CPMAS spectrum and the unit cell of the isotactic polypropene crystal structure determined by X-ray diffraction. This reveals that the unit cell contains pairs of enmeshed left- and right-handed helices and that this packing generates two distinct environments for all three chemical carbon sites in a 2:1 ratio as observed. This small *inter*chain effect was found to be sensitive to sample treatment. Samples quenched from the melt, or those which had not been specially annealed, had splittings in the spectrum which were less resolved and showed a different distribution of intensity within the region where splitting was observed in the well-annealed samples. These observations are understandable in terms of the sensitivity of the shift to the local environment, with the distribution of such environments being dependent on the degree of order induced in the solid by the physical treatment to which it is subjected. Such effects are well known in the liquid state as a dependence of shifts on solvent, concentration and so on. In the solid state, the difference is that these local environments may be frozen in and may thus have a distribution of values leading to line broadening and reduced apparent resolution in the spectra.

Summarising the foregoing two sub-sections we see that, whereas the shifts and chemical structure established for the liquid/solution state are retained in the solid state, there are factors specific to solids which arise principally from the fact that chains are constrained to adopt relatively fixed conformations and chain packing arrangements.

5.3 NMR structural investigations of solid polymers

In the previous sections of this chapter we have outlined some of the factors which can particularly affect high-resolution spectra of ^{13}C spins in solids. In this section we describe some illustrative examples of the way in which NMR can determine the primary structures of solid polymers with emphasis on what can be deduced about such structures and the way that they respond to chemical and physical treatments.

5.3.1 Cellulose and its derivatives

As a naturally occurring polymer, cellulose has a variety of origins and a great deal of work has been devoted to characterising and understanding its microstructure and its reactions. High-resolution ^{13}C NMR has played an important role in these investigations and we look at just three aspects.

Figure 5.6 shows the ^{13}C spectrum of a cellulose and the assignment of the peaks in terms of the basic chemical structure (VanderHart and Atalla, 1984). The main features to note are that the spectrum comprises sharp multiple lines in regions assigned to a single chemical site, for example C1, C4 and C6, as well as broad features on the low-frequency side of these. As shown later in fig. 5.10, these broader lines are associated with short $T_{1\rho}(^1H)$ components and can be assigned to polymer units both on the surface of crystallites and in three-dimensional disordered regions. They are shifted and broadened because they experience both a distribution of environments and motions at frequencies which make decoupling and MAS less effective. Figure 5.7 illustrates the nature of these bands and compares them with the response of a sample known to be amorphous. By comparing the spectra of celluloses from a number of sources (fig. 5.8), VanderHart and Atalla concluded that the variations observed in the sharp resonances associated with the ordered crystalline regions required the existence of two different crystalline forms occurring in different

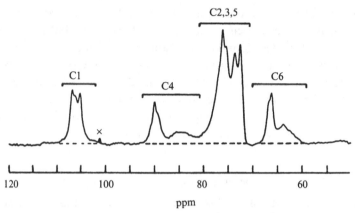

Figure 5.6. ^{13}C CPMAS spectrum of a cellulose (cotton linters). The horizontal bars indicate the ranges of chemical shifts of the carbon sites in the anhydroglucose monomer unit. X marks the first spinning sideband of polyethylene, used as an internal standard. (Reprinted with permission from VanderHart and Atalla (1984). Copyright (1984) American Chemical Society.)

proportions in materials from the different sources. That these effects were possibly due to unit cell inequivalence in a single crystal form could be ruled out because the intensity ratios of lines assigned to the same chemical site (for example, C1) varied with source and were not simple whole numbers. From the available data they deduced the possible forms for the spectra of individual crystalline structures (fig. 5.9).

The nitration and acetylation of cellulose were also studied by high-resolution ^{13}C NMR (Patterson *et al.*, 1985; Doyle *et al.*, 1986). In the case of nitration, the process was followed by monitoring changes in chemical shifts and line intensities as the degree of nitration progressed towards completion as specified by the degree of substitution, DOS = 3. It was found that nitration occurred first at the C6 carbons in the amorphous, disordered regions and that this was complete by DOS = 0.5. A new signal assigned to C1, which was influenced by nitration at C2, appeared at DOS = 0.28 and signals associated with C6 and C4 in the crystalline native cellulose had vanished by DOS = 1.39. The signal from C1 was no longer detectable at a DOS = 2.65. Unlike solution NMR, the solid-state NMR study (i) allowed access to the full range of nitration, (ii) addressed the reaction in terms of the solid structure, including the resolved behaviour of amorphous and crystalline regions and (iii) monitored differential intrinsic reaction rates at the different chemical sites.

In the investigation of acetylation (Doyle *et al.*, 1986), it was shown that for the heterogeneous reaction no differential rates could be ascertained for

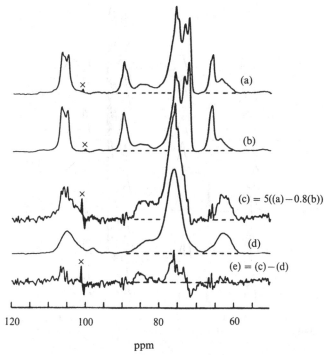

Figure 5.7. ^{13}C CPMAS spectra of (a), cotton linters; (b), acid hydrolysed linters and (d), amorphous cellulose. The difference spectrum, (c), is shown to be very similar to (d) through the difference spectrum (e). This indicates that the hydrolysis attacks chains in the amorphous regions which are suggested to be three-dimensional rather than surface layers. (Reprinted with permission from VanderHart and Atalla (1984). Copyright (1984) American Chemical Society.)

the different chemical sites. Acetylation occurred first in the disordered regions, paralleling observations in solution NMR spectra. Acetylation of the crystalline regions occurred without loss of their integrity. An interesting aspect of this study was the use of the so-called 'delayed contact' CP experiment to discriminate ^{13}C signals associated with the disordered and crystalline regions where the basis of discrimination was the shorter $T_{1\rho}(^1H)$ of the more mobile disordered regions (Aujla *et al.*, 1982) (fig. 5.10).

5.3.2 Tacticity of solid poly(vinyl alcohol)

The use of solution-state high-resolution ^{13}C NMR for elucidating the tacticity in vinyl polymers is well established (Tonelli, 1989). In the solid state, the reduced resolution of spectra, often arising from the con-

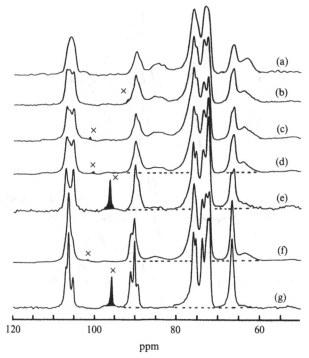

Figure 5.8. ^{13}C CPMAS spectra of celluloses from a variety of sources (a)–(g), showing variation in fine structure, particularly in the C1 and C4 regions (around 105 and 85–90 ppm, respectively). (Reprinted with permission from VanderHart and Atalla (1984). Copyright (1984) American Chemical Society.)

sequences of disorder in chain packing and so on (*vide infra*), means that tacticity effects are not normally observed. Solid poly(vinyl alcohol) (PVA) is an interesting exception which illustrates the subtleties which can be encountered in the solid state (Terao *et al.*, 1983). Figure 5.11 shows the ^{13}C CPMAS spectra of three PVA samples with somewhat differing tacticities as indicated in the solution spectra, shown as line drawings. The three methine lines around 67 ppm in the solution spectra correspond, in order of decreasing frequency, to rr, mr and mm triads, their relative intensities indicating that PVA-A is somewhat isotactic and that PVA-B and PVA-C are close to atactic. Superficially, the spectra of the solids suggest a similar interpretation for the three methine lines. However, the shifts are much larger than those associated directly with tacticity and their relative intensities obtained by deconvolution do not match those of the solution spectra. The *intra-* and *inter*molecular hydrogen bonding discriminates between the different triads, amplifying the shifts and changing

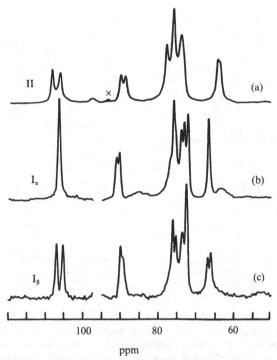

Figure 5.9. ^{13}C CPMAS spectra of a high-crystallinity cellulose II sample, (a) and the two spectra, (b) and (c), derived from spectra (e) and (f) of fig. 5.8 which, in different proportions, account for the variability in the spectra of fig. 5.8. (Reprinted with permission from VanderHart and Atalla (1984). Copyright (1984) American Chemical Society.)

the relative intensities. The methine carbons at the centre of an mm triad can form two *intra*molecular H-bonds but not all of them do so because of other constraints in the crystalline solid and the possibility of *inter*-molecular H-bonds. Thus, the intensity associated with the mm signal in solution is distributed amongst the three peaks observed in the solid-state spectra. Similarly, the methine mr triad contributes to the two lowest frequency peaks whilst the rr triad appears only in the lowest frequency resonance. This study illustrates the detailed information that such measurements can reveal.

5.3.3 *The structure of insoluble polymeric resins*

Many polymeric systems, once formed, are not soluble and must be characterised as solids. Apart from infra-red spectroscopy, solids NMR methods offer the only non-destructive approach for determining the

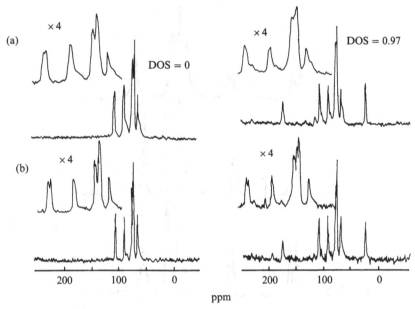

Figure 5.10. ^{13}C CPMAS spectra ((a) normal; (b) delayed contact) for cellulose (left-hand spectra) and acetylated cellulose with a degree of substitution of 0.97 (right-hand spectra). The loss of broad signals associated with the disordered regions is clearly detected and is accompanied by increased resolution. (Doyle *et al.*, *Polymer*, **27**, 19 (1986) by permission of the publishers, Butterworth Heinemann Ltd ©.)

chemical structure and formation mechanisms of these materials. In this section we illustrate the use of NMR for the characterisation of such solids.

Urea–formaldehyde resins. Urea–formaldehyde resins pose a number of challenges due to their high concentration of ^{14}N to which most carbons are directly bonded and, as a result, the spectra are substantially broadened (cf Section 4.4.2.6). In addition, all but the carbonyl carbons have directly bonded protons which means that assignment methods such as dipolar dephasing (Section 4.4.2.3) are not applicable. In their study of a range of urea–formaldehyde resins, Maciel and coworkers (1983) demonstrated the advantages of using higher field spectrometers to reduce the splittings and broadening arising from the ^{14}N nuclei (fig. 5.12). Analysis of the higher field spectra was approached by using a set of chemical shifts of model compounds taken from solution NMR and deconvoluting the solid-state spectra. Spinning sidebands, which can be a problem at higher fields, can be suppressed typically with the TOSS sequence (Dixon, 1981). Whilst

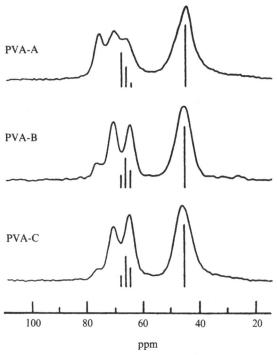

Figure 5.11. ^{13}C CPMAS spectra of solid poly(vinyl alcohol) samples of differing tacticities. The corresponding solution spectra are shown as stick diagrams. (Reprinted with permission from Terao *et al.* (1983). Copyright (1983) American Chemical Society.)

analysis of systems in this way cannot be regarded as quantitative, relative intensities, particularly amongst aliphatic carbon resonances, are well established. Figure 5.13 shows two examples of the reconstruction of experimental spectra from the deconvolution process. Particular aspects of the reaction scheme might be further clarified by using ^{13}C-enriched precursors at different stages in the curing process.

Polystyrylpyridine. The formation of this cross-linked resin from tere-phthalic aldehyde and collidine (2,4,6-trimethylpyridine) demonstrates the versatility of solids high-resolution NMR techniques in examining the formation and characterisation of an intractable polymeric solid (Laupretre *et al.*, 1986a, b). The approach initially involves an understanding of the probable basic reactions – in this case the reaction of the methyl groups of the pyridine with either the aldehyde or the product of this reaction – using the solution-state NMR of model compounds and precursor

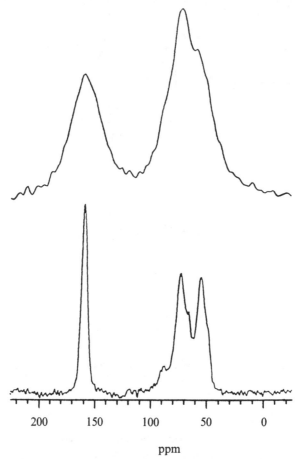

Figure 5.12. The static field (B_0) dependence of the ^{13}C CPMAS spectra of a cured urea–formaldehyde resin. Top, 1.4 T; bottom, 4.7 T. The spectrum at 4.7 T was obtained using sideband suppression techniques. The increased resolution at higher field arises from a reduction in the effectiveness of the ^{14}N broadening of the ^{13}C signals as the ratio of the ^{14}N quadrupole to Zeeman interactions reduces. (Reprinted with permission from Maciel *et al.* (1983). Copyright (1983) American Chemical Society.)

polymers to define the chemical shifts of key sites. This is followed by detailed and careful application of solids NMR techniques to characterise quantitatively the course of the curing process. Figure 5.14 shows the ^{13}C CPMAS spectrum of a typical cured sample of polystyrylpyridine. The resonance frequencies assigned to the designated sites in the model compounds are also shown as stick spectra. Though the solid-state spectrum has limited resolution, it contains sufficient resolved information

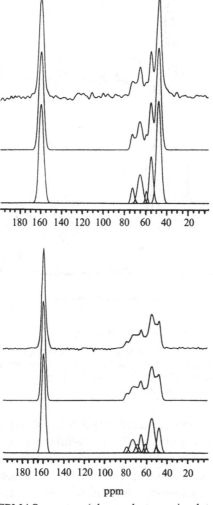

Figure 5.13. ^{13}C CPMAS spectra (observed, top; simulated, middle) of two urea–formaldehyde resins. The simulated spectra were obtained using peaks deconvolved from the observed spectra, these being shown in the bottom spectra. (Reprinted with permission from Maciel *et al.* (1983). Copyright (1983) American Chemical Society.)

to achieve a detailed quantitative analysis. Assignment was facilitated, in this case, by the use of both the dipolar dephasing experiment (Section 4.4.2.3) and spectra taken with very short CP contact times (fig. 5.15). The dipolar dephased spectra identify quaternary and methyl carbon signals whereas spectra taken with short contact times reveal only CH_2 and CH

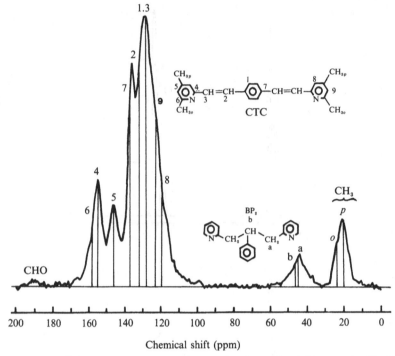

Figure 5.14. ^{13}C CPMAS spectrum of a cured polystyrylpyridine sample. The line spectrum indicates the resonance positions in solution of sites in the model compounds, as indicated. (Data taken from Laupretre *et al.* (1986b). Copyright (1986), Elsevier Science Publishers BV.)

carbons because these cross polarise very quickly. To achieve a quantitative analysis of the structures and the cross-linking process, spectra were determined as a function of the CP contact time. Figure 5.16 illustrates typical results. Fitting these curves to eq 2.44, which describes cross polarisation, yields quantitative information on the extent of reaction at various stages throughout the curing process as well as confirming the basic model for the chemistry involved in cross linking.

Irradiated polyethylene. Irradiation of polyethylene with γ-rays or other forms of ionising radiation causes cross linking, the formation of vinylene double bonds and the creation of methyl groups from vinyl chain ends. Solution-state NMR is applicable to the solvent-extractable portion (the 'sol' fraction) but not to the insoluble 'gel' fraction. The use of solids NMR to study the total system (Pérez and VanderHart, 1988) illustrates both the power and limitations of the approach. A combination of solids

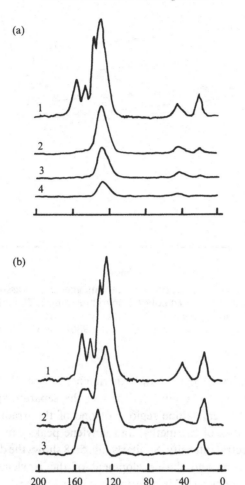

Figure 5.15. Selective observation of (a) protonated carbons and (b) non-protonated and methyl carbons, in a cured polystyrylpyridine sample. The protonated carbons were selected by using short cross polarisation times (1, 1 ms; 2, 40 μs; 3, 20 μs; 4, 10 μs). The quaternary and methyl carbons were selected using the dipolar dephasing (non-quaternary suppression) sequence with different dephasing times (1, 0; 2, 20 μs; 3, 50 μs). (Data taken from Laupretre *et al.*, (1986b). Copyright (1986), Elsevier Science Publishers BV.)

NMR experiments (see below) monitored the disappearance of vinyl end groups, the build up of vinylene and methyl groups and the partitioning of these between the crystalline and non-crystalline regions of the irradiated polymer. Since the concentrations of these various groups are of the order

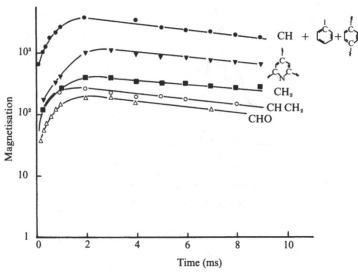

Figure 5.16. The contact time dependence of the magnetisations associated with the different chemical sites in a cured polystyrylpyridine sample. The solid lines are fits to the theory from which the relative intensities of each site can be obtained. (Data taken from Laupretre *et al.* (1986b). Copyright (1986), Elsevier Science Publishers BV.)

of 1 per 10^3 backbone carbons, the experiments are pushing towards the current limits of NMR. Figure 5.17 shows the separate spectra of the crystalline and non-crystalline regions of one of the irradiated samples where the intensities of the methyl and vinylene peaks can be compared with the total spectral intensity. Similarly, fig. 5.18 shows the disappearance of vinyl end groups and the development of the vinylene and methyl resonances as a function of irradiation dose for another of the samples studied. These experiments failed, however, to detect any evidence of cross links which are presumed to be present on the basis of solution-state NMR studies on the 'sol' fractions from samples irradiated at a lower dose level. This was due to a number of factors: spectral overlap with the very intense main peak, uncertainty over the expected shift and the probability that the lines sought would be broad. Notwithstanding these drawbacks, no other approach has come close to revealing the information obtained in this investigation.

The structural features of many other solid polymers have been investigated by ^{13}C high-resolution NMR, typically polyesters (Fyfe *et al.*, 1979), polyimides (Sefcik *et al.*, 1979; Wong *et al.*, 1981; Havens *et al.*, 1981), phenolic resins (Fyfe *et al.*, 1980; Maciel *et al.*, 1984), epoxy resins

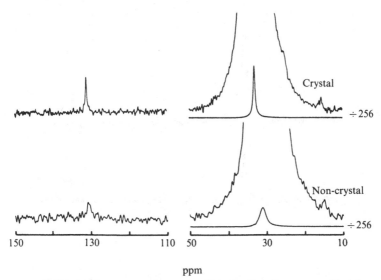

Figure 5.17. ^{13}C NMR spectra of the crystalline (top) and non-crystalline (bottom) regions of an ultra-high molecular weight polyethylene. The sample had been irradiated with γ-rays (167 rad) *in vacuo*. Both spectra are on the same intensity scale and were obtained from spectra generated by three different pulse sequences (normal CPMAS, delayed contact CPMAS and CPMAS preceded by a multiple dipolar echo sequence applied to the protons). (Reprinted with permission from Pérez and VanderHart (1988). Copyright (1988), John Wiley and Sons.)

(Garroway *et al.*, 1979) and many more. The examples given above serve to illustrate the scope of investigations possible.

Although we have discussed the use of ^{13}C exclusively, it is important to note that other spins such as ^{15}N, ^{29}Si, ^{31}P can be exploited in similar fashion. ^{29}Si CPMAS, for example, has been used to characterise some polydimethylsiloxane-based networks (Beshah *et al.*, 1986) (cf Section 8.1.1). The use of other spin-$\frac{1}{2}$ nuclei differs in points of detail, such as shift ranges, sideband problems arising from shift anisotropies, cross-polarisation dynamics and so on, but the general principles remain the same.

5.3.4 *Structural heterogeneity*

Chapters 6 and 7 address the way in which NMR can characterise the organisation of polymer chains on the supramolecular scale. However, for completeness, and to form a link with these chapters, we look at a few examples of how structural heterogeneity can be distinguished and characterised using ^{13}C high-resolution methods. Again it will be clear that structure and motion are intimately linked and that it is this correlation

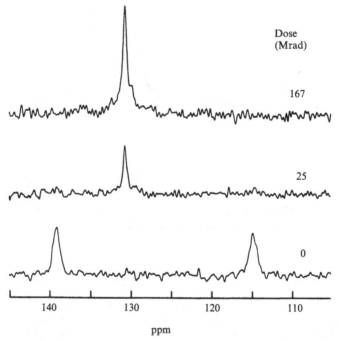

Figure 5.18. ^{13}C CPMAS spectra (100–150 ppm region only) of an irradiated ethylene-1-butene copolymer as a function of irradiation dose. The loss of vinyl end groups (114 and 139 ppm) and the appearance of vinylene groups (130 ppm) is clear. (Reprinted with permission from Pérez and VanderHart (1988). Copyright (1988), John Wiley and Sons.)

which gives NMR its ability to discriminate structural details in spatially heterogeneous systems.

As pointed out in Sections 4.4.2.3 and 4.4.2.4, there are a number of effects which can be used to associate lines in a high-resolution ^{13}C spectrum with other NMR properties of the system, such as $T_{1,1\rho}(^1\text{H})$, ^{13}C/^{1}H dipolar coupling strength and so on. Figure 5.19 (Aujla *et al.*, 1982) shows ^{13}C spectra for two samples of poly(ethylene terephthalate), one amorphous, the other about 35 % crystalline. Two spectra of the partially crystalline sample are shown, one a conventional CPMAS spectrum, the other obtained with delayed CP, which generates the ^{13}C signal from carbons associated with ^{1}H spins with $T_{1\rho}$ values longer than the delay chosen. Close inspection shows that the delayed CP spectrum has lost intensity from the high-frequency sides of the lowest and highest frequency lines and that the group of lines in the centre of the spectrum shows increased resolution. The spectrum of the amorphous sample demonstrates

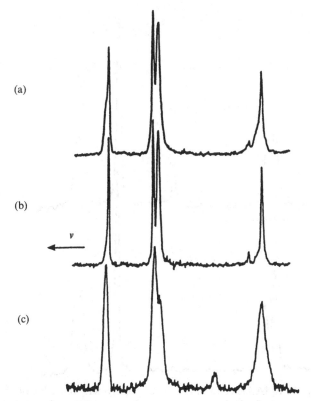

Figure 5.19. ^{13}C CPMAS spectra of PET. (a) normal spectrum of a partially crystalline PET film; (b) the delayed contact (12 ms) spectrum of the same film and (c) the spectrum of a completely amorphous PET. The difference between (a) and (b) corresponds to the loss of signal from the non-crystalline regions which have short $T_{1\rho}(^1H)$. (Reprinted with permission from Aujla *et al.* (1982). Copyright (1982), Springer-Verlag.)

that these changes arise from a loss of the amorphous component of the spectrum. Discrimination is possible in this case because the amorphous regions have considerably faster ^1H rotating frame spin–lattice relaxation than the crystalline part and spin-diffusion between protons in these different regions is not fast enough to average them. The $T_1(^1H)$ values, on the other hand, are not sufficiently different to form the basis for such a discrimination and, indeed, may also be averaged by spin-diffusion because of the longer spin diffusion times available (cf Section 6.2.1).

Figures 5.20 and 5.21 present ^{13}C spectra of some ethylene/propene block copolymers in which such discrimination effects are apparent. Figure 5.20 (Aujla *et al.*, 1982) shows spectra of two such copolymers, both taken

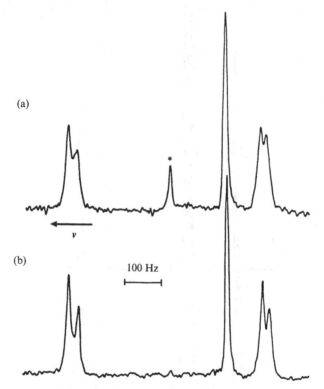

Figure 5.20. ^{13}C CPMAS spectra of two block copolymers. In each case the main block is polypropene which gives the characteristic spectrum of the isotactic form (see fig. 5.5). The end block, about 10 % of the total, is formed using an ethylene feed (a) and a mixed ethylene/propene feed (b), the latter yielding an end block composition with about 40 % ethylene units. Under the CP conditions used only (a) shows the presence of all-*trans* polyethylene chains which have the properties of crystalline polyethylene (peak marked *). (Reprinted with permission from Aujla *et al.* (1982). Copyright (1982), Springer-Verlag.)

under normal CPMAS conditions, but with different contact times. Both copolymers are predominantly polypropene with end blocks which constitute about the same proportion of the total polymer but which, in one case, is mainly polyethylene while the other consists of a 1:1 ratio of ethylene to propene. The spectrum of the former shows a peak at a chemical shift which is characteristic of the all-*trans* conformation of polyethylene. The second sample shows no evidence of any signal which can be ascribed to ethylene-like units, although these must be present. In this case, the difference arises because the end block comprised entirely of ethylene can form crystallites of all-*trans* polyethylene which undergo fast CP and are associated with a relatively long $T_{1\rho}(^1\text{H})$. On the other hand, the

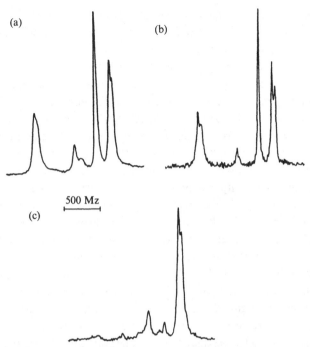

Figure 5.21. ^{13}C high-resolution spectra of block copolymer in which the main block is polypropene with the end block (15% of polymer) being a random copolymer of ethylene (> 85%) with propene. (a) Normal CPMAS spectrum; (b) a delayed contact (27 ms) CPMAS spectrum and (c) a SPE spectrum with a recycle time of 0.3 s. (Data taken from Smith, Thesis, University of East Anglia, 1986.)

end block which is derived from 1:1 ethylene/propene is rubbery, does not cross polarise as rapidly (motion reduces the ^{1}H/^{1}H and ^{1}H/^{13}C dipolar couplings) and is associated with short $T_{1\rho}(^{1}$H) values.

Figure 5.21 (Smith, 1986) furnishes a series of ^{13}C high-resolution spectra for a similar copolymer which has an end block amounting to approximately 15% of the total polymer and which comprises about 85% ethylene. In this case the simple CPMAS spectrum shows the typical isotactic polypropene spectrum together with a typical polyethylene spectrum, the latter having both crystalline and non-crystalline contributions. A delayed CP experiment shows an increased resolution of the 2:1 splittings of the methyl and methylene peaks of the polypropene and the loss of the signal assigned to the non-crystalline regions of the polyethylene component. A SPE experiment, on the other hand, reveals a strong signal from the methyl carbons in the polypropene (figs 4.17 and 5.3) along with weaker signals from the non-crystalline polyethylene region; the faster relaxing disordered regions of the polypropene are reflected in $T_1(^{13}C)$. Other minor signals, probably due to small amounts of more random copolymer structures, can also be seen.

NMR has also addressed another structural issue for ethylene/α-olefin copolymers, namely, which sidechains, if any, in these polymers can be accommodated within the crystalline regions (Laupretre *et al.*, 1986a, b; VanderHart and Pérez, 1986; Pérez *et al.*, 1987). The approach relies on discriminating signals from sidechain groups via their relationship with signals known to originate from, say, the backbone carbons in the crystalline regions. Laupretre and coworkers compared spectra recorded with SPE, for which only carbons with short $T_1(^{13}C) < 3$ s were detected, with delayed contact CP spectra due only to those signals associated with long $T_{1\rho}(^1H)$ arising, in this instance, from carbons in the crystalline regions. The samples used in this study contained relatively high levels of side chains and low crystallinities. As such, there may be some uncertainty over how well the $T_{1\rho}(^1H)$ behaviours for the different regions are decoupled. However, the qualitative conclusion that only methyl side-chains showed any significant propensity for incorporation into the lattice of the crystalline regions is probably correct within the precision of the measurements. Other investigations (VanderHart and Pérez, 1986; Pérez *et al.*, 1987) involved a more detailed and quantitative approach and utilised samples with much lower levels of branching. In these studies, both ethyl side chains and methyl and vinyl end groups were characterised and their distribution between crystalline and non-crystalline regions was quantified.

Again these studies illustrate the fine detail that can be gleaned from NMR methods and the level of understanding of the NMR spin physics which is needed to ensure that defensible information is deduced from the measurements.

5.3.5 *Specialised methods for structure determination*

The examples given above illustrate the use of what have come to be regarded as standard, almost routine, high-resolution NMR techniques for characterising the structure of solid polymers. In this section we briefly describe some rather more specialised methods.

The main indicator of structure in the examples cited thus far is the isotropic chemical shift which, as in solution-state NMR, reflects the local electronic environment of the resonant nuclei. As we have seen, primary chain chemistry, chain conformations and packing are the main factors leading to structural characterisation and discrimination. For static samples, the principal components of the shielding tensor for each site can also be determined, either through selective site enrichment (to avoid problems of spectral overlap) or through the analysis of spinning sideband manifolds from slow spinning experiments in which assignment of lines to each manifold is possible. Knowledge of these shielding tensor components, however, does not add significantly to the characterisation of structure within the context of this discussion. They are important in characterising chain orientation (Chapter 7) and local dynamics but only in specific cases do they address local structural features except in so far that they augment the information accessible through the isotropic shift by providing three numbers to characterise the shielding rather than the one average value.

In crystallographic terms, structure implies a determination of the arrangements of atoms in space and it is in this area that solids NMR offers further possibilities. The interactions which have the potential for giving information of this type are, or course, homonuclear and heteronuclear dipole–dipole coupling. In the early days of NMR it was this possibility which attracted attention. The principal accessible nucleus was ^1H and, as we have seen, the lineshapes of solids containing many protons are generally broad and rather featureless with only the simplest isolated pair or triplet groupings yielding internuclear distances with any confidence. The developments in NMR techniques for solids over the past two decades have allowed this aspect of NMR to be revisited (Tycko and Dabbagh, 1991). The high-resolution capability, made possible by homo- and heteronuclear decoupling methods and MAS, allows the design of experiments which give site-specific measurements of dipolar couplings between, for example, $^{13}C/^1H$, $^{15}N/^1H$, $^{13}C/^{15}N$ and other similar pairs or larger groupings of spins. Before looking at these methods in a little more detail, it is important to make the general point that the dipolar interaction is very short-range and hence, the structural information generated is only local, affecting perhaps atoms within two or three atomic spacings (aside of course from longer range spin diffusion effects which are underpinned by dipolar interactions).

In essence, all the experiments of interest have the form of two-dimensional NMR although they are not always treated in that way (cf Section, 4.5). Once high-resolution methods for dilute spins were developed, the possibility of correlating chemical shifts with dipolar couplings became evident. Separated local field spectroscopy applied to a static sample of a highly oriented polyethylene was one of the first such

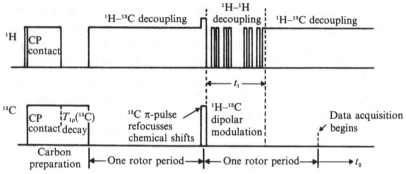

Figure 5.22. The pulse sequence used for the measurement of dipolar rotational spin echo spectra of ^{13}C spins in solids. CP generation is followed by an optional $T_{1\rho}(^{13}$C) decay, useful for separating signals in systems with heterogeneous relaxation characteristics. Data acquisition takes place in time t_2, whilst the dipolar modulation is introduced during the time t_1 which contains an integral number of multiple pulse homonuclear (^{1}H) decoupling cycles. The whole experiment is synchronised to the MAS rotor period, as indicated. (Reprinted with permission from Schaefer *et al.* (1983). Copyright (1984) American Chemical Society.)

investigations (Opella and Waugh, 1977). This two-dimensional ^{13}C NMR measurement involved a CP preparation, followed by an evolution period t_1 during which the protons were subjected to multiple-pulse homonuclear decoupling, and a detection period t_2 during which the ^{13}C signal was recorded under dipolar decoupling conditions. Fourier transformation with respect to both t_1 and t_2 gave a two-dimensional spectrum with the ^{13}C/^{1}H dipolar coupling on the ω_1 axis and the chemical shift along the ω_2 axis. An important point to note is that the multiple-pulse decoupling of the protons from each other leaves the heteronuclear ^{13}C/^{1}H coupling intact while removing the complicating spin-diffusion processes between the protons (cf Section 3.8).

A second such experiment involving MAS was described, *inter alia*, by Schaefer and coworkers (1983) and others (Munowitz *et al.*, 1981; Munowitz and Griffin, 1982). This 'dipolar rotational echo' experiment is illustrated in fig. 5.22. The basic concept is the same as that for the separated local field experiment but with the added effects of MAS. This places a constraint on the number of multiple-pulse cycles which can be fitted into a rotor period as well as requiring that data acquisition be synchronised to the MAS process. The main thrust of this work was to use the dipolar powder lineshapes recovered from the measurements to characterise local motions as described later in this chapter. The methods intrinsically give values for the dipolar coupling which, in the absence of motions sufficient to cause some averaging, can lead to structural information in the form of internuclear distances. However, it should be noted that even in so-called rigid solids, Schaefer and coworkers report reductions in the expected $^{13}C/^1H$ dipolar couplings which, they suggest, arise from high-frequency motion normally present in a solid.

Although these experiments have not been applied widely to the determination of internuclear distances in solid polymers, they have continued to undergo development and refinement, widening the range of dipolar couplings which are accessible (Gullion and Schaefer, 1989; Chirlian and Opella, 1989; Kolbert and Griffin, 1991; Gullion *et al*, 1991). In appropriate systems they could be of great value in determining key features of the local disposition of atoms.

5.4 Molecular motions in solid polymers

As described in Chapter 1, polymeric solids manifest complex distributions of thermally activated molecular motions. These range from high-frequency, modest-amplitude, torsional oscillations of a local nature to very slow movements, often involving longer range phenomena and larger scale displacements. Whilst the spin interactions on which NMR relies for the study of these motions are intrinsically short-range, and thus local in character, they can be sensitive to the full range of motional frequencies present. As temperature is increased, amplitudes and frequency distributions of motion increase and structures involving *inter*chain and *intra*chain distances and conformations will change in parallel. Recognise, too, that an NMR study which aims to characterise motions in solid polymers is likely to be a more extensive investigation than simply the measurement of one or two spectra. Such a study, therefore, requires careful assessment of its aims, together with selection of the relevant NMR experiments, samples and conditions. The reward for these efforts is a detailed and often site-specific view of chain motions, covering such aspects as the precise nature of the motional mechanism (for example, jump or diffusion), symmetries, distributions of rates and their character (homogeneous or inhomogeneous), activation energies and so on. There is no other technique which casts light on these motions with quite such exquisite detail.

As already indicated, the organisation of the second half of the book emphasises polymers rather than NMR phenomena. In consequence, it is not intended to present a simple overview of particular NMR approaches based on the nuclei involved. The aim is to use the collective strengths of NMR to address polymer systems and it will be left to the reader to explore specific aspects of NMR in more depth by referring to the many specialist reviews and books.

Before looking at specific cases it is worth reminding ourselves of the basic NMR features which underpin this insight into molecular motion. These fall into three categories which are not mutually exclusive: lineshapes, spin–lattice relaxation and spin echoes which can be observed in both single- and multidimensional experiments.

Lineshapes. Lineshapes are affected by motions which are comparable to or faster than the frequencies intrinsic to the spectrum of the system in the absence of the motions. For inhomogeneous lineshapes, such as powder lineshapes arising from shielding anisotropy or quadrupole coupling, motions which interchange different frequencies within the spectrum at rates faster than the intrinsic widths of the component lines cause broadening and coalescence of the exchanged lines, often in a manner which additionally reveals information on the precise mechanism of the motion (cf fig. 3.7). Once such motions become fast relative to the frequency differences exchanged by the motion, a new averaged lineshape is achieved which no longer changes as the motional frequency increases. At this point, the relationship of the lineshape to that of the system in the absence of the particular motion can only be used to characterise the type/symmetry of the motion but not its rate, apart, of course, from the trivial observation that it is faster than the characteristic static spectral width which has been averaged. On the other hand, a homogeneously broadened line, such as that arising from homonuclear dipole–dipole coupling between abundant spins (for example, ^1H–^1H), only shows narrowing when the motions modulating the interaction become comparable to the full static linewidth. Again, once any anisotropic motion is fast compared to this width, a reduced linewidth remains which is both independent of the motion and characteristic of the symmetry of the fast motion but not its actual rate.

Spin–lattice relaxation. Once motions have become fast in the sense described above, they can be characterised through measurement of spin–lattice relaxation (SLR). As indicated in Chapters 2–4, spin–lattice relaxation can take a variety of forms, principally laboratory- and rotating-frame SLR which, respectively, respond to motional frequencies of order ω_0 and ω_1, the precession frequencies of the spins in the $\mathbf{B}_0$ and $\mathbf{B}_1(t)$ magnetic fields. SLR rates can be measured in a site-specific manner (for example using ^{13}C CPMAS or ^2H in deuterium-substituted systems) or, for abundant and strongly magnetic species such as ^{1}H and ^{19}F, they can yield the overall relaxation behaviour, often mediated by spin-diffusion, in which motions of one grouping can often dominate. Even for these spin systems, correlation of SLR with lineshapes can be informative and multiple-pulse line-narrowing sequences combined with MAS (CRAMPS) can also yield chemical site-specific characterisation of the spin dynamics and hence SLR and motions.

Spin echoes. One of the distinguishing features of NMR is the ability it confers on the experimenter to manipulate the spin system to reveal information that is not immediately accessible from simple observation of spectra or even SLR measurements. CPMAS is one such case. The use of spin-echo experiments of various kinds extends the rates of motions which can be measured and also allows their nature to be probed in more detail. These and allied multidimensional experiments are of great utility in the study of motions in solid polymers.

As an example of the use of an echo experiment in the characterisation of motion, consider the simplest two-pulse sequence, the P_1–τ–P_2-echo. The formation of an echo implies that the interaction which generates the spread of frequency giving rise to the FID following P_1 can be reversed by P_2. For a powder lineshape determined by shielding anisotropy, a $90°$–τ–$180°$ sequence refocusses the FID to give an echo centred at 2τ. Ideally, this refocussing requires each spin to maintain a constant frequency during the time 2τ. If spins are made to change their frequencies through molecular motions which change the orientation of the spins' shielding tensor then, to the extent that the jumps occur on a timescale comparable to τ, echo formation will be inhibited. Indeed, the degree of dephasing and, hence, loss of echo formation, will be greatest when a spin, on average, accumulates a phase shift of order π during τ, that is, when $\Delta\omega\tau = \pi$. Observation of the echo spectrum obtained by Fourier transformation of the echo reveals information on the nature of the motions and their rates through changes in both spectrum shape and total intensity. Figure 5.23 illustrates some calculated echo shapes and intensities for a solid echo sequence ($90°_x$–τ–$90°_y$) applied to a spin-1 quadrupolar powder pattern of a system undergoing jumps at the rates indicated between orientations separated by the tetrahedral angle (Spiess, 1985a). Both the shape changes and intensity loss and recovery as a function of τ can be used to characterise the motion in detail. The corresponding spectra from the FID are shown and, apart from these not being as sensitive to the slower motions, the echo experiment overcomes to a considerable degree the

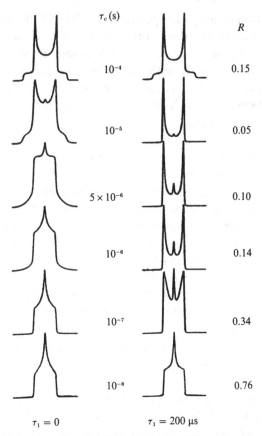

Figure 5.23. Normal (left) and solid echo (right) ^{2}H quadrupolar powder spectra calculated for a model system in which the principal axis of the quadrupole coupling undergoes jumps between two sites related by the tetrahedral angle. τ_c is the correlation time for the jumping process and R is the intensity reduction factor for the solid echo. The distinctive and different spectral shapes for normal and solid echo spectra are illustrated clearly as is the distinctive fast exchange lineshape, characteristic of the tetrahedral two-site model. (Reproduced with permission from Spiess, *Colloid Polymer Science*, **261**, 193 (1983). Copyright (1983), Steinkopff Verlag.)

problem of the deadtime following a pulse (cf Section 4.2). Note also the considerable range of jump times covered by this experiment. We will look in more detail at examples of this and related approaches in due course.

5.4.1 Poly(ethylene terephthalate) (PET) and related polymers

Motions in solid PET, the chemical structure of which is shown in fig. 7.17, have been investigated by a number of groups. English's discussion (1984 and references therein) of extant work at that time indicated the complexities and pitfalls inherent in such investigations arising, at least in part, from the intimate relationship between structure, dynamics and larger scale morphology. PET can be prepared in a range of states from completely amorphous to partially crystalline with crystallinities approaching 60 %. Mechanical relaxation processes, measured at 100 Hz, are observed around 360–400 K (α-relaxation) and 220 K (β-relaxation).

The main thrust of this work was the observation of ^{1}H NMR lineshapes and spin–lattice relaxation behaviour (T_1 and $T_{1\rho}$) over a wide range of temperature for a number of samples of different crystallinity and form (powder, film and fibre) some of which were selectively deuterated (either the methylenes or the aromatics). The ^{1}H NMR lineshape at temperatures below 300 K consisted predominantly of a single line whose width decreased steadily from its rigid lattice value (at about 100 K) to about half of this at about 500 K. The different samples all showed qualitatively similar behaviour in this regard, but with different absolute values of linewidth reflecting different proton concentrations (deuterated samples) or chain packing (partially crystalline, amorphous and so on). Above 300 K, a narrower line was present, the amplitude of which increased to about 20 % of the total signal by 500 K.

Starting at the lowest temperature, both T_1 and $T_{1\rho}$ timescales associated with the broad lineshape component decreased as the temperature was increased. T_1 was a single exponential process and, typically, the aromatic protons (for a sample with deuterated methylenes) showed a broad minimum around 300 K. The fully protonated polymer also exhibited an albeit less-defined minimum at the same temperature with reduced T_1 values, whereas a sample containing only methylene protons showed no evidence of such a minimum at all. $T_{1\rho}$ was non-exponential with a dominant long component which developed a pronounced minimum at about 400 K for a sample containing only methylene protons. There was no such minimum in samples with only aromatic protons or in the fully protonated material.

As English pointed out, a conventional approach to the interpretation of linewidth narrowing or the $T_{1\rho}$ response for the normal, non-deuterated, samples would have assumed that the temperature variation of these quantities was due to changes in rates of motion alone. However, the combined data, described above, are not consistent with such a model. The linewidth narrowing which sets in around 100 K implies that the responsible motions must be occurring at rates comparable to or faster than the rigid lattice linewidth, that is, about 50 kHz. However, motions at rates of this order should give rise to an associated $T_{1\rho}$ minimum at about the same temperature (cf fig. 2.15) since $T_{1\rho}$ is sensitive to frequencies of order $\omega_1/2\pi \sim 10\text{--}100$ kHz. No minimum in $T_{1\rho}$ developed until 400 K which must, therefore, be due to a completely different process to the one responsible for line narrowing at low temperatures. These observations, as well as the temperature dependence of T_1, lead directly to the conclusion that the narrowing of the line from 100 K upwards is brought about through the steadily increasing amplitude of a motion which is occurring at frequencies of the order of gigahertz or higher. The temperature dependence of the linewidth thus reflects an increasing *spatial* rather than *temporal* averaging of the *inter*- and *intra*chain ^1H–^1H dipolar interactions. Typically, such motions will be localised librations which can occur with ever increasing angular amplitudes as the temperature is raised. This can be expressed in the following simple model. When anisotropic motion is fast enough to cause averaging of a spin interaction such as the dipole–dipole coupling as envisaged here, the total second moment of the interaction, $M_{2\text{T}}$, is partitioned into two parts

$$M_{2\text{T}} = M_2(t) + \langle M_2 \rangle \qquad (5.1)$$

where the first term is the zero-average fluctuating part and the second is that part left static. The observed line narrowing in PET is due to a progressive increase in $M_2(t)$ with temperature and the consequent reduction in $\langle M_2 \rangle$, since $M_{2\text{T}}$ is essentially conserved. Only motional frequencies greater than $\langle M_2 \rangle^{\frac{1}{2}}$ will produce this averaging. T_1 and $T_{1\rho}$ behaviour, which is driven by the term $M_2(t)$, shows that, in fact, they are of the order of gigahertz.

The temperature dependence of the linewidth appears to be an activated process but this arises because of the change in 'structure' which involves what might be described as a continuous phase transition, the energy term at any given temperature being the difference in free energy involved in expanding the lattice and allowing larger amplitude librations to occur.

In addition to this high-frequency motion, $T_{1\rho}(^1\text{H})$ detects a motion with a frequency around 100 kHz at about 400 K involving the methylene groups in PET. This correlates with the mechanical α-relaxation and it is presumed to involve specific local motions of these groups. $T_1(^1\text{H})$, in turn, reveals a third motion characterised by a frequency of around 150 MHz at 300 K which correlates with the β-relaxation arising from flips, presumably over 180°, of the aromatic rings in a part of the polymer.

Overall, a model was generated for PET in terms of three different regions, namely, crystalline, less-mobile amorphous (LMA) and more-

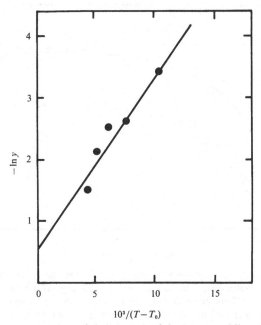

Figure 5.24. The variation of the amount of the more mobile amorphous fraction in PET with inverse temperature and the fit (solid line) to the model for $y(T)$ as described in eq 5.2. (Reprinted with permission from English (1984). Copyright (1984) American Chemical Society.)

mobile amorphous (MMA) regions (cf Section 6.4.2). The high-frequency motion driving line narrowing was shown to be present throughout all regions, whereas the aromatic ring flips associated with β-relaxation were presumed to occur only in the LMA polymer. The evolution of the MMA polymer with increasing temperature, implicit in the growth in intensity of the narrow line, was expressed in terms of a simple free-volume model as follows

$$y(T) = y_0 \exp\left[-\gamma V_c / \alpha V_m (T - T_0)\right] \qquad (5.2)$$

where $y(T)$ is the amount of MMA material at temperature T, y_0 is the total amount of amorphous material, V_c is the critical value of free volume above which LMA becomes MMA in NMR terms, V_m is the volume per chain segment over the temperature range covered, α is the coefficient of thermal expansion and T_0 is the temperature at which MMA material is first detected. γ is an adjustable parameter in the range 0.5–1.0. Figure 5.24 shows the fit of the experimental data to this model.

A model was also generated to describe the librational narrowing of the broad line component. Figures 5.25 and 5.26 portray the variation with

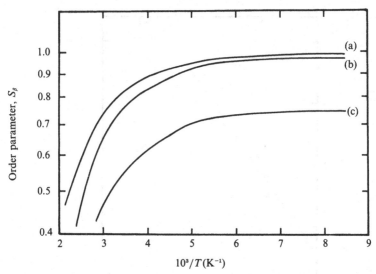

Figure 5.25. The variation with temperature of the ^{1}H dipolar order parameter in PET samples. (a), polycrystalline (about 50 %); (b), crystalline film/fibre; (c), amorphous film/fibre. This shows the rapidly increasing degree of averaging of the dipolar couplings which occurs as the temperature increases. Differences between samples are thought to reflect higher levels of static disorder in the amorphous form of PET. (Reprinted with permission from English (1984). Copyright (1984) American Chemical Society.)

temperature of the order parameter, S_β, and the width, β_0, of an assumed Gaussian distribution of librational angular amplitudes for this motion. In essence S_β^2 measures the degree of spatial averaging, given by

$$S_\beta^2 = \langle M_2 \rangle / M_{2\mathrm{T}} \qquad (5.3)$$

whereas β_0 indicates the spread of libration angles involved at any one temperature.

^{13}C NMR spectra of both static samples and those subjected to MAS were also measured. Whilst they offered support for certain aspects of the interpretation, it was the ^{1}H measurements which, in this study, were crucial. ^{1}H NMR in solids is often more difficult to interpret because of the coupling of spin–lattice relaxation processes through spin-diffusion and the difficulty of separating *intra*chain from *inter*chain couplings. However, this study shows that, used in the right way, considerable insight can be generated into the wide range and nature of motions through ^{1}H NMR.

The ability to observe high-resolution ^{13}C NMR spectra in solid polymers brings with it the promise of being able to study the motions of chemically distinct sites through measurement of spin-relaxation be-

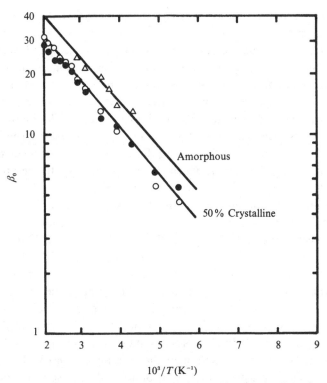

$10^3/T(\mathrm{K}^{-1})$

Figure 5.26. The temperature variation of the width β_0 of a Gaussian distribution of vibrational amplitudes chosen to represent the variation of the ^{1}H spin-lattice relaxation in PET. Open and filled circles refer to partially deuterated polymer with only methylene and aromatic protons, respectively. Open triangles are data for the fully protonated polymer. The activation energy $\Delta E = 1.1$ kcal mol^{-1}. (Reprinted with permission from English (1984). Copyright (1984) American Chemical Society.)

haviour of each line in a spectrum. Early efforts in this direction concentrated on spin–lattice relaxation (SLR), particularly in the rotating frame. An investigation of PET by Sefcik and coworkers (1980) illustrates both the strengths and difficulties of this approach. Interest in $T_{1\rho}$, it is recalled, arises from the fact that the motional frequencies to which this relaxation is sensitive (about 50–100 kHz) are thought to be important in determining the mechanical properties of the solid polymers. Indeed, Schaefer and coworkers (1977) deduced a correlation between the low-frequency SLR behaviour and impact strength for some glassy polymers.

Care must be taken in interpreting $T_{1\rho}$ behaviour of dilute spins such as ^{13}C in solids. Relaxation is usually characterised by spin-locking the ^{13}C magnetisation in the absence of ^{1}H dipolar decoupling followed by acquisition of the high-resolution spectrum with DD and MAS (cf Section 4.4.2.5). Each peak in the spectrum decays as a function of the spin-locking time and this decay can be analysed to give site-specific relaxation behaviour. Note first that this relaxation is often a non-exponential process. The second, and more important, observation is that the relaxation may not be determined entirely, if at all, by the expected mechanism of modulation of the ^{13}C–^{1}H dipolar couplings by thermally activated motions. The reason for this is quite straightforward. The ^{13}C spins only need to experience fluctuating dipolar fields at frequencies of order ω_1 to relax to any reservoir which is at a different spin-temperature. The dilute ^{13}C spins in a typical organic polymer are embedded in a sea of ^{1}H spins. Even in the absence of thermally driven motions, the ^{1}H–^{1}H dipolar spin flip–flop dynamics produce a frequency fluctuation spectrum which can effectively couple the spin-locked ^{13}C spins to the ^{1}H dipolar reservoir which has an effective infinite spin temperature. Schaefer and coworkers (1977, 1984a) devised a scheme which allows the contribution of this spin-dynamics process to the observed ^{13}C rotating frame SLR to be determined. This is achieved by measuring the rate of cross polarisation from a dipolar ordered state of the proton spins to the ^{13}C spins in their resonant rf field. The ^{1}H dipolar ordered state is usually produced by means of adiabatic demagnetisation in the rotating frame (ADRF) (cf Section 2.5). This measures the contribution of the spin dynamic fluctuations to the ^{13}C $T_{1\rho}$ process, but it should be noted that the measurement must be carried out on a static sample since MAS destroys dipolar order very rapidly. In general, the larger the ^{13}C spin-lock field used relative to the ^{1}H dipolar linewidth, the less important is this non-motional contribution. For ^{13}C rf fields of 30–40 kHz, the $T_{1\rho}(^{13}C)$ behaviour of glassy polymers with predominantly aromatic protons or partially motionally averaged aliphatic protons is dominated by motions. The more rigid/crystalline systems, particularly those with aliphatic chains, require much larger ^{13}C rf fields to ensure that motion dominates.

The $T_{1\rho}(^{13}C)$ behaviour of both quenched amorphous and partially crystalline PET samples was also characterised by Schaefer and coworkers (1984a). Not being a single process, relaxation was described in terms of the initial slope giving the average relaxation time $\langle T_{1\rho}(^{13}C) \rangle$ for each resolved line in the MAS spectrum. Apart from the ethylene carbons in the crystalline regions of the partially crystalline samples, the mean cross polarisation rate $\langle T_{IS}(ADRF) \rangle$ indicated that motional contributions dominated $\langle T_{1\rho}(^{13}C) \rangle$ relaxation for each line. For the ethylene carbons in the amorphous regions, $\langle T_{1\rho}(^{13}C) \rangle$ increased approximately as B_1^2, con-

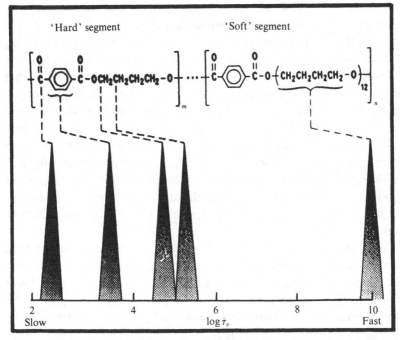

Figure 5.27. A schematic representation of the structure and correlation times for motions of the various carbon sites in the poly(butylene terephthalate-co-tetramethylene-undecakis(oxytetramethylene)terephthalate copolymer. (Reprinted with permission from Jelinski *et al.* (1983a). Copyright (1983) American Chemical Society.)

sistent with motions in the low-to-mid kilohertz region. Annealing at temperatures below and above T_g produced systematic changes in $\langle T_{1\rho}(^{13}C)\rangle$ which were interpreted in terms of changes in chain-packing, free-volume and conformations.

Jelinski and coworkers (1983a, b, c, d) used both ^{13}C and ^2H NMR methods to characterise motions in both poly(butylene terephthalate) (PBT) and its copolymers with tetramethylene-undecakis(oxytetramethylene) terephthalate which comprises hard and soft regions. The structure of these polymers is shown in fig. 5.27. The full range of ^{13}C solids NMR techniques is exploited to produce a view of the motions occurring at each carbon site in the system and their dependence on polymer composition. The high-resolution spectra gave resolved lines for OCH_2 groups in the hard and soft segments, which, together with measurements of $T_1, \langle T_{1\rho}(^{13}C)\rangle$, off-magic-angle sample spinning and static sample difference spectroscopy using specifically and partially deuterated polymers, yielded a complex model of the motions at each site (fig. 5.27). T_1 values and NOEFs indicated motions rich in the megahertz frequency range for the OCH_2 groups in the soft regions which required a broad log χ^2 distribution of correlation times for their interpretation. The same carbons were shown to cross polarise effectively, which also indicated that their fast motions were actually restricted in angular amplitude so that sufficient $^{13}C-^1H$ dipolar coupling remained secular to mediate the CP process. The $\langle T_{1\rho}(^{13}C)\rangle$ values for the soft segment signals were independent of B_1 but varied with composition, whereas those for the hard segments decreased with decreasing B_1. The scenario portrayed in fig. 5.27 was distilled from these measurements in conjunction with shielding tensor principal component data. The terephthalate group undergoes rather slow motions, which are limited to librations and 180° flips of the phenylene rings.

Apart from the specialist capabilities of multidimensional NMR, elucidation of the precise mechanism of a particular motion is often best addressed by observing changes in lineshapes, either directly from FIDs or from appropriate echo experiments. The quadrupolar lineshape for ^2H spins is increasingly used for this purpose: PBT deuterated at the two inner carbons of the butylene group is a good example (Jelinski *et al.*, 1983c). Figure 5.28 compares the ^2H lineshape for this material at different temperatures with spectra simulated on the basis that the C–D bond flips randomly between two equal energy sites which differ by a dihedral angle of 103°. As a measure of the sensitivity of the simulation, note that dihedral angles of 100° or 110° would not fit the data. Several possible molecular mechanisms were identified which could produce the necessary interchange of the C–D bonds needed to fit the ^2H spectra. Of these, only two were consistent with the requirement, supported by ^{13}C measurements, that the terephthalate groups are essentially static: three bond motions involving *gauche* conformation migration ($ttg^{\pm} \leftrightarrow g^{\pm}tt$) and *gauche* pair production ($ttt \rightarrow g^{\pm}tg^{\mp}$).

Underlying such simulations are the generally accepted facts (i) that the ^2H quadrupole interaction dominates the lineshape, (ii) that the interaction is axially symmetric and (iii) that the principal axis lies along the C–D bond. In addition, there are certain experimental considerations to be noted. Unless special care is taken with the design and construction of the ^2H NMR sample probe, it is often difficult to use sufficiently short pulse lengths to ensure uniform coverage of the full spectral width, which typically will be around $+200$ to -200 kHz for ^2H in C–D bonds. Furthermore, because the FID corresponding to the typical Pake powder

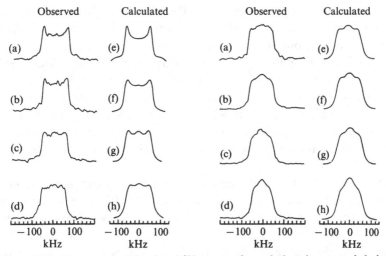

Figure 5.28. Observed and simulated ^{2}H spectra for poly(butylene terephthalate), deuterated at the middle two carbons of the butylene units, as a function of temperature. The temperatures were (a)–(d) (left), -88; -32; -11; 5 °C; and (a)–(d) (right), 21; 44; 63; and 85 °C. The simulated spectra assume a two-site jump process between orientations differing by an angle of 103°. The jump rates for the spectra were (e)–(h) (left), $\ll 10^4$, 1.9×10^4, 4×10^4 and 7.1×10^4 s^{-1}, and (e)–(h) (right), 1.4×10^5; 2.1×10^5; 2.8×10^5 and 5.7×10^5 s^{-1}. (Reprinted with permission from Jelinski *et al.* (1983c). Copyright (1983) American Chemical Society.)

doublet characteristic of the spin-1 quadrupolar NMR lineshape has a very rapid initial decay, the spectra are usually recorded using a $90°_x$–τ–$90°_y$ solid-echo sequence, the echo being Fourier transformed to give the spectrum. In this study of PBT, the spectra shown in fig. 5.28 were corrected for the effects of finite pulse length whereas effects of motions on the timescale of τ were neglected.

^{2}H NMR has developed into one of the most powerful tools for characterising both order and motions in appropriately deuterated polymer systems. A number of reviews discuss the subject in greater detail (Sillescu, 1982; Spiess, 1983, 1985a, b, 1991; Blümich and Spiess, 1988).

5.4.2 Polystyrenes (PS)

There have been a number of detailed studies of motions in solid polystyrenes which afford the opportunity to introduce some of the more advanced diagnostic NMR methods available. Continuing their ^{13}C high-resolution NMR studies, Schaefer and coworkers (1984b) examined both the $T_1(^{13}\text{C})$ and $T_{1\rho}(^{13}\text{C})$ behaviour of polystyrene (PS) and substituted polystyrenes at ambient temperature. They further exploited the ^{13}C–^{1}H dipolar coupling for proton-bonded carbons and the chemical shift lineshape for non-protonated carbons to characterise the amplitudes and nature of specific motions. ^{13}C–^{1}H dipolar couplings were determined using the dipolar rotational spin echo (DRSE) experiment introduced in Section 5.3.5. This experiment generates a discrete sideband representation of the ^{13}C–^{1}H dipolar lineshape which, in the presence of motions that are sufficiently fast to produce some motional averaging, has sideband intensity ratios that can be related to the specific nature, symmetry and amplitudes of the motions.

This range of ^{13}C NMR measurements reveals a complex motional profile in these materials: all observations on PS were at ambient temperature, which is well below its T_g (373 K). Motional heterogeneity was established in as much as a small proportion of the total phenyl ring population, distinguished by short $T_1(^{13}\text{C})$ values, was found to be undergoing fast (MHz) 180° ring flips, possibly with associated ring librations. The DRSE sideband pattern for the aromatic protonated carbons established that the motions of this small fraction were of large amplitude whilst slow spinning spectra, from which the ^{13}C chemical shift anisotropy was determined for the quaternary *ipso* ring carbon, pointed to 180° ring flips rather than large-amplitude rolls of the ring. The distinction between these two mechanisms is that the ring spends negligibly short time in the transition between the two states for 180° flips. Motional heterogeneity implies that these particular rapidly flipping rings are locked into the solid structure on timescales of the order of at least the spin–lattice relaxation times. Indeed, Schaefer and coworkers (1984b) concluded that this fraction of rings is largely unaffected by *intra*chain conformational or configurational defects, tacticity, or physical treatment such as annealing. Rather, they suggested that it arises as an intrinsic part of the *inter*chain packing in the glassy state. Through measurements of $T_1(^{13}\text{C})$ on a PS enriched in ^{13}C at the methine carbons, they associated with the rings undergoing fast flips a similar small fraction of methine carbons undergoing large-amplitude high-frequency motions. This pointed to cooperative motional processes in the polymer.

The longer $T_1(^{13}\text{C})$ signals were associated with relatively short $T_{1\rho}(^{13}\text{C})$ timescales and partially averaged dipolar coupling tensors for the aromatic protonated carbons, from which it was deduced that the main portion of the phenyl rings was undergoing motions of low frequency (100 kHz) with restricted angular displacement (30–40°). Again, motional heterogeneity was invoked to explain the results.

Partially deuterated polystyrenes were examined in great detail by Spiess and his coworkers using powerful techniques which exploit the special NMR properties of the deuteron in solids. The advantages more than outweigh the extra effort required to synthesise the labelled materials. Before considering the results for PS we will look briefly at these techniques.

Deuteron NMR has the advantages, as mentioned above, that the quadrupole interaction in typical chemical environments such as C–D or O–D bonds, is large enough to dominate both the spectrum and relaxation properties but not so large as to make experiments too specialised for regular use. The axially symmetric environment of a C–D bond yield Pake powder doublet lineshapes with a quadrupole coupling constant of 167 kHz (fig. 2.11). The essentially *intra*molecular origin of the electric field gradient which determines the quadrupole coupling strength means that, unambiguously, the motions which affect the NMR properties are reorientations of the C–D bonds. In addition, the changes in lineshape induced by the onset of motion can reflect directly the particular nature, symmetry and mechanism of that motion which, of course, is not a unique feature of ^{2}H NMR. A further advantage lies in the fact that the dynamic range of motional frequencies which can be accessed is considerable, spanning many decades in favourable circumstances. This range is available through various one-dimensional and two-dimensional experiments where the low-frequency limit is usually set by the spin–lattice relaxation processes. In some instances, very slow motions can be gained access to, as described below. Finally, because of the sensitivity to local motions and the absence of efficient spin-diffusion, motional heterogeneity can often be revealed through differential spin–lattice relaxation to separate spectra from regions with characteristically different motions.

Figure 5.23 illustrates the specificity of changes in both solid echo and normal ^{2}H spectra arising from specific mechanisms and rates of C–D reorientation. Each potential symmetry of discrete jump has a different pattern of effects on the spectra and, in the case of the solid echo experiment, on the intensity of the signal obtained. Even in the fast motion limit when spectra become independent of the rate of the motion, they retain characteristics which are distinctive of the symmetry of the motion. Figure 5.29 further illustrates this point for three different types of motion (Spiess, 1983). It may be noted in passing that an axially symmetric coupling such as quadrupolar coupling can give rise to motionally averaged spectra characterised by $\bar{\eta} > 0$.

The picture of motional heterogeneity in polystyrene below T_g based on ^{13}C measurements (Schaefer *et al.*, 1984b) also shows up in the ^{2}H NMR

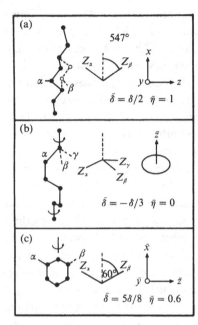 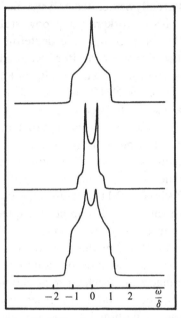

Figure 5.29. Left: three simple motions involving C–D bond reorientation and the resulting averaged (fast exchange limit) quadrupole coupling tensors, and right: their corresponding fast exchange limit ^{2}H spectra. It should be noted that the axially symmetric (static) quadrupole coupling can yield non-axially symmetric tensors and, hence, spectral lineshapes when motions modulate the interaction. The motions illustrated are: (a) a three-bond kink motion; (b) a five-bond crankshaft motion; and (c) a 180° flip of an aromatic ring. (Reproduced with permission from Spiess, *Colloid Polymer Science*, **261**, 193 (1983). Copyright (1983), Steinkopff Verlag.)

response of partially deuterated samples (Spiess 1983). For example, the ^{2}H spectrum of PS-d$_5$ (deuterated phenyl rings) looks somewhat like a rigid Pake powder doublet except that extra peaks are evident and its intensity distribution does not match. The overall ^{2}H spin–lattice relaxation is highly non-exponential and, by selectively monitoring the fast relaxing components, it was shown that they correspond to about 20 % of the total rings undergoing rapid 180° flips, which is in reasonable accord with ^{13}C data. The 180° flips were identified by the fact that the spectral shape matched that shown in fig. 5.29(c).

The solid echo and normal spectra discussed above can give information on rates of C–D bond reorientation characterised by correlation (exchange) times in the approximate range of 10^{-4}–10^{-7} s. The faster motions in this range overlap the range of motions sampled by spin–lattice

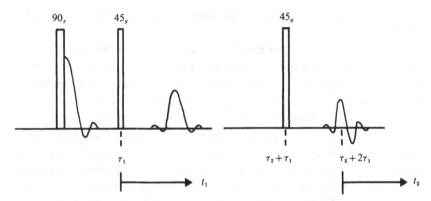

Figure 5.30. A typical pulse sequence for generating spin alignment in ^{2}H spins in a solid. The first two pulses create the spin alignment and must be phase shifted by 90° from each other for on-resonance irradiation. The spin alignment decays between the second and third pulses with a relaxation time of a similar order to T_1. The third pulse produces the spin alignment echo which is the signal in-phase with this last pulse.

relaxation time measurements. The desirable goal of extending the accessible range to slower rates (longer exchange lifetimes) has been achieved by two approaches, namely, spin-alignment echo formation and the two-dimensional exchange experiment.

The spin-alignment echo sequence illustrated in fig. 5.30 is essentially the sequence introduced by Jeener and Broekaert (1967), where the first two pulses produce a transfer of some of the spin order represented by the equilibrium magnetisation M_0 into the secular interactions which determine the shape of the FID. For a spin-1 quadrupolar powder lineshape, this $90°_x-\tau-45°_y$ pulse pair can transfer up to half this spin order into the quadrupole interaction, for which the terms *spin alignment* or *quadrupole order* have been coined (Spiess, 1980; Ahmad and Packer, 1979a, b; Packer, 1980). As the amount of order transferred in this way depends on the quadrupole frequency, the third pulse in fig. 5.30 recalls this as transverse magnetisation which refocusses to form the spin alignment echo at a time τ following the third pulse. The crucial aspects of this sequence are that the spin alignment persists for times of the order of the spin–lattice relaxation time and that motions which occur on timescales up to this extended range affect both the amplitude of the echo and the shape of the spectrum obtained by its Fourier transformation. Full details of these relationships can be found in the review literature (Spiess, 1983, 1985a, b, 1991). The spin-alignment echo is analogous to the Hahn stimulated echo obtained when a train of three 90° pulses is applied to a system whose lineshape is dominated by inhomogeneities in, say, B_0 or the chemical shift (Hahn, 1950).

If it is assumed that reorientations occur on a timescale such that no motion of significance occurs during the period τ_1, the first two pulses can be thought of as simply preparing states of spin alignment, the amplitude of which reflects the value of the quadrupole frequency and hence C–D bond orientation of each bond in the sample. If no motion were to take place in the period t_2 then each transverse magnetisation component recalled by the third pulse evolves with the same quadrupole frequency as during τ_1 and a full refocussing of these signals occurs at a time τ_1 after the third pulse. Changes in the orientation of C–D bonds in the time interval $t_1 = \tau_2$ reduces the degree of refocussing generated by the third pulse and alters the frequency characteristics of the spin-alignment echo in a manner which reflects the statistics of the frequency changes in τ_2. In fact the spin-alignment echo can, under these circumstances, be expressed as a correlation function,

$$E_{sa}(\tau_1, \tau_2, t_2) = \langle \sin[\omega_Q(0)\,\tau_1] \cdot \sin[\omega_Q(\tau_2)\,t_2] \rangle \qquad (5.4)$$

where $\omega_Q(0)$ and $\omega_Q(\tau_2)$ are the values of the quadrupole frequency of a deuteron in a given C–D bond during the evolution period τ_1 and at $t_1 = \tau_2$. The ensemble average $\langle\,\rangle$ accordingly reflects the changes in ω_Q occurring in the time τ_2. As the persistence of spin alignment is governed by spectral densities which are similar to spin–lattice relaxation, timescales for C–D bond reorientations of the order of 1 s can be monitored.

It may be helpful to note that the spin-alignment echo can be thought of as a solid echo which has been stored for the time τ_2 in a state of spin order which is independent of quadrupole frequency: in the limit $\tau_2 \to 0$ the echo becomes a solid echo. In general, account must be taken of motions occurring in the times τ_1 and t_2 as well as t_1. .

Before moving on to describe two-dimensional experiments, it is worth making some general comments about their application to polymers. The discussion thus far has alluded to symmetries, mechanisms and rates of C–D bond reorientations. Polymer systems, as we have already seen, often have complex modes of motion which may, under any given set of conditions, cover a distribution of frequencies, angular amplitudes and mechanisms (for example, rotational diffusion, discrete large-angle jumps and so on). Both the solid echo and spin-alignment echo sequence respond to all of these features. We have already seen, for example, that jumps between orientations separated by different angles produce characteristically different lineshapes both in the intermediate and fast exchange situations (figs 5.23 and 5.29). If for a given motion, there exists a distribution of rates then this will, in turn, affect the response of the echo sequences. A broad distribution, for example, will mean that the reduction in signal intensity which, for a single correlation time process would be maximised when $\Delta\omega_Q\tau \sim \pi$, will never be as great since at any given temperature only a fraction of the spins will satisfy this condition. In essence, such a broad distribution of jump times results in lineshapes which are a complex superposition, particularly in the intermediate exchange regime.

The spin alignment echo sequence forms a natural link to two-dimensional experiments in the sense that it comprises preparation/evolution, mixing and detecting intervals. While fully appreciating the similarities, the way in which the experimental data are treated and thought about is quite distinct (cf Section 4.5). Applications to polymers have been succinctly reviewed (Blümich and Spiess, 1988; Spiess, 1991) and the theory fully described (Wefing and Spiess, 1988; Wefing *et al.*, 1988; Kaufmann *et al.*, 1990; Hagemeyer *et al.*, 1990).

The essence of the two-dimensional approach to characterising bond reorientations is that the NMR frequency is directly related to orientation through the appropriate anisotropic spin interaction. Thus, the preparation/evolution period of a two-dimensional experiment involves labelling the spins' orientations through their different NMR frequencies. The first pulse and period τ_1 of the spin alignment echo sequence played precisely this role for C–D bonds. The start of the mixing or exchange period is defined by a pulse which transfers at least some of the orientation-labelled transverse magnetisation at the time τ_1 into a state in which no evolution under the orientation-dependent spin interaction occurs. Again, in the spin-alignment sequence, the second rf pulse accomplished this transfer. When a further pulse is applied to produce transverse magnetisation at the start of the detection period, the frequencies which particular spins experience during this detection period will reflect the probabilities that they have undergone reorientations during the mixing period t_m. In short, a two-dimensional spectrum denotes the joint probability density of finding spins with frequencies Ω_1 and Ω_2 before and after the mixing time t_m, respectively.

Clearly, this joint probability directly reflects the nature and time evolution of the appropriate bond reorientations. For simple crystalline organic solids, for example, in which the crystallography dictates that well-defined jumps through specific angles are expected to be the mechanism for molecular reorientation, off-diagonal intensity in the two-dimensional

Experiment   Simulation

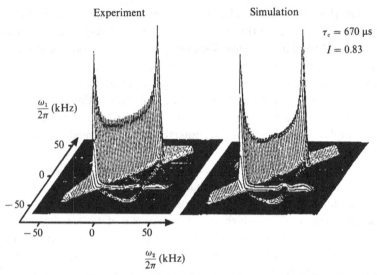

Figure 5.31. The ^{2}H two-dimensional exchange spectrum of polycrystalline dimethylsulphone-d$_4$ at $T = 312$ K. The mixing time during which exchange is monitored was 3 ms. The experimental spectrum (left) compares closely with a simulated spectrum with a jump time for exchange of 670 μs. The spectra show the characteristic elliptical off-diagonal features from which the jump angle of the C–D bond can be measured directly. (Reprinted with permission from Kaufmann *et al.* (1990). Copyright (1990), American Institute of Physics.)

spectrum is predicted and, in fact, appears as elliptical ridges (fig. 5.31). For polymeric solids in which the reorientational motions are more complex, the off-diagonal intensity distribution in a typical ^{2}H two-dimensional spectrum is concomitantly more complex. However, it has been shown (Hagemeyer *et al.*, 1990) that reorientational angular distribution functions (RADs) may be reconstructed directly from such two-dimensional spectra at each value of t_m without recourse to any model of the motion. This is a unique achievement of NMR, which adds considerably to its potential for characterising motions.

Two-dimensional experiments have been used to investigate chain motions in PS-d$_3$ (–(C$_6$H$_5$) CDCD$_2$–)$_n$ in the regime where temperature is decreasing towards T_g (Wefing *et al.*, 1988; Kaufmann *et al.*, 1990). Typical spectra and associated RADs are shown in fig. 5.32. To obtain these RADs, a distribution of correlation times for the chain motions had to be introduced together with the assumption of isotropic reorientation. It was concluded from these measurements that the width of the assumed log-Gaussian distribution of correlation times increased from three decades at $T = 391$ K to five decades at $T = 375$ K with a corresponding substantial

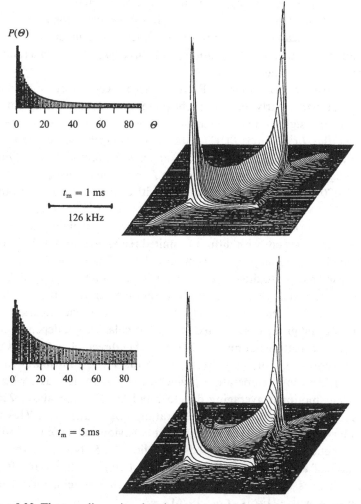

Figure 5.32. The two-dimensional exchange spectra of solid deuterated polystyrene at $T = 391$ K. The mixing times used in the experiments were (top) 1 ms and (bottom) 5 ms. Also shown are the reorientation angle distributions (RADs) derived from the spectra. All such spectra at this temperature could be fitted assuming a log-Gaussian distribution of jump times with a width of about three decades and a mean correlation time of 6 ms. (Reprinted with permission from Kaufmann *et al.* (1990). Copyright (1990), American Institute of Physics.)

increase in the mean correlation time from 6 ms to 30 s. Spectra taken at temperatures close to $T_g = 373$ K were, in fact, detecting chain reorientation over a few degrees on the timescale of seconds. This ability to observe the molecular basis for such motional retardation as the glass transition is approached is again unique to NMR. From this study, and the results of previous measurements of ^{2}H spin–lattice relaxation times, it was found that the mean correlation times as a function of temperature obeyed the Williams–Landel–Ferry equation (Williams *et al.*, 1955) over a range of 11 orders of magnitude.

The relaxation behaviour of PS and, indeed, certain other aromatic polymers, is particularly sensitive to the presence of O_2 at low temperatures where, at first sight, unusual behaviour is observed (Froix *et al.*, 1976). Consider the $T_1(^1H)$ and $T_2(^1H)$ data in fig. 5.33, which compares the response of PS in the presence (curve E) and absence (curve D) of O_2. First, note the coincidence in temperature of the T_1 'minimum' and T_2 'transition' at about 220 K for PS–O_2 which is at odds with conventional interpretation (cf fig. 2.15). Note also the initial $t^{\frac{1}{2}}$ dependence of T_1 recovery below 220 K which reverts to exponential decay at longer times. As explained in Section 3.8.5, this is consistent with diffusion-limited relaxation, in this case to the relatively fixed paramagnetic oxygen sites. Above 220 K, the T_1 relaxation mechanism changes to direct paramagnetic relaxation by rapidly diffusing O_2 molecules which generate a uniform spin temperature in the sample. Thus, in the low-temperature region, T_1 behaviour encompasses spin diffusion to end groups as reflected in the molecular weight dependence of spin–lattice relaxation (Connor, 1970) and either direct relaxation by O_2 or spin diffusion to O_2 or both. By the same token, O_2 contributes to the dipolar field at low temperatures, which causes T_2 to decrease, whereas subsequent motional averaging due to rapid O_2 diffusion above 220 K causes T_2 to increase again thus generating a T_2 'transition'. This is a particularly vivid example of the different behaviour of static and mobile paramagnetic species, as discussed in Section 3.8.5. Note too that the formation of an O_2 complex with the phenyl ring results in an off-axis dipole which is modulated by motion of the ring system, rendering it dielectrically active (Froix *et al.*, 1976).

5.4.3 *Polycarbonates (PC)*

The motions of phenylene rings in solid polycarbonates have been investigated by ^{1}H, ^{2}H and ^{13}C NMR. Whilst detailed interpretations differ, the general deduction is that the rings undergo 180° flips about their 1,4 axes with a wide spatially inhomogeneous distribution of rate. Motions of the main chain are more limited and there is some difference of opinion as to the NMR evidence for them (Schaefer *et al.*, 1984c; Wehrle *et al.*, 1987). The natures of the ring motions are indicated in a number of ways.

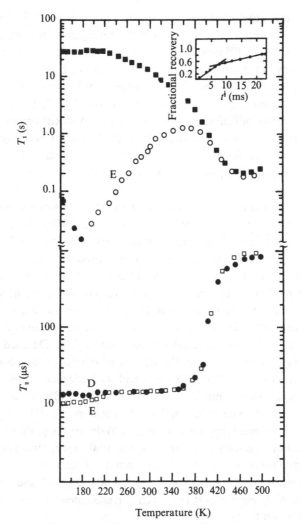

Figure 5.33. $T_1(^1H)$ and $T_2(^1H)$ for degassed (D) and oxygenated (E) polystyrene PS. The insert shows the $t^{\frac{1}{2}}$ dependence of magnetisation recovery (see text). (Adapted with permission from Froix *et al.* (1976). Copyright (1976) American Chemical Society.)

For the case of chloral polycarbonate, protons are confined to the aromatic rings and they thus form somewhat isolated pairs which give rise to a Pake powder dipolar doublet lineshape. The doublet splitting was found to be largely independent of temperature despite motional narrowing conditions (Jones, 1989). This again can only arise if the rings flip by 180° about the 1,4 axis since the dipolar coupling is invariant to this interchange of spins. Dipolar rotational echo measurements of the ^{13}C–^{1}H dipolar coupling of the off-axis carbons in the phenylene rings were also consistent with such a 180° flipping mechanism but required, in addition, a higher frequency libration about the mean positions with an amplitude of about 30° at room temperature (Schaefer *et al.*, 1983, 1985). Through a detailed study of these dipolar sideband spectra and $T_1(^{13}$C) and $T_{1\rho}(^{13}$C) behaviour for a range of different polycarbonates and model compounds, Schaefer and coworkers were able to establish the frequency heterogeneity of these motions which fell within the range 0.3–25 MHz at 300 K. This is similar to the jump-frequency distribution width deduced from ^{2}H measurements (see below).

One method not yet covered is the use of ^{13}C shift anisotropy powder patterns to characterise motions. Roy and coworkers (1986) adopted this approach to characterise phenylene ring motions in bisphenol A polycarbonate. Figure 5.34 shows the observed and fitted lineshapes for this solid which has been ^{13}C-enriched ($> 90\%$) at one of the carbons in the *ortho* position relative to the carbonate residue. The enrichment, which preferentially enhances the signal from this carbon well in excess of the signals in natural abundance from the other sites, allows unequivocal characterisation with high-powered dipolar decoupling. Detailed consideration of motional models and the quality of the fits to the observed spectra led to the conclusion that 180° flips and oscillation about the mean position were occurring simultaneously (Roy *et al.*, 1985, 1986, 1990; O'Gara *et al.*, 1985). The combined CSA lineshape, $T_1(^1$H), $T_{1\rho}(^1$H), $T_{1\rho}(^{13}$C) and $T_1(^{13}$C) anisotropy data were used to derive the parameters of the assumed inhomogeneous stretched exponential correlation function (Williams–Watts) used to describe the motional processes (Connolly *et al.*, 1987). This fit gave a mean activation energy of 464 kJ mol^{-1} and a width exponent of the distribution of 0.15, which corresponded to a very broad distribution of jump times.

A similar view of the ring motions in bisphenol-A polycarbonate was generated by ^{2}H NMR measurements on polymer with deuterium substitution at the ring carbons in the *ortho* position relative to the carbonate residue (Spiess, 1983; Wehrle *et al.*, 1987). Polymer with deuterated methyl groups was also investigated. The object of these measurements was again to understand something of the origin of the mechanical relaxations which are key properties of these materials, typically occurring at about 173 K for a correlation frequency of 1 Hz. One of the aspects of ^{2}H NMR which is brought out very clearly is the way in which the presence of a wide distribution of correlation times for the motions affects the signal intensity

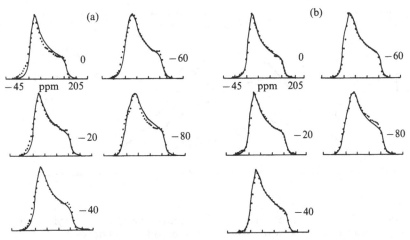

Figure 5.34. The temperature dependence of the ^{13}C powder lineshape of bisphenol A polycarbonate enriched in ^{13}C at an off-axis position in the phenylene ring. (a) 22.6 MHz; (b) 62.9 MHz. The continuous lines are fits to the lineshapes based on a model which allows simultaneous 180° ring flips and ring oscillations. Temperatures are in Centigrade. (Reprinted with permission from Roy *et al.* (1986). Copyright (1986) American Chemical Society).

from echo experiments, such as a solid echo. The important experimental observations were the fully and partially T_1-relaxed ^{2}H lineshapes recorded as a function of temperature which demonstrated clearly that the distribution of correlation times (or frequencies as these workers prefer) is strongly inhomogeneous, particularly at lower temperatures. This was shown by the fact that, although a fully relaxed spectrum appears to be dominated by a rigid-lattice lineshape, the partially relaxed spectra at short recovery times indicated the presence of rings which are flipping through 180° with jump times shorter than 10^{-5} s. These lineshapes and intensity reduction factors could be fitted with either symmetric (log-Gaussian) or asymmetric distributions of correlation times.

There are some points of difference between the ^{2}H study and those carried out using ^{13}C/^{1}H NMR. Whereas the simultaneous flip/oscillation model used to fit the ^{13}C lineshapes in fig. 5.34 is acknowledged to be sensible by Wehrle *et al.*, (1987), they suggest that no allowance is made for the different rates at which different amplitudes of motions occur. $T_1(^2$H) measurements, which should reflect the faster oscillatory motion of the ring, have an activation energy which is considerably less than that ascribed to the flip process itself. A second difference arises from observations relating to motions of the main chain carbons. The ^{2}H-

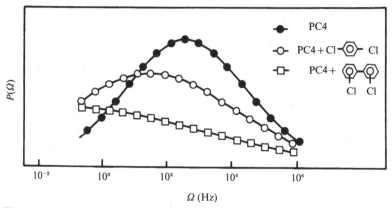

Figure 5.35. Distributions of correlation frequencies $v_c(= \tau_c^{-1})$ for phenylene ring flips in (bisphenol-A) polycarbonate, with and without additives (as shown) derived from the ^{2}H spectra of the ring-deuterated polymer. $T = 180$ K. (Reproduced with permission from Wehrle *et al.*, *Colloid Polymer Science*, **265**, 815 (1987). Copyright (1987), Steinkopff Verlag.)

labelled methyl groups undergo very fast motion about their three-fold axes giving a Pake powder doublet lineshape of reduced width. If the main-chain carbon to which these methyls are bonded were to undergo motions in the appropriate frequency range, the lineshapes would be further averaged. Observations over the temperature range studied ruled out any large-scale conformational changes but could not detect wiggling of the chains with amplitudes up to $\pm 15°$. Schaefer and coworkers (1984c) interpret their ^{13}C data as showing the existence of motions involving the main chain, probably in a cooperative process coupling ring oscillations with movement of the backbone.

The effects of additives on the motions have been investigated specifically to relate events at the molecular level to the bulk mechanical properties. ^{2}H NMR indicates that the distribution of ring-flip rates broadens to lower frequencies in the presence of low molecular weight additives such as *p*-dichlorobenzene and chlorinated biphenyls; that is, more rings move slowly but there are still fast flipping rings present (Wehrle *et al.*, 1987). Figure 5.35 portrays the derived distributions of ring 180° jump frequencies with and without additives. Liu and coworkers (1990) measured the temperature dependence of the $T_{1\rho}(^1$H) behaviour for bisphenol-A polycarbonate with completely deuterated methyl groups in the presence of varying concentrations of perdeuterated di-n-butylphthalate for which relaxation is due solely to ^{1}H–^{1}H dipolar couplings modulated largely by the ring flipping motions. It was concluded that concentrations of the additive in the range 1–5% caused antiplasticisation, rather similar to the observations made by Wehrle *et al.* (1987). Higher concentrations, however, increasingly facilitated motions associated with plasticisation (cf Section 6.9).

5.4.4 Polyethylene (PE)

There have been myriad NMR studies of polyethylenes as shown by the number and scope of the references to polyethylene in the index of Komoroski's (1986) edited compilation on the NMR of solid polymers. The name polyethylene (or polyethene) covers a wide range of materials from mechanically rigid linear high-density material through highly branched, often cross linked, soft and rubbery low-density systems. Use of other olefins as co-monomers with ethylene also generates sidechains, the number, type and distribution of which can have a significant and important influence on the properties of the solids produced. All of these factors can and do influence the character of the motions under various conditions of temperature and physical pre-treatment of the polymer. Because of this, it is important as always to know and specify the origins, characteristics and thermal/physical history of any samples studied. If this is neglected, comparisons can often become meaningless.

It is a recurring theme that molecular motion cannot really be separated from a description of the structural organisation of polymers and polyethylenes are no exception. Broadly speaking, the basic structure we have to keep in mind is that most polyethylenes are partially crystalline, usually with lamellar crystalline regions and other, more disordered and hence more mobile regions. A third, interfacial region is often detected (cf Section 6.4.1).

It is not the intention here to give a complete account of NMR investigations of motions in polyethylenes. Rather, certain features which reinforce the collective capabilities of modern NMR methods will be illustrated. As with most polymers, the main nuclei available are ^{1}H, ^{2}H and ^{13}C. ^{1}H lineshapes and spin–lattice relaxation behaviour have been widely exploited. However, as discussed in Chapter 1, while these measurements are relatively easy to make because of the high sensitivity of ^{1}H NMR, unambiguous interpretation in support of definitive models for motions is difficult. This arises mainly because of the many-particle nature of the dominant ^{1}H–^{1}H dipolar couplings which makes the separation of *intra-* and *inter*chain motions complex and also because of spin diffusional averaging of SLR characteristics. Of course, spin diffusion can be put to good use in probing structural organisation on the scale of the crystalline lamellae (cf Section 6.4.2). Typical effects of motion on ^{1}H lineshapes were illustrated in fig. 2.13.

$T_{1\rho}(^{13}C)$ behaviour of the crystalline regions of polyethylene is generally dominated by ^{1}H spin dynamics unless measurements are made at high temperatures (> 370 K) and with very high dipolar decoupling field strengths (> 80 kHz). Under such conditions VanderHart and Garroway (1979) concluded that the motions responsible for $T_{1\rho}(^{13}C)$ were consistent with the combined rotation (180°) and translation of chains through the crystalline lamellae as postulated earlier on the basis of ^{1}H broadline and $T_{1\rho}$ measurements (McCall and Douglass, 1965; Olf and Peterlin, 1970). There was general agreement on the motional rate involved ($\tau_c \simeq 5$ μs at $T = 373$ K). ^{13}C spectra also indicated that the jump process did not average the shielding tensor – hence the proposed 180° jump mechanism (VanderHart, 1987).

The same motion has been recently characterised by observing the dynamic behaviour of ^{13}C high-resolution spectra of polyethylene under partially saturated conditions (T_1) and by two-dimensional techniques (Schmidt-Rohr and Spiess, 1991b). The basic idea relies on the fact that the ^{13}C SLR in the disordered regions is very much faster than in the crystallites. Previous studies showed that SLR of the crystalline carbons was multi-exponential in character with dominant limiting relaxation times of the order of 10^3–10^4 s for the largest crystallites, which increased approximately as the square of the lamellar crystalline width (Axelson *et al.*, 1983; Cudby *et al.*, 1984; Kitamaru *et al.*, 1986; Colquhoun and Packer, 1987). $T_1(^{13}C)$ values for the disordered regions are usually around 0.5–1.0 s at 300 K.

As well as demonstrating the dependence of the T_1 values on lamellar thickness, Axelson and coworkers (1983) carried out a two-dimensional experiment which confirmed exchange of ^{13}C magnetisation between the two frequencies in the high-resolution MAS spectrum corresponding to the all-*trans* crystalline line and the mixed conformation, amorphous line. These workers suggested that ^{13}C spin diffusion was the mechanism for this exchange and, indeed, other work showed that the quantitative behaviour was consistent with a diffusion process, possibly spin diffusion (Colquhoun and Packer, 1987). However, as explained in Section 3.8.2, a dilute spin system such as ^{13}C cannot readily support such a spin diffusion process over the required distances and the recent study by Schmidt-Rohr and Spiess (1991b) showed quite clearly that the exchange of magnetisation between the disordered and crystalline regions is through actual chain diffusion. Figure 5.36 shows the partially saturated high-resolution SPE

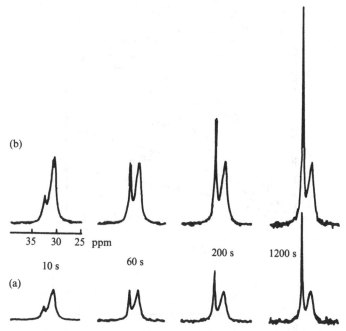

Figure 5.36. (a) High-resolution ^{13}C SPE spectra of a high MW polyethylene, at 305 K, as a function of recycle delay time. The larger intensities of the spectra in the row labelled (b) arise from the nuclear Overhauser enhancement effect of saturation of the ^{1}H spins during the recycle period. This NOE factor is 2 for each peak in each spectrum which, because the intrinsic $T_1(^{13}$C$)$ values of the crystalline and amorphous regions differ by a factor of 10^3, indicates that all the signals arise via the faster spin–lattice relaxation in the amorphous region together with chain transfer between crystalline and disordered regions. (Reprinted with permission from Schmidt-Rohr and Spiess (1991b). Copyright (1991) American Chemical Society.)

MAS spectra of a polyethylene as a function of the recycle time between pulses with and without the nuclear Overhauser effect (NOE). It can be seen firstly that the relative intensity of the crystalline peak grows as the recycle time is increased but that the NOE factor remains constant throughout (about 2). This latter observation confirms that the crystalline peak is produced from polarisation in the short T_1 disordered regions and that chain diffusion exchanges carbons between these regions. This SPE experiment can be thought of as a one-dimensional exchange experiment. Within a few seconds of the ^{13}C pulse, the spins in the disordered regions are fully polarised. Those in the crystallites, however, have very long intrinsic T_1 values and the polarisation which appears at the chemical shift corresponding to the all-*trans* crystalline environment is carried there by

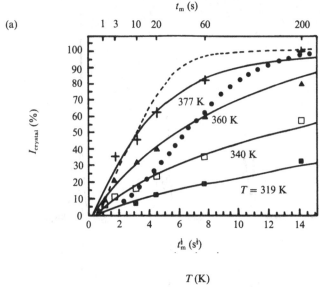

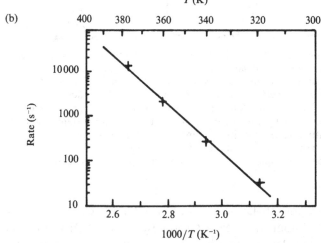

Figure 5.37. Analysis of the variation with recycle time of the intensity of the crystalline ^{13}C signal of the spectra shown in fig. 5.36. In the upper figure, the data are plotted as a function of $t_m^{\frac{1}{2}}$, which is the behaviour expected for a diffusion process at short times. The full lines represent fits generated by a Monte Carlo simulation which allows for constraints to the free chain diffusion (dashed line) arising from, say, chain entanglements in the disordered phase. The lower figure shows the jump rate derived from the fits as a function of temperature. The activation energy obtained from this plot is 105 kJ mol^{-1}, which is close to that for the α-relaxation process in polyethylene. (Reprinted with permission from Schmidt-Rohr and Spiess (1991b). Copyright (1991) American Chemical Society.)

chain diffusion. Two-dimensional versions of the same observation showed this exchange of magnetisation directly and also demonstrated that, if a cross-linked sample were used, exchange was considerably inhibited. Quantitative modelling of the process, as seen in the one-dimensional NMR experiment and incorporating effects of entanglements and other constraints on the totally free diffusion of chains, gave good fits to the experimental data (fig. 5.37). The Arrhénius plot in fig. 5.37 yielded an activation energy of $105 \, \text{kJ} \, \text{mol}^{-1}$ for chain diffusion which implies an association with the mechanical α-relaxation process.

Studies on motions in the disordered regions of polyethylene which have also been widely investigated by NMR encounter the problem of a range of conformations and associated motions which lead to spectra that are often a complex superposition and are partially averaged. Dipolar dephased high-resolution ^{13}C spectra showed that the broad low-frequency line assigned to the more mobile disordered regions has a complex distribution of residual $^{13}\text{C}-^{1}\text{H}$ dipolar couplings. This has been interpreted in terms of both interfacial material with more constrained motions and nearly isotropic chains which are more remote from the crystallite surfaces (Kitamaru *et al.* 1986) (cf Section 6.4.1). Figure 5.38 shows typical spectra and their dipolar dephasing characteristics. Interpretation of these and other observations, such as NOEFs does not generate a wholly acceptable quantitative model of motion, but fitting, say, the temperature and/or frequency dependence of the $T_1(^{13}\text{C})$ values and NOEFs to a typical model with a distribution of correlation times, using the ^{13}C CSA and dipolar coupled lineshapes to calibrate the strengths of the relaxation producing interactions, might be successful.

The main effort to produce such a description of the motions in the disordered regions of polyethylene has focussed on the use of ^2H NMR (Spiess, 1983; Hentschel *et al.*, 1984). These workers carried out a detailed study of a per-deuterated, isothermally melt-crystallised polyethylene using all the various echo and partial relaxation selection methods already mentioned. The fully relaxed ^2H spectra showed the presence of a typical rigid-lattice Pake powder doublet lineshape together with a central narrower signal. The powder doublet spectrum was shown to arise from the crystalline chains and, indeed, as the temperature was raised, it showed subtle changes which were consistent with the increasing frequency of the 180° chain reorientation/translation process discussed above. A detailed investigation of the dynamics in the disordered regions was undertaken utilising the fact that the $T_1(^2\text{H})$ values of these chains were very much smaller than those for the crystalline regions. A combination of solid and

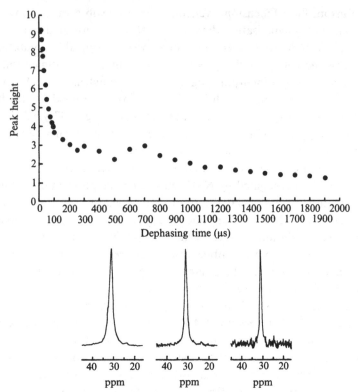

Figure 5.38. Dipolar dephasing behaviour (cf Section 4.4.2.3) of the SPE ^{13}C spectrum of a medium MW polyethylene. The upper figure shows the decay of the intensity of the peak assigned to the disordered region as a function of the dipolar dephasing time. The lower figure shows the SPE spectra at dipolar dephasing times of 10, 100 and 2000 μs (left to right). All SPE spectra were determined with a recycle time of 1 s and, although not apparent from the spectra shown, the frequency of the peak maximum shows a systematic shift with dephasing time. These facts indicate the strongly heterogeneous nature of this, apparently, single peak. (C. J. Groom-bridge, unpublished results).

spin alignment echo experiments was carried out as a function of temperature. Typical solid echo spectra together with their simulations are shown in fig. 5.39. Spin alignment spectra showed that even at 383 K, which is close to the melting temperature, spin alignment persisted in the disordered regions for times of the order of > 20 ms. Even in the melt, solid echo measurements indicated the existence of a residual secular quadrupole coupling. This raises again the question which was addressed briefly at the start of this chapter, namely 'when is a solid a solid?'. From the NMR viewpoint, polyethylene melts show some responses to pulse

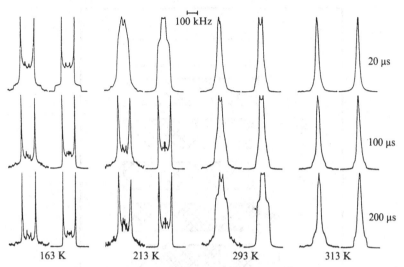

Figure 5.39. Observed (left) and calculated (right) pairs of ^{2}H solid echo spectra for the mobile (short T_1) deuterons in perdeuterated polyethylene. Spectra are shown for different temperatures and solid echo pulse spacings. (Reproduced with permission from Spiess, *Colloid Polymer Science*, **261**, 193 (1983). Copyright (1983), Steinkopff Verlag.)

sequences which would be normally thought of as 'solid-like'. This merely reflects the fact that on the relevant timescales the spins are not able to experience a time-averaged environment which satisfies isotropic symmetry and which leaves some of the anisotropic interactions unaveraged. This property of many polymer solutions and melts is revisited in Section 8.1.

The model developed by Hentschel and coworkers (1984) involved motions in which C–D bonds could occupy increasing numbers of sites through reorientations as the temperature was raised. For example, at lower temperatures it was only necessary to introduce lineshapes corresponding to jumps between two and three equivalent sites, the fast exchange spectra for which are shown in fig. 5.29. As the temperature increased, it was necessary to introduce sites involving averaging between a higher number of locations. Relating the number of sites to the number of bonds required to gain access to these locations produced the response in fig. 5.40 which shows the variation of the fractions of mobile units of different lengths as a function of temperature. The main conclusion from this investigation is that the motions of the chains in the disordered regions of linear polyethylene are localised and constrained, that is, constrained in the sense that their immediate environment of other chains imposes local order which limits the extent and nature of the interconversions of

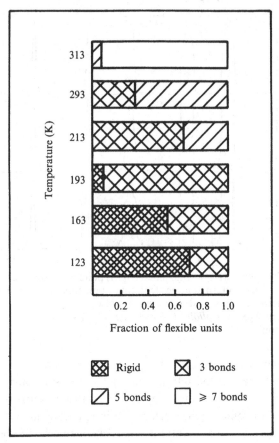

Figure 5.40. The variation with temperature of the populations of mobile flexible units in the amorphous phase of polyethylene derived from the analysis of the ^{2}H solid echo spectra shown in fig. 5.38. (Reproduced with permission from Spiess, *Colloid Polymer Science*, **261**, 193 (1983). Copyright (1983), Steinkopff Verlag.)

conformers that may occur. To some extent this may be thought of as similar to the concept of tube renewal time in the reptation model of polymers (cf Section 1.4), the time in this case being too long for a complete averaging of the deuteron quadrupole coupling to occur. As Hentschel and coworkers point out, their analysis is in terms of the topology of the exchange processes and only gives a lower limit for the lifetimes of the constraints which were deduced to be of greater order than 50 ms.

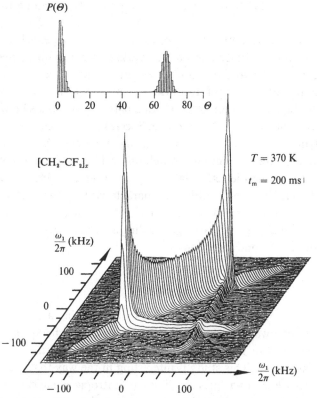

Figure 5.41. Two-dimensional exchange ^{2}H NMR spectra and associated RADs for α-PVF$_2$. (Reprinted with permission from Hirschinger *et al.* (1991). Copyright (1991) American Chemical Society.)

5.4.5 Poly(vinylidene fluoride) (PVF$_2$)

A brief description of the partially crystalline polymer PVF$_2$ illustrates the way in which early intuitive notions of molecular motion are subsequently refined by more sophisticated methodologies. It also serves as a necessary precursor to the treatment of the special electrical properties of PVF$_2$ discussed in Section 8.4. The overall molecular motional profile causing relaxation in PVF$_2$ and its copolymers with polytrifluoroethylene is discussed in detail elsewhere (McBrierty, 1989): here we focus solely upon the crystalline α-relaxation because of its significance in the poling process. ^{1}H and ^{19}F relaxation data, particularly the increase of about 1 μs in T_{2c} at 120 °C, was initially attributed to crystalline chain rotation in the vicinity of defects or rotations of restricted amplitude of all chains about their chain axis (McBrierty *et al.*, 1976). Dielectric data supported by X-ray analysis later pointed to a reversal of the dipole moment along the chain (Miyamoto *et al.*, 1980).

These possible interpretations have since been superceded by the definitive results of Hirschinger and coworkers (1991) who carried out a comprehensive ^{2}H two-dimensional exchange NMR analysis of the α-relaxation process. Without recourse to any model assumptions, they reduced the 16 possible reorientations of a pair of C–D bonds between the four polarisation states in PVF_2 to one. Spectra and associated RADs for the α-polymorph of PVF_2, portrayed in fig. 5.41, specify that the C–D bond directions are reorienting through uniquely defined angles of 67° or 113°. Guided by the dielectric data (Miyamoto *et al.*, 1980), it could be concluded that the chain motion is characterised by an electric dipole moment transition along the molecular direction only, accompanied by a conformational change, $tgt\bar{g} \leftrightarrow \bar{g}tgt$, yielding a unique orientation angle of 113° for the C–D bond directions.

5.4.6 Epoxy polymers

This discussion of motions is concluded by considering a study of epoxy polymers (Garroway *et al.*, 1982). These are the archetypical systems requiring solid-state NMR methods since they are effectively insoluble once they are cured. This study is interesting in the way that the collapse and coalescence of two lines of different isotropic chemical shift as a function of temperature is used to gain access to information on the rates of the exchange process between two sites. This should be familiar to users of high-resolution NMR in the liquid state where this effect is commonly invoked to study dynamic processes of *inter-* or *intra*molecular exchange.

Figure 5.42 shows the temperature variation of the high-resolution CPMAS ^{13}C spectrum of the polymer obtained when diglycidyl ether bisphenol-A (DGEBA) reacts with the cross-linking agent, piperidine. Because piperidine has only aliphatic carbons, the aromatic carbons of the DGEBA component give clearly resolved and assignable signals. In particular, the two lines at lower frequencies in the aromatic region arise from the two sets of two off-axis carbons. These are the *ortho* (lower frequency) and *meta* (higher frequency) carbons relative to oxygen. At low temperatures these lines both show doublet structure which collapses as the temperature is raised. This process takes place over a wide range of temperature and is still incomplete at the highest temperature studied. It is suggested that the origin of the low-temperature splittings is due to inequivalence of the two *ortho* carbons arising from the local symmetry.

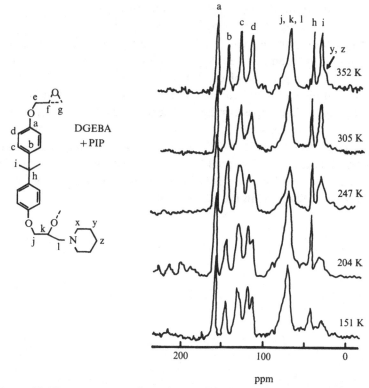

Figure 5.42. The temperature dependence of the ^{13}C CPMAS spectra of an epoxy polymer based on piperdine curing of bisphenol-A bisphenol ether. Assignments of peaks are indicated in the molecular structure. The coalescence of the protonated aromatic peaks, to yield peaks c and d at higher temperatures, arises from the reorientation of the phenyl rings. (Reprinted with permission from Garroway *et al.* (1982). Copyright (1982) American Chemical Society.)

This type of inequivalence was discussed in Section 5.2.1. The corresponding spectrum of the crystalline DGEBA material exhibits such crystallographic splittings which are, however, invariant to change in temperature. This contrasts with other similar compounds which do show temperature-dependent averaging of such line splittings (Harris *et al.* 1987). The averaging process is assigned to ring-flipping at rates which are comparable to the modest chemical shift difference of 7 ppm = 105 Hz.

The temperature range over which line collapse occurs indicates that there is a broad distribution of jump times although a fit to a single correlation-time model is possible, albeit with the addition of a second process at the lower temperatures: the experimental data are treated in terms of a Williams–Watts (1970) type of correlation function. It is shown however, that quite different values of the mean correlation times are obtained depending on whether the distribution is assumed to be inhomogeneous or homogeneous. For the former, each process may be described locally by an exponential correlation function but the correlation time varies from place-to-place. When this distribution is broad, the behaviour of the lineshape can be described to a good approximation as a superposition of lineshapes corresponding to the slow and fast exchange limits. This is reminiscent of the 'apparent phase transition' ideas introduced by Resing (1965, 1968). Unlike the ring flips in polycarbonates and polystyrene, it was suggested that the ring motion in these materials would have to involve more rotational diffusion character to achieve acceptable correlation with the mechanical process.

6

Structural heterogeneity in polymers

The discussion in the previous chapter on structure at the molecular level is now extended to include an examination of the more macroscopic features of polymers. This is important because most useful commercial polymers are heterogeneous with properties that depend sensitively upon the dimensional scale of the different component structures in the material. Such is the case for partially crystalline polymers, blends and composites, segregated block copolymers, filled and plasticised systems. Various processing steps can radically alter the scale of heterogeneity, notably thermal treatment. Blending component polymers to achieve desirable properties also introduces many factors which influence the degree of miscibility in the system; method of mixing, solvents, molecular weight and polydispersity, tacticity, the weight fractions of polymer components and the presence or absence of specific chemical entities which act as compatibility promoters.

Clearly, dimensional scale is all important in defining structural heterogeneity. Before discussing the specific contribution of NMR as such, a brief digression is in order to clarify the way in which different experimental and theoretical approaches relate to one another.

6.1 Experimental probes of heterogeneity: an overview

The sensitivity of different experimental probes to dimensional scale spans many orders of magnitude. Criteria such as experimental procedure and inherent detection limits of the measuring equipment can lead to differences even within a given technique. The fact that component polymers in a blend, for example, may be deemed to be compatible by one method and incompatible by another underlines the confusion that can arise and the

191

consequent need for care when drawing comparisons between the results of different experiments.

The analysis of blends by differential scanning calorimetry (DSC) (Olabisi *et al.*, 1979) is purported to evaluate miscibility on a 10–30 nm scale in terms of the replacement of T_g features characteristic of component homopolymers (assuming they are well separated) by a single transition for the blend at some intermediate temperature. The extent to which the dependence of this T_g on blend composition follows the ideal Gordon–Taylor (1952), Fox (1956), Kelley–Bueche (1961), Pochan–Beatty–Pochan (1979) or Couchman (1980) analysis offers additional clues to the degree of miscibility achieved. Rim and Orler (1987) argue that it is better to use Couchman's expression to estimate T_g for the blend when there are significant differences in ΔC_P and/or $T_g \Delta C_P$ between the component polymers. Non-ideal behaviour and the appearance of a broad, diffuse thermal transition at an intermediate temperature points to a distribution of molecular environments or local compositional fluctuations and therefore to incomplete mixing at a molecular level in the blend (Nishi and Wang, 1975; Smyth *et al.*, 1988). Similar broadening occurs in dynamic mechanical and dielectric spectra for the same reason.

Scattering methods are powerful structural probes which extend down to molecular dimensions. Generally, theoretical estimates at the lower limits of spatial resolution, of the order of a few nanometres, require specification of shape parameters for the scattering entity that are often ill-defined and speculative (Kim, 1972). Similar difficulties can arise in the theoretical interpretation of thermal analysis data which are presumed to probe comparable dimensions (Krause, 1978). Woodward (1989) argues that the resolution of electron microscopy is about 20 nm. For small-angle neutron scattering (SANS), interpretation is only rigorous for mixing rules based upon Gaussian coils for which the distance probed in real space is typically the radius of gyration. SANS and X-ray diffraction can, however, yield interaction parameters (Warner *et al.*, 1983; Hadziioanou and Stein, 1984). On a much coarser scale, the scattering of light which underpins criteria such as visual clarity requires that different structural phases have dissimilar refractive indices and, broadly speaking, the wavelength of light defines the scale of sensitivity (see, for example, Anavi *et al.* (1979)). In this case, extraction of precise quantitative information is difficult. On the other hand, the frequency of the Raman longitudinal acoustic mode (LAM) of laser scattered light, with suitable corrections, is related to lamellar dimensions down to about 10 nm, which is comparable to the sensitivity of small-angle X-ray scattering (SAXS). The degree of polarisation of the longitudinal Brillouin peaks exhibits similar sensitivity (Patterson *et al.*, 1976).

Significant developments have taken place within the dimensional scales probed by optical (and electron) microscopy through exploitation of image enhancement and recognition techniques. Digital image analysis (DIA) probes the size, spatial distribution, boundary shape and cross-sectional profiles of regular structural entities in heterogeneous polymers (Tanaka *et al.*, 1986). Temporal information is also available on phase separation, crystallisation dynamics, and changes induced by external means (as in the electrical poling of piezoelectric polymers).

Spectroscopic techniques, of which NMR is an example, are generally sensitive to the near-neighbour environment. In Fourier transform infrared (FTIR), for example, the position, intensity and shape of selected vibrational bands monitor the local environment and the degree of miscibility achieved (Lu *et al.*, 1983). Other novel procedures utilise the trapping of charge carriers at the interface of discrete domains in dielectric experiments (Wetton *et al.*, 1973, 1978), thermally stimulated depolarisation currents (TSDC) which reveal a MWS effect (Maxwell, 1892; Wagner, 1914; Sillars, 1939) when there is phase separation (Vanderschueren *et al.*, 1982; Albert *et al.*, 1985) and non-radiative energy transfer (NRET) between fluorescent donor and acceptor chromophores attached to the respective component polymers in a binary blend (Mikeš *et al.*, 1980; Semerak and Frank, 1981; Gashgari and Frank, 1981; Albert *et al.*, 1985).

What then are the minimum spatial dimensions necessary to qualify as a structure or, alternatively stated, what are the limits below which the integrity of a given polymer phase is questionable? Wetton and coworkers (1978) argue that molecular relaxation must be dominated by interfacial material which comprises the bulk of the polymer when the linear dimension of the dispersed phase is less than about 5 nm. In this situation, it is risky to interpret data in terms of discrete phases with assignable structural integrity. Indeed, in the limit of very small size, it is reasonable to suppose that techniques such as DSC and mechanical loss may fail to detect the dispersed phase at all.

The emerging picture is one in which fundamentally different molecular characteristics are used to probe spatial dimensions in polymers. Thus, small-angle X-ray scattering (SAXS) or electron microscopy (EM) rely on differences in the electron density of phases or domains whereas NMR often defines a region by virtue of its characteristic molecular motion. NMR parameters sensitive to the effects of spin diffusion, unlike other probes, monitor variations along the shortest path between discrete

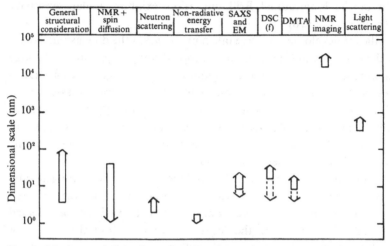

Figure 6.1. Approximate dimensional scales probed by a range of experimental techniques: SAXS, small-angle X-ray scattering; EM, electron microscopy; DSC, differential scanning calorimetry; DMTA, dynamic mechanical thermal analysis.

structural entities. Significant advances in NMR imaging (Veeman and Cory, 1989; Mansfield and Hahn, 1991), which currently identifies quite coarse dimensions will, in time, provide graphic information on spatial heterogeneity on much shorter dimensional scales (see, for example, Nieminen and Koenig (1988); Cory *et al.* (1989); and Gunther *et al.* (1990)). A means to detect a single proton has been proposed (Sidles, 1992).

The futility in discussing structural heterogeneity or degree of miscibility without also specifying dimensional scale is clear from fig. 6.1 which summarises, albeit approximately, the dimensions probed by those techniques which are generally available.

6.2 NMR as a structural probe

As clarified earlier, it is often the case that characteristic features in composite NMR spectra which derive from specific *motions* as well as specific environments can label different constituent structures in the polymer system. Early analysis of partially crystalline polymers in terms of a two-phase model exploited differences in NMR linewidth for near-rigid crystalline and motionally narrowed amorphous spectral components first to identify and then to quantify the relative amounts of material in each phase. With enhanced spectrometer sensitivity and improved data analysis, a third component became evident in some partially crystalline polymers which was generally assigned to material at the boundary between crystalline and amorphous regions (fig. 1.2). Models that describe the chain dynamics and small-scale morphology in this boundary region are applicable to diverse other systems, notably filled elastomers, ionomers and block copolymers. NMR, unlike dielectric or dynamic mechanical measurements, can directly resolve different morphological entities in this way.

In general, useful information on heterogeneity in polymers is available from (i) the short-range, near-neighbour character of the dipole–dipole interaction in solids, (ii) the site-specific detail of high-resolution spectra, and (iii) the diffusive path lengths associated with the transport of energy through the spin system by means of spin diffusion.

A useful refinement of (i) exploits deuteration techniques which open up additional opportunities to examine structural heterogeneity by NMR. Two approaches are especially interesting. In the first, spectrally resolved ^{13}C resonances characteristic of a selectively deuterated component in a composite can only be observed in a cross-polarisation experiment if mixing is sufficiently intimate to guarantee the required ^{13}C–^{1}H dipolar interactions which are short range (0.5 nm) (Schaefer *et al.*, 1981; Belfiore, 1986; Parmer *et al.*, 1987; Gobbi *et al.*, 1987) (cf Section 4.2). The dynamics of the CP process also distinguishes between rigid and mobile domains, as in certain segmented block copolymers. In the second approach, the residual proton signal (typically, 1–2%) in the deuterated component, recorded under MAS, depends on whether the immediate environment is ^{1}H- or ^{2}H-rich which, in turn, is sensitive to the presence of protonated chains introduced by blending or by other means (Vander-Hart *et al.*, 1987).

6.2.1 Spin diffusion and structural heterogeneity

The use of spin diffusion as a structural probe was discussed in a preliminary way in Section 3.8.3. Recall that spin diffusion proceeds in a homonuclear system by means of energy conserving, pairwise, near-neighbour dipolar interactions and, most often, it is the energy which diffuses within the spin system rather than the spin itself. Notable exceptions have been proposed in polyethylene (VanderHart, 1987; Schmidt-Rohr and Spiess, 1991b) and in certain other polymers above T_g (McBrierty *et al.*, 1987).

The effects of spin diffusion on T_1 and $T_{1\rho}$ relaxation reveal the way in which an efficiently relaxing entity such as a methyl group can either fully or partially relax other entities in the polymer system. Useful information can be obtained from a straight forward application of eq 3.42, $\langle r^2 \rangle = 6D_s\tau$ and equating τ with T_1 or $T_{1\rho}$ provides reasonable order of magnitude estimates of the spatial dimensions involved. In a dense rigid proton system, D_s is of the order of 10^{-16} m^2 s^{-1} in the laboratory frame and half this magnitude in the rotating frame (cf Section 3.8.1). Since T_1 is typically of the order of 7×10^{-1} s and $T_{1\rho}$ of the order of 10^{-2} s in the temperature region of interest, the respective communication distances or maximum diffusive path lengths are about 20 nm and about 2 nm. This implies that an efficient relaxing site or region (sink) can relax neighbouring nuclei within this dimensional path length and is largely ineffective for more remote nuclei. For this reason, T_1 decay in polymers is generally exponential whereas, often, $T_{1\rho}$ decay is not.

The lower limit of sensitivity is established when spatial inhomogeneities are comparable to the correlation length of the local dipolar field. In such cases NMR cannot distinguish between spatial fluctuations in the local field and variations in material properties. In this respect, the correlation length of the local dipolar field defines the lower limit for spatial discrimination – of the order of a few near-neighbour distances – which is not unduly severe since spatial heterogeneity is fully expected on the Ångström scale in any event.

By way of illustration, consider the example of *intra*molecular interactions in PMMA (fig. 6.2(a)). The α-methyl protons communicate with and relax the remaining ester methyl and methylene protons which, at best, are only weakly coupled to the lattice at temperatures where the α-methyl T_1 and $T_{1\rho}$ minima respectively occur. This spin diffusion mechanism is shown schematically in fig. 6.2(b). Efficient communication, in this instance, results in exponential T_1 and $T_{1\rho}$ proton decay.

Anticipating the impending discussion on binary blends, this mechanism can be extended to the case where PMMA (polymer A) is blended with a compatible polymer, in this case PVC (polymer B). If the sphere of influence around the α-methyl group also embraces inefficiently relaxing protons in polymer B (fig. 6.2(c)), single rather than multicomponent relaxation is observed, thus defining intimate mixing on a dimensional scale of about 20 nm. This is illustrated in fig. 6.2(d) which contrasts experimental T_1 decays for (i) a simple physical mixture of PMMA and PVC homopolymers and (ii) a blend of the two components mixed on a molecular scale. For case (i) a two-component decay is observed, where

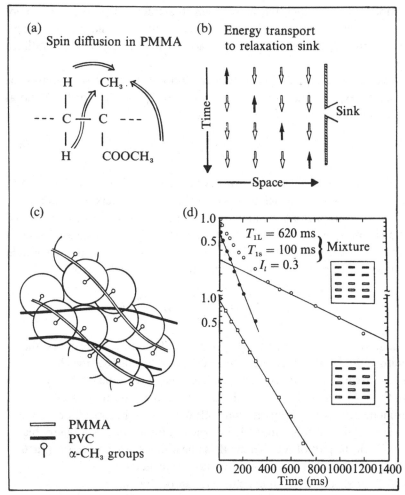

Figure 6.2. (a) Schematic representation of ester methyl and methylene protons relaxed by α-methyl protons *via* spin diffusion in MMA. (b) Energy transport *via* successive flip–flop transitions to the methyl groups (the sink) which are tightly coupled to the lattice, as described in Section 2.6.1. The filled arrows denote the higher energy spins and each row represents successive 'snapshots' in time as the spin energy progresses to the sink. (c) Schematic representation of the spheres of influence surrounding the efficiently relaxing α-methyl groups on the MMA chain. All the protons in MMA and the PVC protons in a compatible PMMA/PVC blend are within the spheres of influence and therefore can relax efficiently *via* spin diffusion to the α-methyl groups. (d) Comparison of T_1 relaxation for a physical mixture of PMMA and PVC (○, ●) with the response for a compatible blend (□).

component intensities reflect precisely the fractions of the two homo-polymers in the mixture. For the intimately mixed blend case (ii), single exponential T_1 decay is observed in accord with the arguments outlined above.

Quantitative support for these ideas follows from a description of the observed exponential rate of relaxation $k(= T_1^{-1}$ or $T_{1\rho}^{-1})$ for the blend as follows (McBrierty *et al.*, 1978):

$$k = \frac{N_M}{N_T}k^\circ_{\,m} + \frac{N_R}{N_T}k^\circ_{\,r} \qquad (6.1)$$

In this context, the relaxation rate is denoted k rather than R in keeping with the original nomenclature. $k^\circ_{\,m}$ is the intrinsic relaxation rate of the $N_M = 3$ α-CH$_3$ protons; $k^\circ_{\,r}$ allows for a contribution to the observed relaxation from N_R protons in polymer B; and $N_T = N_A + N_B$ is the total number of protons in the blend. Equation 6.1 may be restated in terms of the weight fraction W of polymer A containing the relaxing methyl groups:

$$N_T k = \left(\frac{3k^\circ_{\,m}}{M_A} - \frac{N_R k^\circ_{\,r}}{M_B}\right) W + \frac{N_R}{M_B}k^\circ_{\,r} \qquad (6.2)$$

where $N_T = [N_A W/M_A + N_B(1 - W)/M_B]$. The model presumes that a uniform spin temperature has been established throughout the blend, and if all the assumptions are correct, predicts a linear relationship between $N_T k$ and W. Alternatively stated, the observation of linearity is evidence of intimate mixing on a dimensional scale defined by the time T_1 or $T_{1\rho} = k^{-1}$.

The validity of this approach is borne out for a number of binary blends for which the plot of $N_T k$ versus W is linear in accord with eq 6.2 (fig. 6.3). Data are presented for three resonance frequencies ω_0.

A number of pertinent conclusions can be drawn.

- The observation of linearity in all cases indicates intimate mixing on a dimensional scale of about 20 nm or less.
- The small intercepts on the $N_T k$ axis reinforce the assumption that methyl groups are the dominant source of relaxation.
- The computed intrinsic relaxation time for CH$_3$ groups at the lower ω_0 frequency is the same within experimental error for PVME/PS and PSAN/PMMA: $T_1^m(30 \text{ MHz}) = (k^\circ_{\,m})^{-1} = 36.6 \pm 0.5$ ms, which is an interesting observation in the light of the different locations of the methyl groups on the polymer chain in PVME and PMMA. Both sets of 40 MHz data yield $T_1^m(40 \text{ MHz}) = 48.5 \pm 1.0$ ms. The ratio $T_1^m(40 \text{ MHz})/T_1^m(30 \text{ MHz}) = 1.33$, in accord with NMR theory.

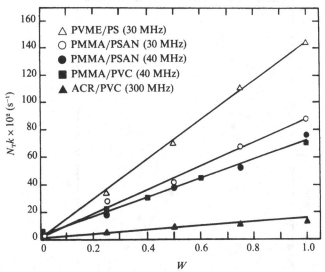

Figure 6.3. Plot of $N_\mathrm{T}k$ versus the weight fraction W of the polymer component containing the relaxing methyl groups (eq 6.2).

• The estimated $T_1(300\ \mathrm{MHz}) = 290 \pm 55$ ms for methyl groups in the $1:1$ blend of PVC and poly(methyl methacrylate-co-methacrylate) (ACR) falls short of the 366 ms anticipated from an extrapolation of the 30 MHz data. The minimum in ACR at 300 MHz is also exceedingly broad (Zhang *et al.*, 1988), which is consistent with a distribution of superimposed main chain motions (see below) rather than the isolated motion of methyl groups with reasonably fixed orientations in space. The lower relaxation time reflects the distribution.

Consider now the reverse situation where the effects of spin diffusion are suppressed, specifically by examining the results of an experiment which uncouples the α-methyl, ester methyl and methylene protons in PMMA (Douglass and McBrierty, 1979, unpublished results). Recall the discussion of Section 3.7.3 which explained how this could be achieved by placing the effective field in the rotating frame, B_e, at the magic angle, $\Theta = 54.7°$, to the laboratory field B_0 (fig. 2.4). Thus, decoupling transforms the observed exponential $T_{1\rho}$ decay for coupled protons into tricomponent decay (fig. 6.4), reflecting the three types of proton in PMMA (table 6.1). Note that the intensity of the shortest $T_{1\rho}$ component due to the 'isolated' α-CH$_3$ groups agrees reasonably well with the theoretical value. More meaningful comparison of component magnitudes requires consideration of additional

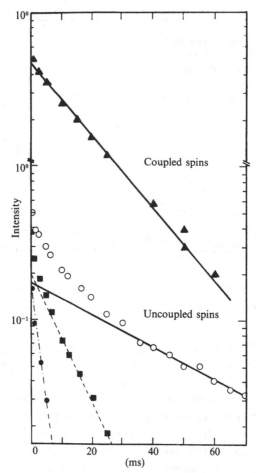

Figure 6.4. Rotating frame $T_{1\rho}$ ($B_1 = 28$ G) decay in PMMA at -100 °C for coupled and uncoupled protons (see text).

factors such as spatial distributions of α-methyl groups (McBrierty *et al.*, 1972) and superimposed main-chain motions (Gabrys *et al.*, 1987) which is beyond the scope of this discussion.

These observations suggest a second approach for investigating structural heterogeneity which exploits new methods of controlling and, on occasion, suppressing spin diffusion. The essence of the procedure is to (i) pulse the spin system into a predetermined non-equilibrium state by establishing magnetisation gradients among the different domains; (ii) allow mixing to occur by means of spin diffusion between different spin systems thereby relaxing the gradients in a controlled way and, (iii)

Table 6.1. *Rotating frame* $(B_1 = 28\ G)$ *component* $T_{1\rho}$ *relaxation times and intensities, I, in PMMA at* $-100\ °C$ *for coupled and uncoupled proton spins* (*see text*)

Spin state	α-methyl		Ester methyl		Methylene	
	$T_{1\rho}$	I	$T_{1\rho}$	I	$T_{1\rho}$	I
Coupled	← --------------------- 4.0 ms --------------------- →					
Uncoupled theoretical	—	0.375	—	0.375	—	0.25
Experimental[a]	3 ms	0.34	46 ms	0.34	11 ms	0.32

[a] Assignments to ester methyl and methylene groups are arbitrarily based on the marginally better accord between experimental and theoretical component intensities.

monitor the magnetisations of the domains and the effects of mixing by a suitable detecting pulse sequence. The Goldman–Shen experiment (1966) and subsequent modifications (Assink, 1978; Cheung *et al.*, 1980; Cudby *et al.*, 1984), all of which rely on differences in $T_{1\rho}$ or T_2, have been widely used. Limitations arise mainly from an inability to label regions unambiguously in the preparative step. More elegant approaches utilise pulse sequences that beneficially suppress spin diffusion in the preparation and detection regimes (Havens and VanderHart, 1985).

As pointed out in Section 3.8.2, spin diffusion rates and sensitivities are high for protons but identification of specific relaxation pathways between spin systems can only be inferred. Monitoring spin diffusion between ^{13}C nuclei in ^{13}C-enriched polymers obviates many of these difficulties and provides new means to probe exceedingly short pathways to specifically labelled sites in polymers, on a scale of the order of Ångström units (Linder *et al.*, 1985). Rotating-frame relaxation of protons attached to specific carbons in the CPMAS experiment also affords a route to site-specific information (Parmer *et al.*, 1988).

In phonon-assisted spin diffusion, energy transfer between dissimilar nuclei by mutual spin flips must take account of the energy imbalance arising from the different magnetogyric ratios of the two nuclei (Solomon, 1955). Thermal motions can make up this energy deficit, as demonstrated in studies on ^{19}F–^1H energy transfer in polymers containing ^1H and ^{19}F nuclei. Experiments described by eqs 2.45 and 2.46 have been carried out on neat PVF_2 and in blends containing PVF_2 (Douglass and McBrierty, 1978). By initially perturbing the ^1H spins and subsequently monitoring the magnetisation of the ^{19}F spins in such blends, information on the degree of mixing of component polymers may be obtained.

Multidimensional NMR exploits the direct resolution of spectral lines in polymer blends where communication between resolved moieties by means of spin diffusion is revealed through generation of off-diagonal elements in the spectrum (cf Section 4.5) (Caravatti *et al.*, 1985, 1986). Typically, two-dimensional nuclear Overhauser effect spectroscopy can be used in certain situations to examine polymer–polymer interactions (Ernst *et al.*, 1987). The observation of cross peaks in the two-dimensional NOE spectrum depends on (i) r_k^{-6} for the interacting protons; (ii) the motional behaviour of the polymers; and (iii) the time available for the dipolar interactions to evolve. Specifically, cross peaks can only be observed when the time-averaged distance r_k is in the range 0.2–0.4 nm (Mirau *et al.*, 1988). These experiments provide graphic evidence of communication between component polymers in a blend and do not rely upon differences in T_2 or $T_{1\rho}$ for the participating spins.

Before proceeding to a systematic discussion of specific examples, it is helpful to understand the nature of the interface between prominent structural regions in polymers.

6.3 Nature of the interface

The interface which is unusually sensitive to composition is important in determining the ultimate properties of multiphase polymer systems (Helfand and Wassermann, 1982). It is estimated that the interface is typically 1.0–1.2 nm thick and that about 70% of the chains reverse back into the crystal rather than proceeding to the amorphous regions as such (Lacher *et al.*, 1986; Schmidt-Rohr and Spiess, 1991b). Schmidt-Rohr and coworkers (1990) deduce that chains in the interfacial regions of PE are conformationally disordered, akin to the amorphous regions, but with restricted mobility (cf Section 5.4.4). Not surprisingly, much effort has been expended in more fully understanding interfacial chain configurations and dynamics (Douglass and McBrierty 1979; Cosgrove *et al.*, 1981; Jelinski *et al.*, 1982; Spiess, 1983, 1985b). Multinuclear NMR assisted, in some cases, by selective deuteration has been particularly revealing (Voelkel and Sillescu, 1979). Examples include (i) discrete non-interacting phases as in the ^{13}C and ^{2}H study of the tetramethylene terephthalate/ tetramethyleneoxy terephthalate copolymer by Jelinski and coworkers (1982); (ii) phase separation with substantial mixing at the boundary as in the ^{1}H study of the styrene/ethylene oxide copolymer (Kaplan and

O'Malley, 1979) and (iii) wholly unstructured and random mixing as revealed in a styrene/butadiene copolymer by ^{13}C solid-state NMR (Schaefer *et al.*, 1981).

An interesting phenomenon is observed in systems containing discrete domains of glassy and rubbery polymer in the way that cooperative molecular motional effects occur between the two phases (Wardell *et al.*, 1974; Morèse-Séguéla *et al.*, 1980; Krause *et al.*, 1982; Tanaka and Nishi 1986). Typically, in a styrene–butadiene–styrene (SBS) triblock copolymer motion is conferred on the glassy styrene chains, through mixing or otherwise, by mobile butadiene polymer above its glass transition and, conversely, the polystyrene restricts the mobility of the rubbery chains in the interfacial region.

The motivation to understand the dynamics of interfacial molecules extends beyond purely polymeric systems. Heterogeneity in elastomers filled with carbon black particles, for example, arises from molecular motions and configurations that are in many ways similar to the non-crystalline component of a partially crystalline polymer (O'Brien *et al.*, 1976) (cf Section 8.1.1). These ideas similarly apply to interfacial effects in block copolymers (Tanaka and Nishi, 1985) and to the chain dynamics at the boundary of clusters in ionomers (McBrierty *et al.*, 1987).

Historically, the observation of a third, intermediate, component in NMR spectra for a number of partially crystalline polymers (Fujimoto *et al.*, 1972; Bergmann, 1978) prompted the idea that the boundary between crystalline and amorphous polymer was not sharp but that interfacial material constituted a separate phase. Indeed, it was necessary, albeit indirectly, to invoke a third phase to describe the temperature dependence of T_{2L} assigned to mobile chains in some two-component T_2 decays where the formation of a plateau in the T_{2L}-temperature curve above T_g implied a weighted average of intermediate and long T_2 components arising from constrained amorphous material at the crystal interfaces and more remote mobile amorphous polymer (McBrierty *et al.*, 1976). In practice it is difficult to resolve this third, interfacial, component when the amount of material in the interface is small or its T_2 does not differ greatly from those of either of the dominant phases. NMR linewidth data for oriented polychlorotrifluoroethylene (PCTFE) fibres (cf Chapter 7) rather graphic-ally illustrate constrained amorphous material at the crystal surface where the longer T_2 in the two-component decay does not reflect rapid, nearly isotropic, motion as hitherto observed in partially crystalline polymers but depends upon fibre orientation, β_1, in the magnetic field B_0 (fig. 6.5(a)). The degree of motional vigour of polymer molecules in the

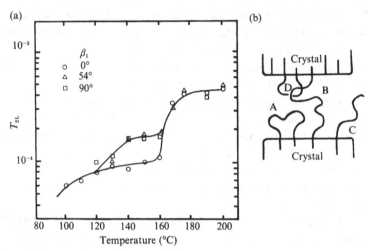

Figure 6.5. (a) Long T_2 component in PCTFE fibres as a function of fibre orientation β_1 in B_0. Note that this nominally amorphous component is not isotropic. (b) Folds (A), tie molecules (B), cilia, (C) and entanglements (D) at the interface of a crystal in a partially crystalline polymer. (Reprinted with permission from Douglass *et al.* (1976). Copyright (1976), American Institute of Physics.)

boundary region lies between the restricted motion of molecules in the crystalline regions and the near liquid-like motions of amorphous material above T_g. This observation coupled with the directional dependence of the motion implicit in the fibre data would suggest that folds, unterminated chains anchored at one end (cilia) and, perhaps, inter-crystal tie molecules are the principal morphological entities participating in such constrained motions (fig. 6.5(b)). Of course, other options whereby any point of constraint, not necessarily confined to the crystal surface, can give rise to anisotropic motions are not ruled out. An added feature emerged from computer simulation studies which predicted topological entanglements between loops on neighbouring crystals. Whether or not the return of chain folds to the crystal can be described by a random or adjacent re-entry model has been widely debated (see, for example, Natarajan *et al.* (1978); Schilling *et al.* (1983)).

Model calculations based on the dynamics of chains constrained in this fashion are consistent with experimental observation (Douglass *et al.*, 1976, 1977). Consider the comparatively simple and readily visualised motion of a cilium comprising N connected chain segments where the polar axis of the local coordinate frame is constrained to lie along the chain axis. The orientation of a typical internuclear vector in the Nth frame, S_0, relative to its orientation, S_N, in a coordinate frame XYZ defined by the

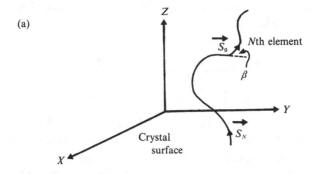

(a)

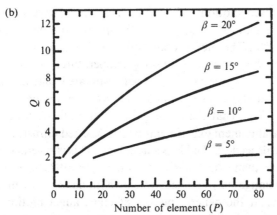

(b)

Figure 6.6. (a) Visualisation of a cilium on a crystal surface in terms of N connected chain segments; (b) Dependence of the parameter Q in eq 6.4 upon P and β. (Reprinted with permission from Douglass *et al.* (1976). Copyright (1976), American Institute of Physics.)

normal to the crystal surface, is shown in fig. 6.6(a). For a chain which is rotating about its own axis, the spin–spin relaxation time T_2 for the Nth element is (Douglass *et al.*, 1977)

$$T_2^{-2} = e^{-3N\beta^2} \cdot T_{2RL}^{-2} \qquad (6.3)$$

It is reasonable to assume that the chains are rotating about their own axes at temperatures where interfacial effects are manifest. Also, under these conditions, T_{2RL} neglects *inter*chain contributions which are generally negligible. A powder-averaged T_2 for the $(P-m)$ elements of the cilium as a whole may be computed as the sum over individual elements: T_2 for the first m elements which are within a factor of two and therefore experimentally indistinguishable from the $N = 0$ element (for which $T_2 \approx T_{2\mathrm{RL}}$) are excluded. The result is

$$T_2 = \left(\frac{(P-m)(1-e^{-3\beta^2})}{e^{-3(m+1)\beta^2}[1-e^{-3(P-m)\beta^2}]} \right)^{\frac{1}{2}} \cdot T_{2\mathrm{RL}}$$

$$= Q \cdot T_{2\mathrm{RL}} \tag{6.4}$$

Equation 6.4 is portrayed as a function of P and β in fig. 6.6(b). Observe that Q, the ratio of the motionally averaged T_2 (rapid rotation about the chain axis) to the *intra*-rigid lattice T_2 for the powder case, increases sharply with increasing P and β, indicating a progressive increase in the range of solid angle available for motion and a concomitant increase in T_2 as one proceeds along the chain away from the point of constraint.

Restricted localised motions of flexible chain units were invoked to explain the temperature dependence of ^{2}H spectra for deuterated PE (Spiess, 1983, 1985b). The model visualised a decrease in the number of constraints on mobility with increasing temperature, paralleling the concomitant increase in free volume. It was estimated that the average length of a flexible unit increased from 3–5 bonds at room temperature to 10–15 bonds at 380 K. That the topological constraints are long-lived is evident in ^{2}H spin-alignment experiments which detected ultraslow motion on a 50 ms timescale as described in Section 5.4.4. This is several orders of magnitude slower than the correlation time $\tau_c < 10^{-9}$ s for a single conformational change as predicted by conventional T_1 measurements. These conclusions fit the general picture of constrained motion at the crystalline/amorphous interface and, again, are independent of the details of the specific models used (cf Section 1.4).

With these fundamental concepts in mind, let us now turn to a number of illustrative examples of the way in which NMR probes structural heterogeneity in polymers.

6.4 Heterogeneity in homopolymers

All polymers, by their nature, are structurally heterogeneous when perceived on scales approaching molecular dimensions. Even for somewhat longer dimensions, glassy polymers, historically viewed as amorphous and isotropic, exhibit local variations in chain packing that give rise to structural heterogeneity of a short-range and quite subtle nature. Nor does simple chemical structure guarantee concomitantly simple morphology, a fact which is borne out for ostensibly one of the simplest polymers, polyethylene (PE), which has been exceedingly difficult to characterise despite unprecedented effort.

6.4.1 Polyethylene (PE)

Recall that early low-resolution NMR lineshape analysis of PE furnished resolvable spectral components assigned respectively to crystalline, interfacial and amorphous regions in the polymer (Fujimoto *et al.*, 1972; Bergmann, 1981; McBrierty *et al.*, 1980; Doskocilova *et al.*, 1986). The relative amounts of material and the character of the molecular motions deduced for each phase were intuitively reasonable and in accord with model calculations such as those described in Section 6.2. Cheung and Gerstein (1981) used the Goldman–Shen experiment to examine the size and shape of amorphous regions in PE by monitoring the diffusion of magnetisation to the crystalline domains.

High-resolution spectra have subsequently given much detailed insight through exploitation of chemical shifts and relaxation rates which are preponderantly sensitive to differences in chain conformation and dynamics as described in Chapter 5 (Pembleton *et al.*, 1977; Earl and VanderHart, 1979; Cudby *et al.*, 1984; Colquhoun and Packer, 1987; Kitamaru *et al.*, 1986). Recall, too, that the ^{13}C resonance at 33 ppm, assigned to methylene backbone carbons in the all-*trans* conformation reflects behaviour in the ordered, crystalline lamellar regions whereas resonances near 31 ppm arise from methylene protons in less ordered configurations with a high *gauche* conformational content (fig. 6.7). As many as four structures have been invoked to describe ^{13}C spectra (Kitamaru *et al.*, 1986) in contrast to $T_1(^1H)$ decay which is single-component by virtue of efficient spin diffusion between protons at room temperature. Site-specific features attributed to methyl groups, branches and cross links in appropriately prepared solid polyethylene samples have also been identified, mimicking, albeit in crude fashion, the rich ^{13}C spectral detail of PE in solution (Bovey *et al.*, 1976; Cheng and Smith, 1986; Pérez and VanderHart, 1988).

Selective relaxation of component spectra is exceedingly useful in delineating contributions from different regions in PE but care is necessary in quantifying the relative amounts of material present (cf Section 4.4.2.7). The pulse sequence described in fig. 6.7(a) illustrates the selectivity which is possible when recording ^{13}C FIDs under ^1H dipolar decoupling conditions following spin–spin relaxation for a time τ_1 without dipolar decoupling (Kitamaru *et al.*, 1986) (cf Section 5.4.4). Proper adjustment of τ_t fully

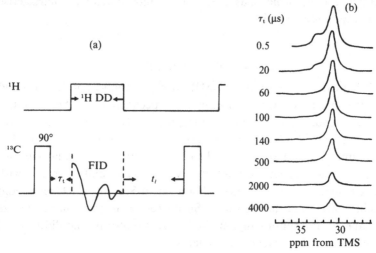

Figure 6.7. Pulse sequence used to observe the ^{13}C FID under DD conditions after relaxing the transverse magnetisation with ^{1}H decoupling for a given time period τ_t. Appropriate choice of τ_t allows suppression of the less mobile component such as the shorter crystalline T_{2c}. Appropriately short values of t_l discriminate against the long T_1 components. The recorded ^{13}C chemical shifts in PE are portrayed on the right. $t_l = 3.5$ s suppresses the longer $T_{1c} = 263$ s and 2560 s components. Progressive suppression of the crystalline and non-crystalline components with short T_2 is essentially complete at $\tau_t = 100$ µs leaving the amorphous peak at 31 ppm (Reprinted with permission from Kitamaru *et al.* (1986). Copyright (1986) American Chemical Society.) (cf fig. 5.38).

suppresses the shorter $T_2(^{13}C)$ components as observed, for example, in fig. 6.7(b) where only the amorphous peak at 31 ppm remains when $\tau_t >$ 100 ms; choice of an appropriately short cycle time $t_l = 3.5$ s discriminates against the longer, crystalline $T_1(^{13}C)$ components in these spectra.

A series of studies on polyethylenes and other polyolefins (Cudby *et al.*, 1984; Packer *et al.*, 1984; Kenwright *et al.*, 1986; Colquhoun and Packer, 1987; Packer *et al.*, 1988) exploit the high sensitivity of ^{1}H *and* ^{13}C relaxation to sample history and its effect on morphology. Two general conclusions are drawn: (i) earlier more intuitive assignments of low-resolution multicomponent relaxation in PE to discrete morphological regions are confirmed, ruling out other possible interpretations such as intrinsic non-exponential NMR relaxation from one or more of the presumed discrete phases, and (ii) ^{1}H *and* ^{13}C relaxation appear to have a common origin controlled, as expected, by a combination of intrinsic relaxation and magnetisation transport; the latter is clearly sensitive to sample morphology and domain size.

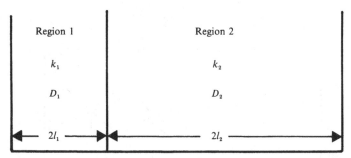

Figure 6.8. Parameters for the one-dimensional, two-region spin diffusion model for relaxation in a partially crystalline polymer. k_1 and k_2 are the intrinsic relaxation rates in regions 1 and 2; D_1 and D_2 are the corresponding spin diffusion coefficients.

The relationship between NMR relaxation times and crystallite lamellar thickness has been quantitatively explored in these and other studies by resorting to spin diffusion models as described in Sections 3.8 and 5.4.4. Consider the simplest, one-dimensional, two-region spin diffusion model, which, among the assumptions made, ignores a third, interfacial region (fig. 6.8). A full theoretical treatment of this problem (Booth and Packer, 1987) showed that the relaxation behaviour of the total magnetisation is described by a superposition of exponential processes, the rates and amplitudes of which are determined by three dimensionless parameters:

$$\alpha = (D_1/l_1^2 \Delta k); \qquad \beta = (D_2/l_2^2 \Delta k); \qquad \lambda = (D_1 l_2/D_2 l_1)$$

where $\Delta k = (k_2 - k_1)$. In the slow diffusion limit ($\alpha, \beta \ll 1$) and with $k_1 \gg k_2$, the amplitudes and rates are given by

$$\left. \begin{aligned} A_n &= 8[(2n-1)^2 \pi^2 (1 + l_1/l_2)]^{-1} \\ R_n &= k_2 + \left(\frac{2n-1}{2}\right)^2 \pi^2 (D_2/l_2^2) \end{aligned} \right\} \tag{6.5}$$

with $n = 1, 2, 3 \ldots$.

It can be seen that if the diffusion term dominates, the rates are all simple multiples of the diffusion rate (D_2/l_2^2). In addition, successive rates and amplitudes are proportional to $(2n-1)^2$ and $(2n-1)^{-2}$, respectively.

Values of the spin diffusion coefficients for proton rotating frame relaxation, $D_2[T_{1\rho}(^1\text{H})]$, have been estimated by using Raman longitudinal acoustic mode measurements to provide independent estimates of l_2. Figure 6.9 illustrates the quadratic relationship observed. Typically, $D_2[T_{1\rho}(^1\text{H})] = 2 \times 10^{-16} \text{ m}^2 \text{ s}^{-1}$, consistent with estimates based on the dipole–dipole coupling strength. Also shown in fig. 6.9 are data which

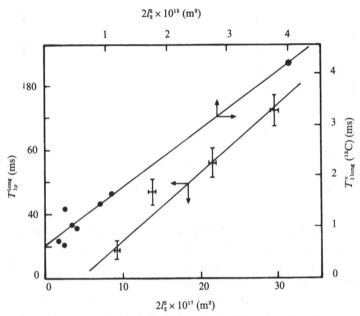

Figure 6.9. Long-time, limiting $T_{1\rho}(^1\text{H})$ and $T_1^x(^{13}\text{C})$ relaxation time constants versus the square of the lamellar thickness $2l_2^2$ for two sets of PE samples. (Reprinted from Cudby *et al.*, *Polymer Comm.*, **25**, 303 (1984) by permission of the publishers Butterworth Heinemann Ltd ©.)

indicate that a similar quadratic relationship may exist for $T_1(^{13}\text{C})$ values. The corresponding diffusion coefficient has a value of the order of $10^{-20} \text{ m}^2 \text{ s}^{-1}$.

An experimental study (Packer *et al.*, 1988) of $T_1(^1\text{H})$ and $T_{1\rho}(^1\text{H})$ behaviour in a series of samples of the same high-density, linear PE, conditioned so as to have a range of crystal thicknesses, densities and so on, demonstrated close agreement of derived l_1, l_2 values with those deduced from other techniques. Whilst $T_{1\rho}(^1\text{H})$ relaxation was multi-exponential, $T_1(^1\text{H})$ was found to be single-component and, hence, was in the fast diffusion limit. Under these conditions, the simple two-region model gives

$$R_1 = f_2 \Delta k + k_1 \qquad (6.6)$$

where f_2 is the fraction of the system in region 2. The observed linear behaviour of R_1 with fractional crystallinity (from DSC) was in accord with the above relationship. However, this gave negative values for k_1. Magnitudes of f_2 calculated from the values of l_1, l_2 derived from the $T_{1\rho}(^1\text{H})$ data also gave a linear relationship with R_1 but with positive and realistic values for k_1. It was suggested that this was evidence for the existence of a third, intermediate region between the crystalline and disordered phases.

$T_1(^{13}C)$ behaviour of the crystalline resonance signal in PE is also multi-exponential and exchange of magnetisation between the crystalline and disordered regions was suggested as the cause. Axelson and coworkers (1983) demonstrated this using a two-dimensional exchange experiment whereas Colquhoun and Packer (1987) suggested that the multiple component $T_1(^{13}C)$ relaxation showed quantitative characteristics consistent with a slow diffusion limit for the two-region outlined above. Whilst $^{13}C-^{13}C$ spin diffusion was considered as a possibility, the observation of significant temperature dependence for these processes (VanderHart, 1987; Schmidt-Rohr and Spiess, 1991b) indicated that the primary mechanism for magnetisation transport was physical diffusion of the chains (cf Section 5.4.4 and fig. 5.37). The diffusion coefficient for this process was estimated as 2×10^{-19} m^2 s^{-1} by VanderHart (1987) and 3×10^{-20} m^2 s^{-1} by Colquhoun and Packer (1987). Schmidt-Rohr and Spiess (1991b) derived a range of diffusivities in the range $10^{-17}-10^{-21}$ m^2 s^{-1} for temperatures up to 380 K.

The consistency and reasonableness of these findings is gratifying in light of the approximations used in the model: (i) the inevitable distribution of lamellar widths, (ii) the neglect of an interfacial component, (iii) the possibility that crystalline defects within lamellae can act as sinks for spin diffusion, (iv) the fact that intrinsic relaxation may itself be a function of lamellar width and (v) the conceptual difficulties in dealing with spin diffusion effects for the essentially random though significant distributions of $^{13}C-^{13}C$ distances at natural abundance.

6.4.2 *Poly(ethylene terephthalate)* *(PET)*

Structural features are often more easily analysed when a polymer is oriented (McBrierty 1974a; Smith *et al.*, 1975) (Chapter 7). It helps too when the polymer in question is amenable to modelling. This is the case with PET (Havens and VanderHart, 1985) where (i) morphology can be controlled, (ii) $T_1(^1H)$ is much longer than the time, τ, required for spin diffusion to attain equilibrium thus ensuring that the total magnetisation is virtually constant over the diffusion period ($T_1(RT; 200$ MHz$) \approx 10^2\tau$) and (iii) as is implicit from the observation of a single broad line in PET, spin diffusion coefficients are similar in the different morphological regions (unlike for PE), thus permitting the assignment of a single diffusion

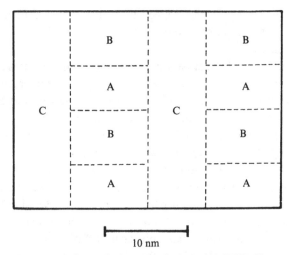

10 nm

Figure 6.10. Representation of the morphology of PET fibres. Results are calculated from a two-dimensional solution to the diffusion equation. Both vertical and horizontal directions are drawn to the indicated 10-nm scale. A, mobile amorphous, B, constrained amorphous, C, crystalline. (Reprinted with permission from Havens and VanderHart (1985). Copyright (1985) American Chemical Society.)

coefficient to all domains. Using a pulse sequence which suppresses spin diffusion in the preparation and detection periods and monitoring the rates of polarisation redistribution, information is obtained on the shape, interfacial area and size of the three identified regions in oriented PET, namely crystalline, constrained non-crystalline and mobile non-crystalline. This contrasts with the two-domain structure suggested by $T_2(^1H)$ data (Ito *et al.*, 1985) and expands upon the earlier observations of Cheung and coworkers (1980). In interpreting multicomponent decays in terms of characteristically different structural regions, Havens and VanderHart rule out their possible attribution to differences in relaxation of aromatic and aliphatic protons as discussed in Section 5.4.1 (English, 1984).

From their analysis of spin diffusion effects, these authors arrive at a mean square diffusive path length of $\langle r^2 \rangle = (\frac{4}{3}) D_s t$ and not the more commonly encountered estimate of $\langle r^2 \rangle = 2D_s t$ for diffusion in one dimension (eq 3.42). The morphology of PET fibres, visualised in terms of the scale model in fig. 6.10, comprises crystalline regions along with two distinctly different non-crystalline domains separated from one another by a few nanometres. This rather more complex morphology is in keeping with the conclusions of Murthy and coworkers (1991) who find that the amorphous regions in PET are not wholly random even in the bulk.

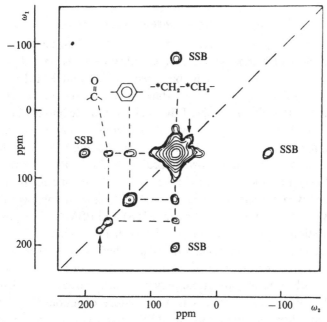

Figure 6.11. Two-dimensional spin-diffusion spectrum of ^{13}C-enriched PET obtained with a mixing time of 0.4 s. Spinning sidebands in both dimensions from the doubly enriched (90 %) CH$_2$ signal are denoted SSB, and the two arrows mark residual ^{13}C signals from the deuterated PMMA spinner. (Reprinted with permission from Linder *et al.* (1985). Copyright (1985), American Institute of Physics.)

Linder and coworkers (1985) observed the effects of spin diffusion between carbons in ^{13}C-enriched isotropic PET as shown in the two-dimensional NMR spectrum of fig. 6.11. The appearance of off-diagonal peaks clearly implies diffusion between carbonyl (165 ppm) and ring (133 ppm) carbons and the enriched methylene carbons at 63 ppm as described in Section 4.5.

6.5 Amorphous–amorphous polymer blends

Generally, miscibility is favoured when the polymer–polymer interaction parameter χ_{12} is small or negative. For blends cast from solvent, the two polymer–solvent interaction parameters should ideally be equal and greater than χ_{12}. Compatibility is enhanced too when specific moieties on the different parent chains have an affinity for one another, as with proton donor/acceptor groups. A number of specific examples bear out these points.

6.5.1 Modified polystyrene–methacrylate systems

Pairs of polymers comprising styrene and methacrylate chains give rise to an important class of blends for which miscibility depends on a diverse range of parameters. This arises in part because PS and PMMA are normally incompatible and various procedures are required to render them miscible (Schultz and Young, 1980; Löwenhaupt and Hellmann, 1990). Dependence on molecular weight is evident in blends of polystyrene and poly(n-butyl methacrylate) (PS/PBMA) which are miscible only for polystyrene components of rather low molecular weight (Di Paola-Baranyi and Degré, 1981). Modification of the PS component with specific proton donor moieties even at concentrations as low as 2 % greatly enhances the compatibility of the modified polystyrene (MPS) and PBMA. This poses intriguing questions as to the manner in which such a low concentration of hydroxyl or other proton-donating moieties, attached to the styrene chain and separated by of the order of 50 normal styrene groups, can render compatible an otherwise incompatible polymer pair (Pearce *et al.*, 1984).

Modification by chemical means is a procedure commonly used to achieve compatibility. Copolymerisation of PS with the correct proportion of acrylonitrile to form poly(styrene-co-acrylonitrile) (PSAN), for example, leads to a compatible blend with PMMA, as revealed by a number of experimental indicators (fig. 6.12). The enthalpy, entropy and excess volume of mixing are small, which is consistent with blend properties that approach those of a 'regular' solution (Naito *et al.*, 1978; McBrierty *et al.*, 1978); T_g follows the Gordon–Taylor (1952) and Kelley–Bueche (1961) predictions; the dependence of density on composition obeys the volume additivity rule; and dielectric and NMR data confirm the cooperative nature of large-scale main-chain motion of the component polymers commensurate with extensive mixing. The linearity of $N_T k$ versus W (fig. 6.3) determined from the $T_{1\min}$ data at about 0 °C (fig. 6.12(d)) confirms mixing on a scale of about 20 nm or less. However, a similar analysis for the corresponding $T_{1\rho\min}$ data at about -100 °C is not linear (fig. 6.13): the observed rate falls short of the magnitude expected for a tightly coupled system which implies that the two polymers are not fully compatible on the shorter scale of 2 nm. Thus, the two component polymers are segregated on a scale between 2 nm and the upper limit of 20 nm probed by T_1 where compatibility is definitely established.

Natansohn and Eisenberg (1987) report compatibility in a blend in which both the MMA and the PS are modified, namely, poly(methyl methacrylate-co-4-vinylpyridine) (PMMA-4VP) and poly(styrene sul-

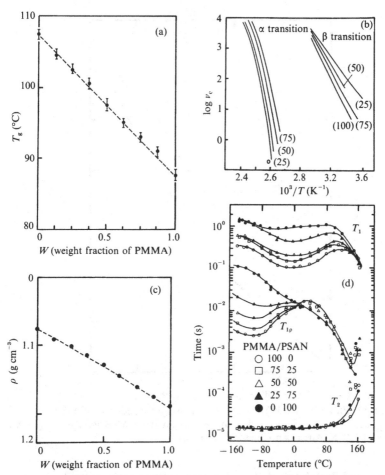

Figure 6.12. Data for PSAN/PMMA blends: (a) T_g versus W; (b) dielectric loss versus inverse temperature. The numbers in parenthesis denote the weight fraction of PMMA; (c) density versus W; and (d) NMR T_1, $T_{1\rho}$ and T_2 versus temperature. (Reprinted with permission from Naito *et al.* (1978) and McBrierty *et al.* (1978). Copyright (1978) American Chemical Society.)

phonic acid) (PS-SSA). Dipolar coupling over spatial distances of a few Ångström units between the methoxy protons of PMMA-4VP and the aromatic protons of PS-SSA is revealed in one- and two-dimensional nuclear Overhauser effect correlated (NOESY) spectra. The role of *inter*molecular hydrogen bonding as a compatibility promoter has been studied by Zhang and coworkers (1991a, b) in terms of the downfield shifts of the ^{13}C carbonyl resonance, when observed in the methacrylate component of a blend.

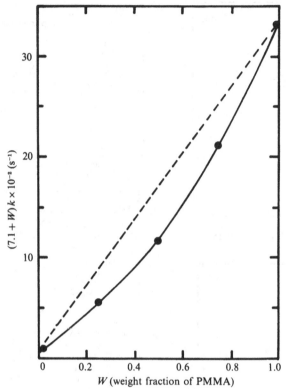

Figure 6.13. Plot of $(7.1 + W)k$ versus W, the weight fraction of PMMA (eq 6.2). $k = T_{1\rho}^{-1}$ at $-100\,°C$. (Reprinted with permission from McBrierty *et al.* (1978). Copyright (1978) American Chemical Society.)

6.5.2 *Poly(vinyl chloride)–poly(methyl methacrylate)*

A diverse range of investigations on PVC/PMMA blends illustrates the complex dependence of compatibility and, by implication, microhetero-geneity, on the method of mixing, molecular weight, polydispersity, tacticity and composition (Razinskaya *et al.*, 1972; Schurer *et al.*, 1975; Vanderschueren *et al.*, 1982; Vorenkamp *et al.*, 1985; Parmer *et al.*, 1988). Vorenkamp and coworkers (1985), for example, showed that an increase in the PMMA isotactic triad content renders blends with PVC increasingly immiscible. For atactic poly(vinyl chloride) (a-PVC) and syndiotactic poly(methyl methacrylate) (s-PMMA), differences in solubilisaton of a-PVC for s-PMMA and s-PMMA for a-PVC strongly influence the mixing achieved; the 1:1 monomer ratio corresponding to the 40/60 PVC/PMMA blend composition affords the highest degree of miscibility. The two

Table 6.2. *Description of PMMA and PVC samples discussed in the text*

Sample	M_w	M_n	Tacticity (%)[a]			Reference
			s	h	a	
PMMA	102 000	95 000	—	—	100	Parmer *et al.* (1988)
PMMA	190 000	150 000	89	11	—	Albert *et al.* (1985)
PVC	62 000	57 000	—	—	100	Parmer *et al.* (1988)
PVC	80 000	43 000	—	—	100	Albert *et al.* (1985)

[a] s ≡ syndiotactic; h ≡ heterotactic; a ≡ atactic.

polymers are therefore complementary in the sense that the modestly basic carbonyl groups in PMMA have an affinity for the weakly acidic α-protons in PVC. T_1 and $T_{1\rho}$ versus temperature data are broadly similar to those for PSAN/PMMA (fig. 6.12(d)) and spin diffusion analysis again confirms miscibility on a 20 nm scale but not at 2 nm (Albert *et al.*, 1985).

The fact that the blend is inhomogeneous on a dimensional scale of a few nanometres is not particularly surprising in the light of the probable microheterogeneity in the PVC homopolymer itself (Section 6.9.1) but, interestingly, it is at odds with the findings of Parmer and coworkers (1987, 1988). In their CPMAS ^{13}C experiment, where one of the blend components is fully deuterated, the observation of a ^{13}C resonance in the deuterated component, it is recalled, depends upon intimate contact (1–2 nm) with the proton-containing component of the polymer pair to achieve the necessary cross polarisation. The fact that the ^{13}C resonance from the quarternary carbon of PMMA is observed in a 35/65 PVC/d_8–PMMA blend but not in the neat d_8-PMMA homopolymer implies that the PVC protons are communicating with the deuterated PMMA carbons and, consequently, that there is intimate mixing of a substantial part of PMMA and PVC.

In support of this conclusion, exponential $T_{1\rho}(^1H)$ decay, determined from the ^{13}C carbonyl resonance in MMA by the method of Stejskal and coworkers (1981), is observed in the 40/60 and 70/30 PVC/PMMA blends at 50 °C. This contrasts with non-exponential $T_{1\rho}(^1H)$ decay measured directly by spin-locking procedures in the vicinity of the α-CH$_3$ minimum at -80 °C (Albert *et al.*, 1985). These differences in the extent of short-scale miscibility in the two studies, in all probability, reflect a sensitivity to differences in PMMA tacticity (table 6.2). Miscibility can also depend on molecular motion (Zhang *et al.*, 1988; 1991a); linewidth data reveal significant differences in the motional vigour of molecules at the tempera-

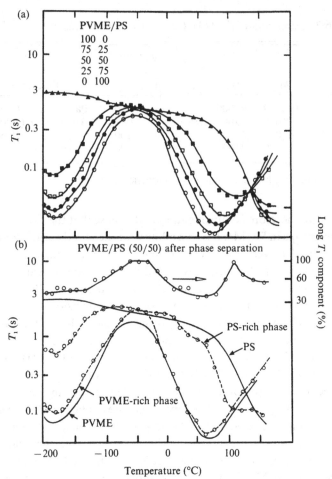

Figure 6.14. (a) T_1 versus temperature for PVME/PS blends; (b) T_1 versus temperature for the 50/50 PVME/PS blend after quenching from 160 °C in an ice–water mixture. (Reprinted with permission from Kwei *et al.* (1974). Copyright (1974) American Chemical Society.)

tures where the respective estimates were made, that is, -80 °C and $+50$ °C.

6.5.3 *Polystyrene–poly(vinyl methyl ether)*

Solvent is important in the preparation of compatible blends cast from solution (Robard *et al.*, 1977). For the PS–PVME pair, the choice of solvent critically determines the miscibility of the ensuing blend: aromatic hydrocarbons such as toluene or benzene facilitate miscibility: solvents such as chloroform or trichloroethylene do not (Bank *et al.*, 1971).

Kwei and coworkers (1974) deduced from proton NMR T_1 and T_2 measurements that PS/PVME blends cast from toluene are microhetero-geneous where extensive, though incomplete, mixing has occurred. Analysis of the T_1 data for the visually compatible blend (fig. 6.14(a)) again defines an upper limit of microheterogeneity of about 20 nm (fig. 6.3). Similar analysis of the relaxation minimum at high temperatures in the vicinity of the PVME glass transition also yields a linear plot comparable to those in fig. 6.3. This is interesting for the following reason. The significant molecular motions associated with the glass transition of PVME are characterised by a higher T_2 and therefore a lower spin diffusion coefficient since $D_s \propto T_2^{-1}$. Taken with the observed T_1 values for the blend, which are somewhat lower, it is clear that heterogeneity is probed on a reduced dimensional scale in this case. This reasoning is consistent with the view, proposed earlier, that molecular motion can enhance compatibility. There are differences, however, between T_2 components for the blend and T_2 for the parent homopolymers which can only be rationalised in terms of cooperative interactions between blend components and the presence of PS-rich and PVME-rich phases. Thus, while T_1 data signify extensive mixing, T_2 data reveal some microheterogeneity in the blend.

Figure 6.14(b) shows the temperature dependence of T_1 after phase separation by spinodal decomposition induced by quenching the blend in an ice–water mixture from 160 °C. Comparison of component T_1 and T_2 magnitudes with those of the homopolymers again indicates separation into PS-rich and PVME-rich phases.

Selective deuteration procedures, described earlier, add to the picture of microstructure in PS/PVME blends (Parmer *et al.*, 1987; Gobbi *et al.*, 1987). Compare the weak aromatic ^{13}C signal due to residual phenyl ring protons in deuterated polystyrene, d-PS, with the much enhanced signal from the d-PS/PVME blend (fig. 6.15). Clearly there is communication on a molecular scale between the aromatic carbons in the d-PS component and protons in PVME, of the order of, at most, a few nanometres. On a cautionary note, other studies have shown that deuteration of the PS component raises the lower critical solution temperature (LCST) – the temperature at which phase separation occurs in the blend – by about 40 °C. As such, the deuterated and undeuterated systems are not strictly comparable (Yang *et al.*, 1986).

Different, though complementary, techniques yield information on the nature of specific interactions between PS and PVME. In FTIR spec-troscopy, for example, vibrations associated with phenyl rings in PS and $COCH_3$ groups in PVME are sensitive to the degree of molecular mixing

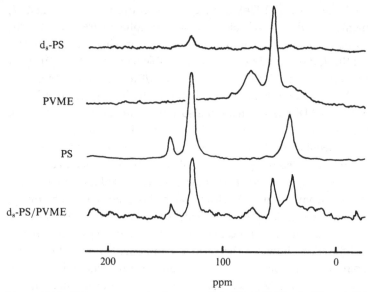

Figure 6.15. ^{13}C resonance in deuterated d_8-PS, PVME, d_8-PS and PS/PVME. (Reprinted with permission from Parmer *et al.* (1987). Copyright (1987) American Chemical Society.)

(Lu *et al.*, 1983). Two-dimensional NMR provides another perspective as shown in fig. 6.16 (Caravatti *et al.*, 1985). Generation of off-diagonal peaks, symptomatic of communication between phenyl protons in PS and methine/methoxyl protons in PVME, reflects extensive mixing in blends cast from toluene whereas their absence confirms immiscibility when chloroform is the solvent. The mixed phase constitutes 60–80% of the compatible blend. Mirau and coworkers (1988) confirmed this interaction in a 40% wt/vol solution of the blend but not in a 25% wt/vol solution, indicating that the interproton separations exceed 0.4 nm in the latter.

Quantitative aspects of mixing are explored further in an experiment which selectively saturates (inverts) a group of isochronous proton spins. The transfer of magnetisation *via* spin diffusion with other coupled spins is then monitored during the variable mixing time, τ_m (Caravatti *et al.*, 1986). High-resolution spectra are generated in the detection mode by means of the combined MREV-8 pulse sequence and MAS from which integrated peak intensities are determined. In the parent PS and PVME homopolymers, the aliphatic protons are selectively inverted and magnetisation redistribution is monitored as in fig. 6.17(a), (b). The short-time behaviour shows the effects of spin diffusion whereas T_1 relaxation accounts for the

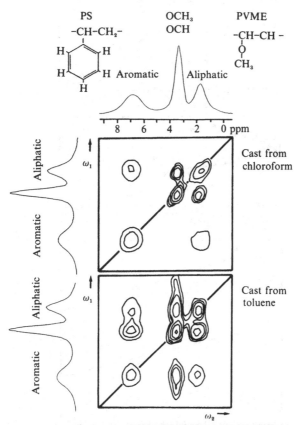

Figure 6.16. Two-dimensional proton spectra of PS/PVME blends cast from chloroform and cast from toluene. MAS spinning frequency = 2.8 kHz; T = 328 K and mixing time τ_m = 100 ms. (Reprinted with permission from Caravatti *et al*. (1985). Copyright (1985) American Chemical Society.)

observed decay at long τ_m. Note the slower rate of spin diffusion in PVME due to the extensive mobility of this polymer at temperatures well above its glass transition. For the compatible blend, the selective inversion of the PS aromatic peak and the subsequent growth of the methine/methoxyl peak intensity in PVME by means of spin diffusion is demonstrated in fig. 6.17(c). In contrast, the blend cast from chloroform displayed only T_1 decay without the initial rise observed in the compatible blend, which is characteristic of interpolymer spin diffusion within the mixed phase. The relative proportions of material in the proposed three-phase model for the compatible 64/36 PS/PVME blend are estimated as PS (8.5%), PS/PSVME (79%) and PVME (12.5%).

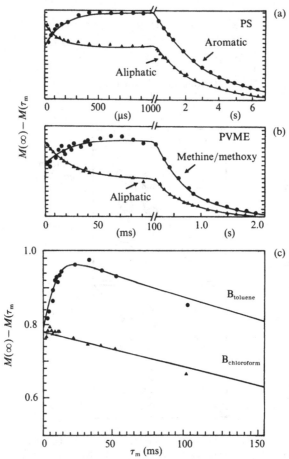

Figure 6.17. *Z*-magnetisation for individual spectral lines plotted as a function of the mixing time τ_m (see text). (a) PS; (b) PVME; (c) Blend cast from toluene ($B_{toluene}$) and from chloroform ($B_{chloroform}$). The data were recorded at 340 K. (Reprinted with permission from Caravatti *et al.* (1986). Copyright (1986) American Chemical Society.)

Takegoshi and Hikichi (1991) used ^{13}C NMR to examine the effects of blending on molecular motion and concluded that, as the PS content is increased, the characteristic transition temperature, T_0, increases, the activation energy of the glass transition decreases and motional anisotropy increases.

6.5.4 Poly(ethylene terephthalate)–bisphenol A polycarbonate

Enhancement of carbon–carbon spin diffusion by ^{13}C enrichment has been used successfully to probe miscibility in blends of poly(ethylene terephthalate) (PET) and bisphenol A polycarbonate (BPAPC) (Linder *et al.*, 1985). Prior thermal and dynamic mechanical analysis showed that blends were amorphous and intimately mixed when the PET content exceeded 70 wt % (Nassar *et al.*, 1979; Cruz *et al.*, 1979).

Carbon–carbon spin diffusion is not observed in non-enriched blends of PET/BPAPC. Recall, however, the results for neat PET doubly enriched at the methylene groups, where spin diffusion occurs between the labelled groups and the unlabelled aromatic and carbonyl carbons (Section 6.4.2). Linder and coworkers (1985) showed that ^{13}C enrichment of methylene groups in PET or carbonyl groups in BPAPC greatly enhances spin diffusion rates in the blend and allows detection of interpolymer spin diffusion along identifiable pathways involving, in all cases, the labelled sites. The use of an interrupted decoupling delay before ^{13}C FID data acquisition which suppresses signals from proton-bearing carbons greatly simplifies interpretation. One- and two-dimensional NMR reveal interpolymer spin diffusion between the two blend components, which clearly points to intimate mixing on a dimensional scale of less than 1 nm. Spin diffusion from labelled carbonyl groups in BPAPC to carbonyl, non-protonated aromatic and methylene carbons in PET at rates of the order of 0.5 s^{-1}, confirms these conclusions.

6.6 Amorphous–crystalline polymer blends

Generally speaking, when compatible pairs of amorphous and partially crystalline polymers are blended together, mixing occurs between the amorphous polymer and the amorphous part of the crystalline polymer. Crystallisation kinetics are usually altered. The following exemplify the contribution of NMR in unravelling these phenomena.

6.6.1 Poly(vinyl chloride)–poly(ε-caprolactone)

In a sense, PVC/PCL is a convenient blend to bridge the amorphous–amorphous and amorphous–crystalline categories since the development of a PCL crystalline phase in the blend depends sensitively upon composition; the crystallinity of PCL decreases towards zero as the PVC content approaches 70 wt % (Brode and Koleske, 1972). This is also

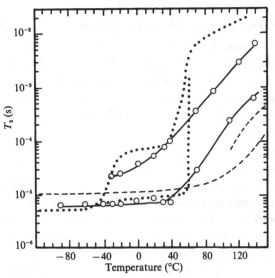

Figure 6.18. NMR proton T_2 data for PCL (...), PVC (– – –) and a 30/70 PCL/PVC blend (O–O–O).

reflected in the shift to lower temperatures of the secondary transition in PCL, attributed to local CH_2 motions, when PVC becomes the major component. This relaxation is akin to the γ-relaxation in PE (Illers, 1969; Vanderschueren *et al.*, 1980).

The propensity for PCL to mix intimately with a variety of polymers has been attributed to hydrogen bonding between the polar oxygen in PCL and specific chemical moieties in its complementary partner (Olabisi 1975; Coleman and Zarian, 1979; Mirau *et al.*, 1988). Interactions of C=O with α-hydrogens in PVC approach saturation for PVC concentrations of 60 wt %. Also, a single T_g is observed for a wide range of blend compositions, which increases systematically with increasing PVC concentration. Agreement with the predictions of Gordon–Taylor (1952), Fox (1956) and Kelley–Bueche (1961) implies extensive mixing.

NMR data for PCL/PVC blends confirm a marked sensitivity to the structural first-order transition in PCL which heralds the onset of melting at about 60 °C (Albert *et al.*, 1984). The resulting highly mobile PCL polymer, which is present at quite low temperatures, induces a response in the blends which closely resembles the behaviour of plasticised PVC (cf Section 6.9.1). Drawing the obvious analogy, mobile PCL is presumed to plasticise at least part of the PVC in much the same fashion as conventional plasticisers. It is evident, too, from the proton T_2-temperature dependence

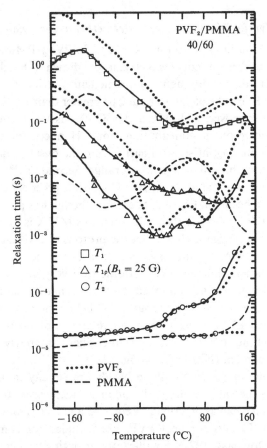

Figure 6.19. ^{19}F (30 MHz) T_1, T_2 and $T_{1\rho}$ data for the 40/60 PVF$_2$/PMMA blend: (...) and (– – –) denote the PVF$_2$ and PMMA response, respectively. (Reprinted with permission from Douglass and McBrierty (1978). Copyright (1978) American Chemical Society.)

(fig. 6.18) that motions in the blend are cooperative; note the premature increase in T_{2s} for the more rigid material at about the same temperature as the onset of general motions in the mobile plasticised phase.

Based upon a comprehensive range of data, it is generally concluded that PCL forms an intimate blend with part of the PVC component and, once mobile, acts as a plasticiser for PVC: excess PVC remains unplasticised and relatively immobile up to its melting point.

6.6.2 Poly(vinylidene fluoride)–poly(methyl methacrylate)

Consider the ^{19}F data in fig. 6.19 for a 40/60 wt % PVF$_2$–PMMA blend, from which a number of conclusions can be drawn (Douglass and McBrierty, 1978). First, although not immediately evident from the semilog plot, $T_2(^{19}$F) for the blend is about 2 μs greater than $T_2(^{19}$F) for the PVF$_2$ homopolymer between -140 °C and 0 °C. This is consistent with an increase in the off-resonant contribution from ^{1}H nuclei (cf eq 3.22), assuming intimate mixing, arising primarily from the onset of molecular motion in PMMA within this temperature range. Second, $T_1(^{19}$F) for the blend exhibits a minimum at about -200 °C which is precisely where the *proton* T_1 minimum occurs in PMMA. A similar, though less pronounced, sensitivity of $T_{1\rho s}(^{19}$F) to the $T_{1\rho}(^1$H) minimum in PMMA is observed at about -100 °C. This suggests that either the protons of PMMA and the fluorines of PVF$_2$ are interacting in a solid solution or there is a phase-separated system where spin energy is diffusing over very short distances. In either case intimate mixing is implied. Note too, that the short, rigid-like, T_2 disappears in the blend at about 50 °C below the corresponding temperature in neat PVF$_2$, indicating extensive crystal premelting in the blend: PVF$_2$ melting point depression intensifies progressively with decreasing PVF$_2$ content (Nishi and Wang, 1975).

Cross-relaxation studies on the same blends are equally informative. Figure 6.20 shows the temperature variation of the mean cross-relaxation rate, σ, in the 40/60 blend as described by eqs 2.45 and 2.46 (Douglass and McBrierty, 1978). Results for neat PVF$_2$ are included for comparison (McBrierty and Douglass, 1977). Significantly, σ begins to increase at a lower temperature in the blend than in neat PVF$_2$, indicating premature enhanced cross relaxation. This argues for additional motions in the blend which are directly modulating the ^{1}H–^{19}F interaction near -40 °C at the difference frequency $\omega(^{19}$F$)-\omega(^1$H$) \approx 1.888$ MHz (for $\omega_0(^1$H$) = 30$ MHz), as required in phonon-assisted spin diffusion (cf Section 3.8). The motion of α-methyl groups is the obvious source (McCall, 1969). There are two possibilities: either appreciable numbers of PVF$_2$ and PMMA molecules are at near-neighbour distances or incorporation of PMMA broadens the spectrum of motions responsible for the σ peak at $+10$ °C in PVF$_2$. In either case, intimate mixing of component polymers is implied.

Tékély and coworkers (1985) confirmed and extended these findings by measuring $T_{1\rho}(^1$H) for protons, which is sensitive to induced morphological change *via* its dependence on cross polarisation. The observed

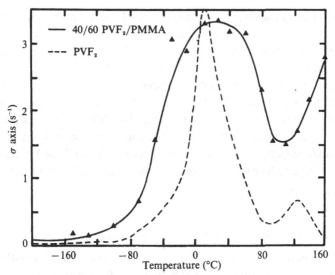

Figure 6.20. Temperature variation of cross relaxation σ for the 40/60 PVF_2/PMMA blend. Corresponding results for neat PVF_2 (dashed curve) are included for comparison. (Reprinted with permission from Douglass and McBrierty (1978). Copyright (1978) American Chemical Society.)

exponentiality of $T_{1\rho}(^1H)$ and its dependence upon PVF_2 content in 40/60 and 50/50 PVF_2/PMMA blends suggested that PMMA protons either participate in direct dipolar interactions with PVF_2 nuclei or have sufficient PVF_2 neighbours to alter appreciably the PMMA molecular motions, in agreement with the conclusions of earlier experiments (Douglass and McBrierty, 1978). The situation differs somewhat at lower PMMA concentrations, for which $T_{1\rho}(^1H)$ decay is non-exponential, implying some PMMA segregation as PVF_2 partially crystallises. Recall the tendency for the PVF_2 component to crystallise as its content increases beyond 50 wt % (Nishi and Wang, 1975). As crystallisation proceeds, the longer $T_{1\rho}(^1H)$ component approaches that of neat PMMA, from which it would appear that PMMA forms an interfacial region of PMMA-rich material.

6.7 Crystalline–crystalline polymer blends

The discussion thus far reflects a general progression towards wholly segregated block copolymers. Certain copolymers with homopolymer constituents that are partially crystalline form a penultimate category where mixing may be intimate or, of more immediate interest, segregation may occur on a scale of intermediate morphological dimensions. In the former case, the copolymer chain itself comprises alternating sequences, often random in character, of the constituent polymers for which important information on structure can be obtained and, on occasion, checks made on synthesis by virtue of the sensitivity of NMR to the local molecular environment.

The random copolymer poly(vinylidene fluoride/trifluoroethylene) (PVF$_2$/TrFE) is especially interesting because of its remarkable piezo-electric, pyroelectric and ferroelectric character (Douglass *et al.*, 1982; McBrierty *et al.*, 1982a, b, 1984). Detailed consideration of its structure/property relationship will be deferred to Sections 8.4.2 and 8.4.3. Consider instead the polyethylene/polypropene polymer pair which has a number of interesting features.

6.7.1 Polyethylene–polypropene

Different preparative procedures lead to different PE/PP copolymer morphologies. For example, ^{13}C NMR indicates the formation of a block copolymer structure when a Zeigler–Natta catalyst is used to copolymerise ethylene and propene (Prabhu *et al.*, 1978, 1980). On the other hand, preparing the PE/PP blend in rather novel fashion using a modified surface growth procedure devised by Zwijnenberg and Pennings (1976) generates a rather different structure. Prior investigation revealed two essentially non-interacting components coexisting as separate crystalline phases with extensive *c* axis orientation (Coombes *et al.*, 1979). The proton T_1, $T_{1\rho}$ and T_2 response of the copolymer is essentially a simple superposition of the component data for neat PE and PP with, perhaps, weak spin diffusion coupling of PE and PP protons. Taken with earlier observations, the NMR data suggest a model for the blend based upon the interpenetration of PE and PP shish-kebab structures and they also afford some insight into the kinetics of crystallisation of the PE/PP blend from solution (McBrierty *et al.*, 1980).

6.8 Block copolymers

A significant proportion of block copolymers tend to be phase-separated systems comprising hard and soft segments with interfaces that may or may not be sharp. The hard, glassy or partially crystalline segments act as physical cross links and reinforcing fillers. Microdomains can be in the form of regular arrays of spheres, lamellae or cylinders where the stability of the hard segments relies on one of a number of possible interactions: hydrogen bonding as in polyurethanes; ion clustering as in ionomer resins; the formation of a network of crystalline lamellar regions as in thermoplastic polyester elastomers; or simply polymers in their glassy state such as polystyrene in styrene–butadiene block copolymers. Domain sizes between 0.5 and 100 nm can be ascertained (Löwenhaupt and Hellmann, 1990). Roe and coworkers (1981) pointed out that structures in block copolymers become unstable in a number of circumstances: when the parent polymers are similar; when block lengths are fairly short; when a common solvent is added; or when the temperature is suitably altered relative to the critical solution temperature. Typically, for styrene–butadiene block copolymers, SAXS measurements show that heating produces a thermally reversible transition involving a gradual disappearance of microdomains due to the progressive intermixing of components.

Solution-state NMR has given much information on physical characteristics such as average block length (Wilkes, 1977), mole fraction (Jelinski *et al.*, 1981), end capping ratio (Williams *et al.*, 1981) and the degree of 'blockiness' in the copolymer (Klesper *et al.*, 1970). Interesting structural studies have exploited the selective swelling of one or other component in diblock and triblock systems (Heatley and Begum, 1977). The following examples illustrate the power of high- and low-resolution solid-state NMR in probing the structure of block copolymers.

6.8.1 *Polystyrene–polybutadiene–polystyrene*

The morphology of SBS triblock copolymers is visualised in terms of PS domains of regular size and geometry embedded in a rubbery PB matrix. SBS has properties that represent a desirable combination of the thermoplastic and elastomeric character of the constituent homopolymers. In many respects the PS domains behave in similar fashion to filler particles or multiple cross links in as much as each domain is the terminal centre of several hundred elastomer chains (Bishop and Davidson, 1969). The interface between the PS domains and the PB matrix is reasonably sharp except in cases where the block copolymer is on the point of macrophase separation (Krause, 1978). Drawing an analogy with carbon black-filled rubber (cf Section 8.1.2), Tanaka and Nishi (1985, 1986) estimate the thickness of the interfacial layer in SBS to be about 2 nm, comprising 80 % rubber molecules. Cooperative motional effects in the block copolymers, described earlier, are intuitively consistent with results expected from what is, in essence, a form of cross-linked system.

Proton T_1 data for an SBS block copolymer are compared with the homopolymer results in fig. 6.21. In the vicinity of the PB minimum, two

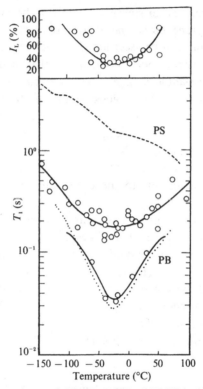

Figure 6.21. Comparison of NMR $T_1(^1H)$ (30 MHz) data for an SBS block copolymer (○) with the response for PS (– – –) and PB (...). The solid lines denote the theoretical fit to the experimental T_1 and I_L data as described by eqs 3.45–3.47. 20 % of protons in the blend are in the PS component. (Reprinted with permission from McBrierty and Douglass (1981). Copyright (1981), John Wiley and Sons, Inc.)

T_1 components are observed, the component magnitudes *and* intensities of which are temperature-dependent. The solid lines in fig. 6.21 denote the theoretical fit to the data based upon the model of partial spin diffusion coupling between the two phases described in Section 3.8.3. Recall that the component T_1 values predicted by the spin diffusion model do not, in general, reflect the response of the two phases, but, rather, are modes of decay of the coupled system: nor do the observed component intensities bear any simple relationship to the proton contents of the two phases and they cannot therefore be used straightforwardly to estimate the proportions of discrete phases in the copolymer.

Comparing T_1, T_2 and $T_{1\rho}$ data for the bulk SBS copolymer and for macroscopic single crystals (Keller *et al.*, 1970) confirms the absence of molecular orientation in either phase of the extruded copolymer and the insensitivity of the molecular motional characteristics up to 150 °C to the *shape* of the PS domains in the copolymer (Wardell *et al.*, 1976). Tanaka and Nishi (1986) also conclude that the form of magnetisation recovery following the application of the so-called Goldman–Shen (1966) sequence to a number of styrene–butadiene block copolymers is independent of domain shape.

The structure of SBS block copolymers may be contrasted with copolymers of PS and PB prepared by random copolymerisation which locates the styrene and butadiene components in close contact with one another. Anderson and Liu (1971) observed a single T_1 component where the T_1 minimum moves progressively to higher temperatures with increasing styrene content. If, in fact, there is phase separation with spin diffusion operating efficiently in these copolymers, the dimensional scale of the microphase regions cannot be greater than about 6 nm.

In interpreting NMR data for block copolymers, the geometry of the interface cannot be neglected in all cases. Notably, Lind (1977) showed that an adequate description of NMR data for a copolymer of polydimethylsiloxane (PDMS) and bisphenol-A polycarbonate (BPAPC) required preferred spin diffusion along the length of the PDMS blocks rather than across a single diffusion barrier between the PDMS and BPAPC components.

6.8.2 Polyurethanes

As with the SBS block copolymers, NMR discriminates between the hard and soft segments that characterise the polyurethanes (PU) by virtue of clear differences in their respective molecular motions. The shorter T_2 of the anticipated two-component FID is assigned to the hard segments which are usually isocyanate-derived, whereas the longer T_2 identifies the soft segments which are usually polyether- or polyester-based. Assink and Wilkes (1977) explored phase mixing in polyurethanes by observing proton spin diffusion effects and predicted a significant interfacial region, in contrast to the SBS systems. The nature of phase separation and consequent domain structure has been the subject of much discussion (Leung and Koberstein, 1985). Two models are especially favoured. In the first, the hard segments are considered to be extended sequences (Bonart and Muller, 1974; Bonart *et al.*, 1974); in the second, a coiled or folded configuration is preferred (Van Bogart *et al.*, 1983; Koberstein and Stein, 1983). Jelinski and coworkers (1981, 1983a–d, 1984) used ^{13}C and [2]H

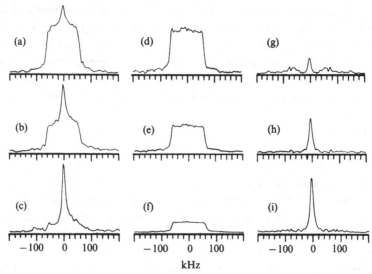

Figure 6.22. Solid-state quadrupole deuterium NMR spectra of hard segment labelled polyurethanes recorded at 22 °C and 55.26 MHz. Spectra are described as follows: (a) 70, (b) 60 and (c) 50 wt % hard segment. Spectra (d)–(f) are of the all-hard segment material reported at different gains. Spectra (g)–(i) were generated by subtracting an appropriate amount of the symmetrised spectra in the centre column from those in the left column. All spectra were recorded with a 2 s recycle delay. (Reprinted with permission from Dumais *et al.* (1985). Copyright (1985) American Chemical Society.)

NMR to explore the relative merits of each model. The attendant site-specific information and the absence of complicating spin diffusion effects simplified data interpretation.

By way of illustration consider the ^{2}H NMR study of PU comprising the following hard and soft segments labelled specifically at the butanediol moeity of the hard segment (Dumais *et al.*, 1985) where, in this representation, $D \equiv {}^2H$.

Hard segment

Soft segment

In this experiment, which expands upon the Assink–Wilkes (1977) study, the diphenylmethane diisocyanate of the hard segment and the soft segment polyol residues remain undetected because of selective deuteration. The source of the signal is therefore unambiguous and the effects of spin diffusion are minimised. Deuterium spectra, recorded for PU with 70, 60 and 50 wt % hard segment (fig. 6.22) reveal a narrow resonance of constant linewidth superimposed on a broad (120 kHz) line. Dumais and coworkers argue that part of the hard segment polymer experiences motion which is nearly isotropic and speculate that some hard segments may be short enough to dissolve in the soft segment-rich domains or a proportion of hard segments may be located in the interfacial region between the hard and soft domains. Quantitative estimates of the relative amounts of each phase are reasonably consistent with estimates from SAXS analysis. These data, in summary, lend credence to, but not unequivocal support for, the Van Bogart–Koberstein–Stein model of coiled or folded rather than extended configurations.

6.9 Polymer–diluent systems

For the purposes of this discussion, polymers with low molecular weight additives are classed as a blend. Additives which lead to a lower glass transition temperature and modulus of elasticity of the host polymer are termed plasticisers. Antiplasticisers, on the other hand, cause the modulus and tensile strength to increase with concomitant enhanced embrittlement (Jackson and Caldwell, 1967). The view that the antiplasticiser molecule acts as a physical cross link between two sites on different chain segments (Makaruk and Polanska, 1981) is borne out in ^{13}C NMR studies on polycarbonate antiplasticised with 2,2′-dinitrobiphenyl (DNB) (Gupta *et al.*, 1983).

The underlying molecular criteria which are important in relating microscopic behaviour to macroscopic properties include the mobility and internal flexibility of the diluent molecules and the nature of their coupling to the polymer matrix. Typically, Zhang and coworkers (1991) used ^{31}P two-dimensional NMR to demonstrate a direct correlation between the onset of diffusive motion of a phosphate ester diluent, tetraxylyl hydroquinone diphosphate (TXHQDP), in a blend of polystyrene and poly-(phenylene oxide) (PS/PPO) and the appearance of a mechanical loss peak. The sensitivity of the chosen two-dimensional experiment to slow motions renders the NMR and dynamic mechanical timescales roughly compatible. The presence of both mobile and immobile diluent molecules in the plasticised polymer is also established.

This section explores further the molecular origins of plasticisation and the way in which concomitant structural heterogeneity develops, often in a subtle way. Such is the case with poly(vinyl chloride) (PVC) where NMR not only reveals these effects but also highlights again the intimate relationship between molecular structure and motion.

6.9.1 Plasticised poly(vinyl chloride)

$T_{1\rho}(^{13}C)$ NMR investigations by Sefcik and coworkers (1983) on PVC plasticised with tricresyl phosphate (TCP) demonstrate the role of TCP (a) as an antiplasticiser impeding cooperative main chain motions at lower concentrations (< 15 wt %), and (b) as a plasticiser inducing more facile main-chain motions thereafter. From a structural point of view, plasticiser expands free volume, thereby facilitating long-range segmental motions at lower temperatures and, more specifically, alters the chain cohesion parameters in a way that has an impact on the mechanical properties of the system as discussed above. Typically, enhanced cohesion due to antiplasticisation suppresses the mechanical β-relaxation and enhances the tensile strength of PVC (Kinjo and Nakagawa, 1973).

Other studies address the structural implications of selective plasticisation, mindful, of course, that local heterogeneities can also arise, for example, from the incorporation of stabilisers into PVC (McBrierty, 1979; Douglass, 1980; Gevert and Svanson, 1985) or from ordered 'paracrystalline' regions associated with short segments of syndiotactic polymer (Davis and Slichter, 1973). The fact that the plasticiser is not completely dissolved in PVC shows up most clearly in multicomponent linewidth data (fig. 6.23). At room temperature, incorporation of the plasticiser diisodecylphthalate (DIDP) develops a long exponential tail, characteristic of mobile material, on the rigid-like Gaussian FID for unplasticised PVC. At 75 °C, three components are resolved: the short T_2 denotes rigid polymer, the long T_2 can be unequivocally assigned to the plasticiser itself and the intermediate T_2 manifests plasticised polymer. There are therefore three distinguishable types of material present, one of which is insensitive to the presence of plasticiser. The Goldman–Shen experiment reveals that the scale of structural heterogeneity is small, with linear dimensions of no more than a few nanometres. Similar complexity arises in polystyrene–toluene systems (Rossler *et al.*, 1985) and in PS/PPO plasticised with TXHQDP described above (Zhang *et al.*, 1991).

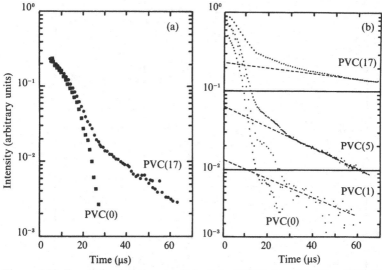

Figure 6.23. Free induction signals of plasticised PVC: (a) at 25 °C and (b) at 75 °C. The numbers in parentheses are wt % diisodecylphthalate. (Reprinted with permission from Douglass (1980). Copyright (1980) American Chemical Society.)

6.9.2 *Plasticised d_8-poly(methyl methacrylate)*

The 70/30 polymer–diluent blend of perdeuterated atactic PMMA and DNB provides a graphic example of the way in which the dynamics of cross polarisation affords detailed information on miscibility (Belfiore, 1986). Recall that the rate-determining step for *inter*-segment cross polarisation is the proximity of $^{13}C_{polymer}$ and $^1H_{diluent}$ dipolar coupled nuclei. The evolution of ^{13}C magnetisation with contact time τ is given by eq 2.44 which reduces to

$$S(\tau) = S_0\{1 - T_{CH}^{(SL)}/T_{1\rho}(^1H)\}^{-1} \cdot [e^{-\tau/T_{1\rho}(^1H)} - e^{-\tau/T_{CH}^{(SL)}}] \qquad (6.7)$$

when $T_{1\rho}(^{13}C) \gg T_{CH}^{(SL)}$. Recall too that $T_{CH}^{(SL)}$ will be short for a rigid system and for short ^1H–^{13}C dipolar distances.

Consider the CP contact-time curves for ^{13}C resonances in both the polymer and the diluent where $T_{CH}^{(SL)}$ and $T_{1\rho}(^1H)$ govern the short and long contact-time behaviour, respectively (fig. 6.24). A number of observations can be drawn.

- As expected, $T_{CH}^{(SL)}$ is short for DNB carbons because of their relative immobility and ready access to protons.
- The build-up of ^{13}C magnetisation is appreciably slower for d_8-PMMA carbons for which direct contact with protons is less probable.

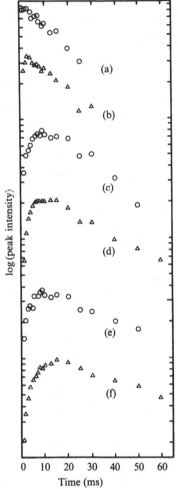

Curve	$T_{CH}^{(SL)}$ (ms)	$T_{1\rho}(^1H)$
(a)	0.6	23
(b)	1.0	26
(c)	8.2	20
(d)	3.7	37
(e)	4.0	48
(f)	6.8	50

^{13}C spectrum for the blend: peaks A and B correspond to the diluent; C, D, E and F correspond to d$_8$-PMMA.

Figure 6.24. ^{1}H–^{13}C cross polarisation contact-time curves for the ^{13}C resonances of both polymer and diluent in a 70/30 blend of d$_8$-PMMA and DNB. The curves have been shifted vertically for clarity: (a) protonated aromatic carbons, DNB (peak A, 133 ppm); (b) CNO$_2$, DNB (peak B, 148 ppm); (c) C=O, PMMA (peak F); (d) OCD$_3$, PMMA (peak E); (e) α-CD$_3$, PMMA (peak C); (f) quaternary, PMMA (peak D). (Reproduced from Belfiore, *Polymer*, **27**, 80 (1986) by permission of the publishers, Butterworth Heinemann Ltd ©.)

- d$_8$-PMMA methoxy and α-CD$_3$ carbons have the shortest $T_{CH}^{(SL)}$ values, indicating that these polymer carbons in particular are in a near-neighbour environment with the protonated diluent molecules.
- Spin diffusion is operative in DNB since $T_{1\rho}(^1H)$ values evaluated at the protonated and nitrogenated aromatic carbon sites are comparable. It is

not efficient enough, however, to average all the $T_{1\rho}(^1H)$ values for the blend to a common value.

- Differences in $T_{1\rho}(^1H)$ reveal that the CP process involves a broad distribution of ^1H–^{13}C intermolecular distances which is not surprising since the modest DNB content (30 wt %) precludes the location of all d_8-PMMA chain segments near diluent molecules.

In summary, Belfiore concludes that there is intermolecular CP transfer from diluent ^1H nuclei to polymer ^{13}C nuclei, which is indicative of intimate mixing between d_8-PMMA and DNB.

7

Oriented polymers

7.1 Introductory remarks

Preferred orientation can arise, either by accident or design, when polymers are subjected to particular fabrication procedures (Ward, 1982, 1985). Mechanical deformation typically induced by drawing, compression or rolling, can lead to anisotropy at the macroscopic level where, in effect, the orientational dependence that characterises the structural units at the molecular level is retrieved in part in the bulk: ultra-high-modulus fibres are notable examples. In other instances, molecular orientation can be achieved by suitable application of electric and magnetic fields and preferred orientation can occur at inter-phase surfaces in heterogeneous polymer systems, as discussed in Section 6.3. On the other hand, unwanted orientation may be induced by rapid spinning as in MAS (Maciel *et al.*, 1985).

In materials as morphologically complex as polymers, one is inevitably concerned with macroscopic averages, or moments, of the distribution of structural units. Clearly it is important to determine such moments if the overall properties of the polymer are to be understood and quantified. The natural inclination is to resort to X-ray analysis which, indeed, has been predictably successful. However, a number of other theoretical and experimental tools have collectively given greater insight into the structure of non-isotropic polymers. Results from a variety of NMR experiments correlate well with each other (Kretz *et al.*, 1988) and with alternative approaches including X-ray analysis (McBrierty *et al.*, 1971b; Windle, 1982; Pietralla *et al.*, 1985; Ward, 1985). It is the use of NMR to extract information on such distributions that forms the central theme of this chapter. Multidimensional NMR applied to highly oriented polymers which correlate molecular order and dynamics (Yang *et al.*, 1988) and

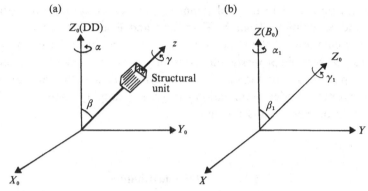

Figure 7.1. (a) Orientation of a typical structural unit in the sample coordinate frame $X_0 Y_0 Z_0$ where Z_0 is the draw direction (DD) of a uniaxially stretched polymer; (b) orientation of the sample in the laboratory frame XYZ where Z denotes the direction of B_0 and Z_0 is the sample draw direction.

which generate more definitive information on motional mechanisms (Yang *et al.*, 1990; Hirschinger *et al.*, 1991) are alluded to elsewhere.

The model introduced in Section 1.3 to account for the macroscopic properties of a polymer visualises the material as an assembly of structural units (single crystals, crystallites, chain segments and so on) with tensor properties that are directionally dependent. As pointed out in Sections 1.3 and 3.2, computation of their aggregate contribution involves the rotational transformation of tensors from a molecular coordinate frame in a typical structural unit (xyz) to a sample coordinate frame ($X_0 Y_0 Z_0$) (figs 3.1 and 7.1). This step introduces the statistical distribution of units into the analysis. The Wigner rotation matrix formalism (Appendix 4) can be used to model data from a diverse range of experiments in a unified way (McBrierty, 1974b). The fact that the Euler angles involved in the rotation transformations are individually indexed (eq A2.1) allows symmetry to be treated in a reasonably straightforward manner, often by inspection. This point is illustrated below.

A practical consequence of this generalised approach is that distribution moments derived in one experiment can be used directly to model diverse other tensor properties. Typically, moments obtained from anisotropic NMR linewidth data have been used to predict the mechanical moduli of uniaxially oriented polyethylene (McBrierty and Ward, 1968; McBrierty *et al.*, 1971b) and the piezoelectric coefficients of poly(vinylidene fluoride) (cf Section 8.4.2).

As might be anticipated, the transformation of tensors between coordinate frames is sensitive to the crystallographic symmetry of the structural units and to textural symmetry introduced during processing (Roe and Krigbaum, 1964a, b; Krigbaum and Roe, 1964). In the limit where orientations of the structural units are averaged uniformly over a sphere, macroscopic properties are isotropic, as is commonly encountered in bulk polymers. The manner in which symmetry and the rank of the tensor property under examination can greatly simplify analysis will be clarified in the ensuing discussion.

7.2 Nature of the distribution

7.2.1 Symmetry considerations

Full characterisation of the distribution of structural units in the polymer is generally only required in studies on the nature of the plastic deformation mechanism itself. In modelling specific properties, only a limited number of non-zero moments are required as specified by the rank of the tensor under examination and by whatever symmetry elements apply. Thus, spin–spin relaxation times, T_2, or equivalently M_2, involve moments to order 4 whereas measurements based upon linear dichroism, for example, only introduce moments to order 2. All odd moments vanish for distributions with a centre of symmetry since $\mathbf{P}(\beta) = \mathbf{P}(\pi - \beta)$ in which case the subscript l in the general analysis (Appendix 4) is even throughout.

Now consider the specific geometry shown in fig. 7.1. Textural fibre symmetry characteristic of uniaxially drawn polymers implies a uniform distribution of the polar axes of structural units about the draw direction or fibre axis (z about Z_0 in fig. 7.1(a)): all values of the angle α are equally probable and therefore terms in α are averaged uniformly over 2π such that the subscript m in eqs A4.1 and A4.4 is zero. Expressions for the distribution function and corresponding moments may then be written

$$\mathbf{P}(\beta\gamma) = \sum_{ln} \left(\frac{4\pi}{2l+1}\right)^{\frac{1}{2}} P_{l0n}\, Y_{l-n}(\beta\gamma) \qquad (7.1)$$

and

$$P_{l0n} = \frac{(-1)^n}{2\pi} \left(\frac{2l+1}{4\pi}\right)^{\frac{1}{2}} \langle Y_{ln}(\beta\gamma)\rangle \qquad (7.2)$$

If, in addition, the structural units are randomly disposed about their own polar axes or they possess the appropriate crystallographic symmetry (*vide infra*), γ is likewise averaged and the subscript n is also zero in which case

$$\mathbf{P}(\beta) = \sum_{l \text{ even}} P_{l00} \cdot P_l(\cos \beta) \qquad (7.3)$$

and

$$P_{l00} = \frac{2l+1}{8\pi^2} \langle P_l(\cos \beta) \rangle \qquad (7.4)$$

The angle brackets denote the average over the distribution. This is equivalent to Hermans' orientation function or Saupe's order parameter S when $l = 2$ (Hermans *et al.*, 1946; Maier and Saupe, 1960). Note that $\mathbf{P}(\beta)\,d(\cos \beta)$ is the fraction of structural units oriented between β and $\beta + d\beta$ with respect to the draw axis. Frequently encountered averages over $\cos^2 \beta$ and $\cos^4 \beta$ may be written

$$\left.\begin{aligned}
\langle \cos^2 \beta \rangle &= \frac{16\pi^2}{15} P_{200} + \tfrac{1}{3} \\[2mm]
\langle \cos^4 \beta \rangle &= \frac{32\pi^2}{105} \{\tfrac{2}{3}P_{400} + 3P_{200}\} + \tfrac{1}{5}
\end{aligned}\right\} \qquad (7.5)$$

For isotropic distributions

$$P_{lmn} = \delta_{l0}\,\delta_{m0}\,\delta_{n0}/8\pi^2 \qquad (7.6)$$

where δ is the Dirac delta function.

Developing further the consequences of symmetry, consider briefly those classes of crystallographic symmetry that are formally equivalent to transverse isotropy of the structural unit ($n = 0$) with specific reference to NMR linewidth data for PTFE and PE. Hexagonal symmetry with a six-fold rotation axis ($n = 0,6$) and pseudo-orthorhombic symmetry ($n = 0,4$) characterise the crystal structure of PTFE and PE, respectively. Since T_2 or M_2 involves even moments to order 4, n cannot be greater than 4. This precludes $n = 6$ terms and therefore, in the analysis of NMR second moments and T_2 values, hexagonal symmetry is formally equivalent to transverse isotropy ($n = 0$). In contrast, the assumption of transverse isotropy in PE is only approximate from the standpoint of crystallographic symmetry since the contribution from $n = 4$ terms should not be ignored. More detailed insight into the effects of crystallographic symmetry is to be found in original papers by Roe and Krigbaum (1964a, b), Roe (1965, 1966, 1970) and McBrierty (1974b).

Expressions for the distribution functions in eqs A4.1, 7.1 and 7.3 do not converge particularly rapidly and many moments are required to describe adequately the distribution in highly oriented polymers. The problem is

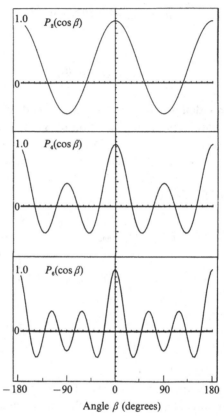

Figure 7.2. Angular dependence of the first three even Legendre polynomials $P_l(\cos \beta)$.

less severe for more modest orientations where the moments themselves converge rapidly. As might be expected, the amount of information on the distribution increases with the number of moments that can be accurately specified. However, as will be evident in due course, the extent to which certain moments add to the overall knowledge of the distribution can also depend very much on the molecular characteristics of the polymer and on the technique used.

To illustrate some of the general features of distributions, consider the angular dependence of the first three even Legendre polynomials (fig. 7.2). Typically, for systems approaching near perfect alignment along the draw axis Z_0 in the polymer ($\beta \approx 0$), all three functions are positive and approach unity for perfect alignment. Where there is a preponderance of structural unit orientations orthogonal to the draw axis ($\beta \approx \pi/2$), both

$\langle P_2(\cos\beta)\rangle$ and $\langle P_6(\cos\beta)\rangle$ are negative whereas $\langle P_4(\cos\beta)\rangle$ is positive and so on. Thus, casual inspection of the sign of the moments can often generate a reasonable picture of the distribution of structural units in the polymer.

7.2.2 Theoretical distribution functions

The pseudo-affine deformation model (Kratky, 1933; Kuhn and Grün, 1942; Ward, 1962) is used on occasion to describe statistical distributions in partially ordered polymers. Upon drawing, this model envisages the unique axes of structural units undergoing the same change in direction as lines connecting pairs of material points in the polymer without change in volume. The moments of the distribution are related to the draw ratio λ according to

$$
\left.
\begin{aligned}
\langle \cos^2\beta\rangle &= \frac{\lambda^3}{\lambda^3-1}\left\{1-\frac{\tan^{-1}(\lambda^3-1)^{\frac{1}{2}}}{(\lambda^3-1)^{\frac{1}{2}}}\right\} \\
\langle \cos^k\beta\rangle &= \frac{\lambda^3}{(k-2)(\lambda^3-1)}\left\{(k-1)\langle\cos^{k-2}\beta\rangle-1)\right\}
\end{aligned}
\right\} \tag{7.7}
$$

$\lambda = L/L_0$ where L_0 and L are, respectively, the length before and after drawing. The predicted dependence of distribution moments upon λ is portrayed in fig. 7.3. More complicated expressions for P_{lmn} have been worked out for biaxial orientation (l, m and/or $n = 0$) (Richardson and Ward, 1970; Cunningham, 1974). An interesting comparison has been drawn between the predictions of pseudo-affine deformation and the affine deformation of a rubbery network (Ward, 1985) which showed that the second moment of the distribution in the latter case grows at a rate which increases with increasing λ, in contrast to the behaviour of the pseudo-affine scheme portrayed in fig. 7.3.

Several investigations have used the Gaussian distribution expressed as follows.

$$
\mathbf{P}(\beta) = C e^{-\sin^2\beta/2\sin^2\bar{\beta}} \tag{7.8}
$$

where C is a normalisation constant. The dependence of the predicted moments $\langle P_l(\cos\beta)\rangle$ on the width of the Gaussian distribution, $\bar{\beta}$, determined from eqs 7.3 and 7.8, has been worked out by Hentschel and coworkers (1981). For modest orientations, eq 7.3 converges rapidly and a few moments are sufficient to describe the distribution adequately ($l \leq 8$ for $\bar{\beta} > 20°$). As expected, the number of moments required progressively increases as orientation improves and $\bar{\beta}$ decreases.

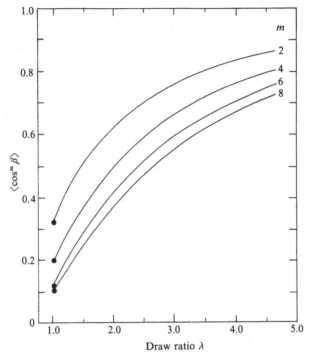

Figure 7.3. Dependence of the orientation distribution functions $\langle \cos^m \beta \rangle$ on draw ratio based upon the pseudo-affine deformation model (eq 7.7). The points ($\bullet$) denote the moments for the isotropic material corresponding to $\lambda = 1$.

7.3 NMR of oriented polymers

The sensitivity of NMR to molecular orientation is evident in a wide range of deformed polymers. The additional information contained in the linewidth dependence on sample orientation β_1 in B_0 (fig. 7.1(b)) has two important consequences: first, it provides a quantitative probe of the distribution of structural units and, second, it offers a more rigorous test of molecular motional assignments in the polymer. In principle, it is possible to determine all moments; in practice, constraints arise from experimental factors such as signal-to-noise ratio or from fortuitously inappropriate molecular chain configurations. Measurements at an arbitrary number of sample orientations β_1 relative to B_0 overdetermine the accessible moments (McCall and Hamming, 1959). Procedurally, experience dictates that the lower order moments of the distribution are evaluated first and these are used in successive fits to the experimental data to determine moments of higher order.

For oriented partially crystalline polymers, it is often assumed at least initially, that there is no preferred orientation in the amorphous regions. While this is not always true (see, for example, Murthy *et al.* (1991)), it offers a reasonable first approximation. The moments of the composite spectrum then reflect a superposition of crystalline and amorphous contributions described as follows:

$$M_N(\beta_1) = f_x M_{Nc}(\beta_1) + (1 - f_x) M_{Na} \qquad (7.9)$$

where $N = 2, 4$, f_x is the degree of crystallinity and the subscripts c and a denote, respectively, the crystalline and amorphous contributions. N signifies the order of the NMR moment. When fitting experimental data, M_{Na} may be viewed as a disposable parameter in the absence of quantitative information on the linewidth of the amorphous regions. A formal expression for the NMR moments of spectra is given in eq 3.18. The situation, of course, is somewhat more complicated for samples which are doubly oriented rather than axially symmetric. Rolled nylon-6,6 (Olf and Peterlin, 1971) is a case in question for which eqs 7.1 and 7.2 are required.

T_1 and $T_{1\rho}$ relaxation times are less informative principally because correlation frequency distributions and spin diffusion tend to smear out spatial anisotropy (McBrierty *et al.*, 1970, 1971a). On the other hand, chemical shift and deuterium NMR have been richly informative largely due to their site-specific character and to the fact that only single-spin interactions are involved. The ability of synchronous two-dimensional MAS NMR to probe orientation both in the crystalline and amorphous phases of partially crystalline polymers is finding increasing application (Harbison and Spiess, 1986; Harbison *et al.*, 1987).

7.3.1 Dipolar linewidth studies

The dependence of second and fourth moments of dipolar broadened lines on β_1, the sample orientation in B_0, is a direct consequence of molecular distributions in partially ordered polymers. The analysis exploits eq 3.18 and, where appropriate, eq 7.9. For polymers with fibre symmetry and transversely isotropic structural units, M_{2c} and M_{4c} may be written in abbreviated form as follows (McBrierty *et al.*, 1971b).

$$M_{Nc}(\beta_1) = \sum_{l=0}^{2N} C_N(\beta_1, l) \langle P_l(\cos\beta) \rangle \qquad (7.10)$$

The $C_N(\beta_1, l)$ coefficients are determined from known positions of resonant nuclei (lattice sums) and fundamental constants.

Figure 7.4 portrays the manner in which anisotropy in $M_2(\beta_1)$ develops

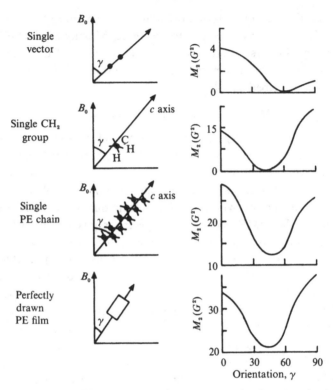

Figure 7.4. Diagram illustrating the development of anisotropy in the second moment M_2 for a highly uniaxially oriented sample of a crystalline PE. (Reproduced from McBrierty, *Polymer*, **15**, 503 (1974a) by permission of the publishers, Butterworth Heinemann Ltd ©.)

in oriented PE fibres: note the dominant influence of the methylene internuclear vectors on the dependence of M_2 on β_1 for the sample as a whole, once again reflecting the short-range nature of the dipolar interaction. A contrasting situation arises with polyoxymethylene (POM) fibres (fig. 7.5) where the behaviour of $M_2(\beta_1)$ can best be understood by considering the tabulated $C_N(\beta_1, l)$ coefficients for the POM rigid lattice (table 7.1) and, in particular, the negligible contributions made by $l = 2$ terms to the moments (McBrierty and McDonald, 1973). This is because the angle between the internuclear vector in the methylene group and the c-axis of the main chain is, fortuitously, almost exactly equal to the magic angle $\cos^{-1} 1/\sqrt{3}$ and therefore the contribution of this, usually dominant, pair of protons is almost zero. Lattice sums for $l = 2$ are therefore small. These contrasting examples illustrate the extent to which $M_N(\beta_1)$ can be sensitive to specific crystal structures in a polymer.

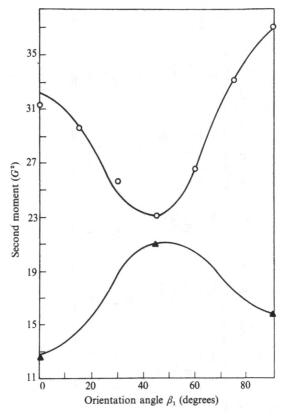

Figure 7.5. Comparison of anisotropy in uniaxially drawn low-density PE (○) $[\langle P_2(\cos\beta)\rangle = 0.78; -196\ °C]$ and POM (▲) $[\langle P_2(\cos\beta)\rangle = 0.81; -172\ °C]$. (Data taken from McBrierty and Douglass (1980). Copyright (1980) Elsevier Science Publications BV.)

7.3.2 *Chemical shift anisotropy*

Analysis of the orientational dependence of the chemical shift tensor in terms of structural unit distributions requires (i) unobscured spectra and (ii) tensors that are well defined in the principal axis system. Ideally, chemical shifts that span a wide range of shielding and exhibit appreciable anisotropies are desirable. As is evident from tables A1.2 and A1.3, the range and scale of anisotropy for ^1H nuclei is small. ^{13}C chemical shifts, on the other hand, span a broad range and are therefore particularly useful. ^{19}F nuclei are also amenable both in regard to range and scale of anisotropy. ^{15}N nuclei also have favourably large shielding anisotropies.

Unwanted contributions to the spectrum, such as dipolar broadening, can be selectively suppressed by suitable pulse sequences, thus facilitating

Table 7.1. $C_N(\beta_1, l)(G^2)$ Coefficients in eq 7.10 for oriented POM (McBrierty and McDonald, 1973)

	l	β_1 0°	15°	30°	45°	60°	75°	90°
$C_2(\beta_1, l)$	0	19.54	19.54	19.54	19.54	19.54	19.54	19.54
	2	-0.01	-0.01	-0.06	0.00	0.00	0.00	0.01
	4	-13.93	-9.53	-0.32	5.66	4.03	-2.00	-5.22
$C_4(\beta_1, l)$	0	837	837	837	837	837	837	837
	2	-9	-9	-6	-2	1	4	5
	4	-963	-659	-23	391	278	-138	-361
	6	75	30	-28	-11	24	3	-23
	8	130	12	-44	39	-10	-22	36

requirement (i) above. Regarding (ii), interpretation is greatly eased if the chemical shift tensor is axially symmetric, which can, on occasion, be achieved by judicious choice of temperature, as in PTFE (Vega and English, 1979, 1980).

The most general orientational dependence of the chemical shift tensor is specified in eq 3.2 where the angular information is contained in $\sigma(\beta_1) - \tilde{\sigma}$. Expressions for the first and second moment (eqs 3.16 and 3.17) greatly simplify for axial symmetry ($\eta = 0$). Also, when the PAS is coincident with the molecular coordinate frame (xyz) (fig. 3.1), the angle β_2 is zero and the Nth moment of the line assumes the simpler form of eq 3.18, which may be alternatively expressed as

$$\Delta\sigma_N = (\Delta\sigma)^N \sum_{i=0}^{N} \sin^{2i}\beta_1 \sum_{j=0}^{N} C_{ji}^{N} \langle\cos^{2j}\beta\rangle \qquad (7.11)$$

where $\Delta\sigma = \sigma_{\parallel} - \sigma_{\perp}$. The relevant coefficients C_{ji}^{N} are listed in table 7.2. The first moment is related to the order parameter S as follows

$$\Delta\sigma_1 = \Delta\sigma\, S(\tfrac{2}{3} - \sin^2\beta_1) \qquad (7.12)$$

As in the analysis of dipolar broadened lines, fits to the experimental data recorded as a function of the angle β_1 yield moments of the distribution from which an approximation to the distribution function $\mathbf{P}(\beta)$ can be constructed from eq 7.3.

7.3.3 Quadrupole lineshape analysis

Spectra of nuclei with non-zero quadrupole moments are almost entirely dominated by *intra*molecular nuclear quadrupole interactions between the quadrupole moment and the electric field gradient (EFG) tensor. Deuterium is a particularly useful probe because the EFG is almost axially symmetric about the C–^{2}H bond in aliphatic chains and is approximately so in other cases. Data are therefore relatively easy to interpret and are amenable to graphic visualisation. Basically, the sensitivity of ^{2}H NMR to the direction of the C–^{2}H bond relative to B_0, as described by eq 3.2 with $\eta = 0$, provides the probe of local order in a non-isotropic polymer. In consequence, different structural unit distributions lead to different lineshapes. Experimental techniques for coping with the large spectral linewidth, up to 250 kHz, are now routine and selective deuteration adds a further important dimension.

Table 7.2. *Coefficients C_{ji}^N in eq 7.11 (Brandolini et al., 1983)*

$j \backslash i$	N=1		N=2			N=3				N=4				
	0	1	0	1	2	0	1	2	3	0	1	2	3	4
0	-1/3	1/2	1/9	-1/3	3/8	-1/27	1/6	-3/8	5/16	1/81	-2/27	1/4	-5/12	35/128
1	1	-3/2	-2/3	4	-15/4	1/3	-7/2	75/8	-105/16	-4/27	20/9	-10	35/2	-315/32
2	—	—	1	-5	35/8	-1	25/2	-245/8	315/16	2/3	-40/3	-385/6	-105	3465/64
3	—	—	—	—	—	1	-21/2	189/8	-231/16	-4/3	28	-126	385/2	-3003/32
4	—	—	—	—	—	—	—	—	—	1	-18	297/4	-429/4	6435/128

7.3.4 Analysis of two-dimensional MAS spectra

The discussion in Section 3.6.2 referred to the detailed information contained in rotational sidebands of spectra recorded under MAS in the slow spinning regime. Two-dimensional MAS spectra for oriented polymers are equally informative where, for example, sensitivity (i) to modest orientation in partially drawn polycarbonate (Pietralla *et al.*, 1985), (ii) to differences in orientation for different structural elements of a polymer chain (Blümich *et al.*, 1987) and (iii) to different regions in an oriented polymer (Harbison and Spiess, 1986) have been demonstrated.

As described in Appendices 6 and 7, the two-dimensional MAS spectrum $S(\Omega_1, \Omega_2)$ is expanded in a converging series of subspectra $S_L(\Omega_1, \Omega_2)$ the corresponding sideband intensities I_{LMN} of which can be used to determine the moments of the distribution of distinguishable structural moieties on the polymer chain (Harbison *et al.*, 1987).

7.4 Orientation in polyethylene

The use of NMR to study statistical distributions in partially drawn polymers was first applied to PE, exploiting the structural model described in Section 1.3 (McBrierty and Ward, 1968; McBrierty *et al.*, 1971b). Briefly, it was assumed that linewidth anisotropy as a function of sample orientation in B_0 is due to the reorientation of structural units upon drawing, that the structural units are randomly disposed about the draw axis (fibre symmetry), that the structural units themselves are transversely isotropic and that the amorphous regions are isotropic. In short, eqs 7.3 and 7.9 apply. In retrospect, these approximations turned out not to be overly severe since the resulting moments of the distribution agreed reasonably well with estimates by other methods such as X-ray analysis and optical birefringence and their use allowed adequate prediction of mechanical moduli as discussed below.

Consider the NMR fourth-moment data for uniaxially oriented low-density polyethylene (LDPE) recorded as a function of sample orientation β_1 in B_0 (fig 7.6) (McBrierty *et al*, 1971b). Moments of the distribution to order 6 listed in table 7.3 were obtained by fitting eqs 7.9 and 7.10 to the experimental data using the coefficients $C_N(\beta_1, l)$ for PE in table 7.4. $\langle P_8(\cos\beta)\rangle$ could not be determined directly and $\langle P_6(\cos\beta)\rangle$ only marginally so, because of the relatively small values of the coefficients $C_4(\beta_1, 8)$ and $C_4(\beta_1, 6)$. The higher order moments reported in table 7.3 were estimated using a smoothing procedure which suggested itself from X-ray

Table 7.3. *Orientation distribution coefficients* $\langle P_l(\cos\beta)\rangle$ *derived from smoothed distribution functions (see text). The figures in the final column give the rms deviations between calculated and experimental fourth moments of the absorption line. Data taken from McBrierty et al. (1971b)*

	$\langle P_l(\cos\beta)\rangle$					M_{4a}	rms
Draw ratio	$l=2$	$l=4$	$l=6$	$l=8$	$l=10$	(G^4)	deviation (G^4)
1.3	0.19	−0.08	0	0	0	2000	17.8
2.3	0.73	0.47	0.15	0	0	1650	20.2
3.7	0.80	0.84	0.70	0.25	0.15	1650	10.3

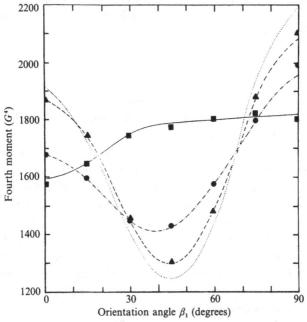

Figure 7.6. Comparison of experimental and theoretical variation of the ^{1}H NMR fourth moments plotted against orientation angle β_1. The lines are theoretical and the points experimental data. The dotted curve represents the theoretical result for perfect orientation. Draw ratios: (▲), 3.7; (●), 2.3; (■), 1.3. (Reprinted from McBrierty *et al.*, *J. Phys. D*, **4**, 88 (1971) by permission of IOP Publishing Ltd.)

spectra which indicated negligible orientation transverse to the draw direction for the more highly drawn samples. Accordingly, the distribution moments of orders 6, 8 and 10 were chosen such that $\mathbf{P}(\beta)$ for the higher draw ratios was constrained to approach zero at $\beta = 90°$. The polar plots

Table 7.4. *Values of* $C_4(\beta_1, l)$ *for low-density polyethylene* (G^4). *The sum* $\sum_l C_4(\beta_1, l)$ *applies to the perfectly oriented polymer. Data taken from McBrierty* et al. (*1971b*)

l	β_1						
	0°	15°	30°	45°	60°	75°	90°
0	1629	1629	1629	1629	1629	1629	1629
2	−756	−680	−473	−189	95	302	378
4	1252	857	29	−509	−362	179	469
6	−173	−69	65	26	−56	−7	54
8	87	8	−29	26	−6	−15	24
$\sum_l C_4(\beta_1, l)$	2039	1745	1221	983	1300	2088	2554

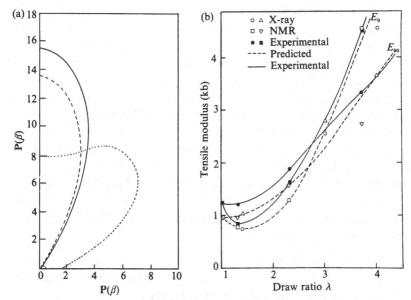

Figure 7.7. (a) Polar plots of $\mathbf{P}(\beta)$ versus β in oriented drawn PE for the draw ratios indicated. $\lambda = 3.7$ (——); $\lambda = 2.3$ (- - -); $\lambda = 1.3$ (----). (b) Comparison of predicted and experimental extensional (E_0) and transverse (E_{90}) tensile moduli for oriented LDPE as a function of draw ratio. (Reprinted from McBrierty *et al.*, *J. Phys. D*, **4**, 88 (1971) by permission of IOP Publishing Ltd.)

of fig. 7.7(a) are approximations to $\mathbf{P}(\beta)$ determined from eq 7.3 for each λ. Figure 7.7(b) compares experimental extensional (E_0) and transverse (E_{90}) moduli with their theoretical counterparts using moments determined from NMR and X-ray analysis of the statistical distributions in the drawn polymers. The minimum in E_0 is predicted together with the cross-over pattern of E_0 and E_{90}. Neither the NMR nor the X-ray data predict the E_{90} minimum, which has been explained in terms of reversible mechanical twinning (Frank *et al.*, 1970).

Three components are usually resolved in room temperature $M_2(\beta_1)$ data for ultra-drawn partially crystalline polymers (Smith *et al.*, 1975; Ito *et al.*, 1983). The spectrum of ultradrawn PE ($\lambda = 30$) is made up of a broad component which reflects highly oriented crystalline regions; a weak isotropic narrow component which is assigned to low-molecular-weight material that does not crystallise; and an intermediate component which is attributed to ordered chains rotating about their own axis. The overall character of $M_2(\beta_1)$ for the intermediate component supports its assignment to 'tie' molecules between crystal regions in the highly ordered polymer (Smith *et al.*, 1975) (cf Section 6.3).

[13]C chemical shifts of the methylene group have been used as a probe of orientation in highly drawn linear PE (Opella and Waugh, 1977; VanderHart, 1979). The fact that the chemical shift spectrum is centred at 12.7 ppm when $\beta_1 = 0°$ offers clear evidence of near perfect alignment along the draw axis since it is presumed that $\sigma_{33} = 12.7 \pm 1.3$ ppm is parallel to the chain axis (table A1.2). The latter assumption may not be fully justified, due to a reported slight misalignment of the principal axis of the chemical shift tensor and the chain axis (Nakai *et al.*, 1988b). Measurements on an oriented sample of ultra-high molecular weight PE ($M_W = 4 \times 10^6$) revealed a second, monoclinic, phase which is induced in this material upon drawing (VanderHart and Khoury, 1984). The monoclinic peak is located at 35.0 ppm, which is downfield from the usual orthorhombic peak at 33.6 ppm in the oriented sample (fig. 7.8). A number of observations can be drawn from these spectra.

- The crystallinity of the drawn and undrawn samples are, respectively, $f_x = 0.46 \pm 0.03$ and 0.59 ± 0.20. The latter determination agrees with the estimate based upon density measurements.
- The relative intensities of the orthorhombic and monoclinic peaks resolved under CPMAS conditions do, in fact, denote the wt% of the phases; the monoclinic phase constitutes 8.5 ± 1.0 wt% of the total crystalline content.
- Spin diffusion considerations indicate that both crystalline phases have the same average proximity to the non-crystalline regions.

Orientation in the non-crystalline regions is clearly evident in the non-spinning ^{13}C spectra recorded as a function of β_1 (fig. 7.9). Analysis of the

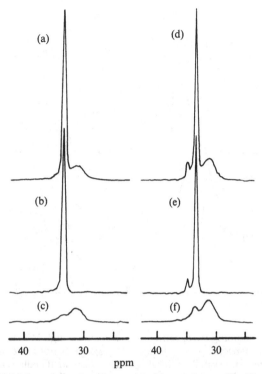

Figure 7.8. ^{13}C MAS spectra of undrawn (a)–(c) and drawn (d)–(f) ultra-high MW PE. Spectrum (b) = (a) minus (c) and (e) = (d) minus (f). Spectra (a) and (d) are normalised to the same intensity and spectra (c) and (f) are adjusted to eliminate the non-crystalline signals in spectra (b) and (e). (Reproduced from VanderHart and Khoury, *Polymer*, **25**, 1589 (1984) by permission of the publishers, Butterworth Heinemann Ltd ©.)

first moment of the lineshape yields values of $\langle P_2(\cos \beta) \rangle = 0.66 \pm 0.06$ for the combined orthorhombic and monoclinic crystalline regions and 0.23 ± 0.04 for the non-crystalline regions.

^{2}H NMR data for perdeuterated uniaxially oriented d_4-PE ($\lambda = 9$) and d_4-PE single crystal mats illustrate the sensitivity of ^{2}H lineshapes to local order in deformed polymers (Hentschel *et al.*, 1981). In this study, experimental spectra were analysed by assuming at the outset a Gaussian distribution (eq 7.8) of structural units (chains) about the preferred axis in oriented PE. The analysis of the single-crystal mats proceeds as follows. Moments P_{l00} ($l \leqslant 8$) determined from eq 7.8 are used to compute theoretical lineshapes which are then fitted to the experimental spectra for the crystalline regions recorded as a function of β_1 (the amorphous contribution is selectively suppressed). The optimum fit is achieved by

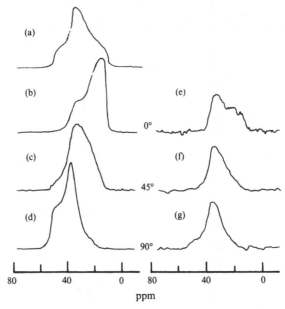

Figure 7.9. Non-spinning ^{13}C spectra of undrawn ultra-high MW PE (spectrum (a)) and drawn PE (spectra (b)–(g)). Spectra (e)–(g) denote the non-crystalline component. The non-crystalline lineshapes have been artificially scaled to 50 % of the spectra (a)–(d). Note the preferred orientation in the non-crystalline spectra. (Reproduced from VanderHart and Khoury, *Polymer*, **25**, 1589 (1984) by permission of the publishers, Butterworth Heinemann Ltd ©.)

adjusting the width of the Gaussian distribution to $\bar{\beta} = 12 \pm 1°$ and by including an isotropic contribution equivalent to 20 % of the total signal intensity (fig. 7.10). The agreement is excellent for an analysis which only invokes two adjustable parameters and is in reasonable accord with predictions of $\bar{\beta} = 15 \pm 2°$ from X-ray analysis.

The amorphous contribution can be examined preferentially by exploiting differences in T_1 to delineate crystalline and amorphous contributions. Both planar and conical distributions (Spiess, 1992) are invoked to allow for the *gauche* conformers in the amorphous regions. Analysis of the deuterium spectra reveals that 25 % of the amorphous regions are wholly disordered whereas the remaining part is ordered but to a significantly lower degree than the crystalline regions: the second moment of the distribution is 0.66 compared to 0.99 for the crystalline regions (Spiess, 1985b).

Biaxially oriented PE was first used to test the applicability of two-

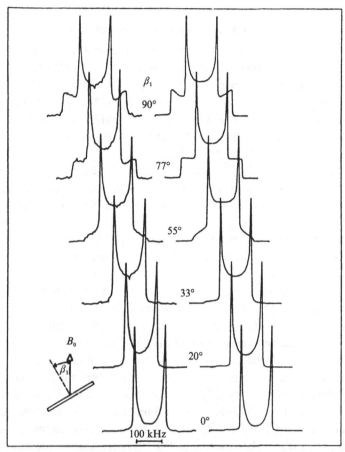

Figure 7.10. Observed (left) and calculated (right) deuterium spectra of PE single crystal mats for various orientations in B_0. The calculated spectra were convoluted with a Gaussian of variance 0.9 kHz to account for the inter-deuteron dipolar coupling. (Reproduced from Hentschel *et al.*, *Polymer*, **22**, 1516 (1981) by permission of the publishers, Butterworth Heinemann Ltd ©.)

dimensional MAS NMR as a probe of partially ordered polymers (Harbison and Spiess, 1986). Results are in general accord with the consolidated understanding of molecular structure and order in partially drawn PE and serve as a useful precursor to the more complete analysis of orientation in drawn poly(ethylene terephthalate) (PET) discussed in Section 7.6.

7.5 Orientation in polytetrafluoroethylene

An early study on uniaxially drawn PTFE compared the results of three different approaches involving the analysis of (i) X-ray diffraction patterns recorded on a flat plate (Bacon, 1956; Gupta and Ward, 1970), (ii) X-ray pole figures (Wecker *et al.*, 1972, 1974) and (iii) NMR $T_2(\beta_1)$ data (O'Brien and McBrierty, 1975). Results correlate reasonably well but the T_2 data could only confidently predict the difference between the second and fourth moments, $\langle \cos^2 \beta \rangle - \langle \cos^4 \beta \rangle$ (fig. 7.11).

Analysis of chemical shift spectra are more revealing (Brandolini *et al.*, 1982, 1983, 1984). For spectra recorded above 35 °C, the lineshape is axially symmetric with the unique axis coincident with the helical axis of the TFE chain (Garroway *et al.*, 1975). This choice of temperature and the use of the MREV-8 pulse sequence (cf table 4.1) to suppress complicating dipolar contributions greatly simplifies interpretation. Values of $\langle P_i(\cos \beta) \rangle$ obtained from a moments analysis of spectra to progressively higher order using eqs 7.9 and 7.11 along with data listed in table 7.2 were used to construct approximations to the distribution functions $\mathbf{P}(\beta)$ (eq 7.3) for each draw ratio. Orientation develops less rapidly than predicted by the pseudo-affine deformation model and falls short of that reported for PE. This, perhaps, reflects the extensive annealing of the oriented samples and also differences in sample crystallinity which become important if the amorphous regions are initially assumed to be random. As in the case of PE, tension tends to align the structural units along the draw direction at higher draw ratios. For $\lambda = 1.4$, however, $\mathbf{P}(\beta)$ exhibits a maximum around $\beta = 40°$, paralleling observations in PE. The deformation mechanisms in the two polymers are clearly similar. Orientation transverse to the draw direction is favoured when PTFE is modestly compressed ($\lambda = 0.82$).

A related experiment examined the shrinkage and accompanying disorientation induced in unconstrained drawn samples annealed at 300 °C (Brandolini *et al.*, 1984). By extrapolating the dependence of $\mathbf{P}(\beta)$ on the draw ratio λ' back to $\lambda' = 1$, a value of $\mathbf{P}(\beta)$ is predicted which is close to that for a random distribution (fig. 7.12). Thus, annealing of the unconstrained samples induces shrinkage and disorientation but a totally random distribution is not retrieved: there is a modest residual dependence of lineshape on β_1, which is indicative of remnant orientation, even in the most heavily annealed samples. $\langle P_2(\cos \beta) \rangle$ is strongly dependent on shrinkage whereas $\langle P_4(\cos \beta) \rangle$ is not.

Kasuboski (1988) examined orientation in the amorphous regions of oriented PTFE ($\lambda = 2.4$) using differences in $T_{1\rho}$ for crystalline and

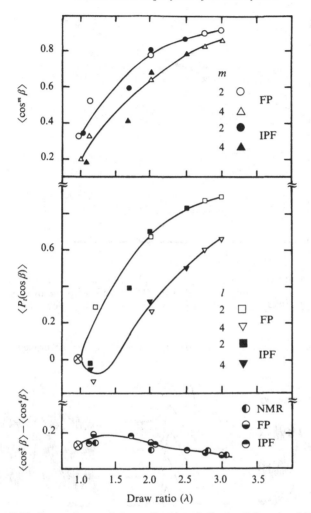

Figure 7.11. Dependence of $\langle \cos^m \beta \rangle$, $(m = 2,4)$, $\langle \cos^2 \beta \rangle - \langle \cos^4 \beta \rangle$, and $\langle P_l(\cos \beta) \rangle$ $(l = 2,4)$ on draw ratio of uniaxially oriented PTFE: flat plate X-ray analysis (FP); data compiled from X-ray inverse pole figures (Wecker *et al.*, 1972, 1974) (IPF); and $T_2(\beta_1)$ analysis (NMR) (see text). The points $\otimes$ denote the theoretical values for a random sample $(\lambda = 1)$. (Reprinted from O'Brien and McBrierty (1975) by permission of the Royal Irish Academy.)

amorphous regions to delineate the two contributions to the spectrum. The spin-locking pulse prefixing the MREV-8 sequence was set at $\tau = 30$ ms which is long compared with the amorphous $T_{1\rho a} = 8$ ms component but insufficient to relax fully the $T_{1\rho c} = 25$ ms crystalline component: in this way, 99% of the amorphous component was suppressed while 50% of the

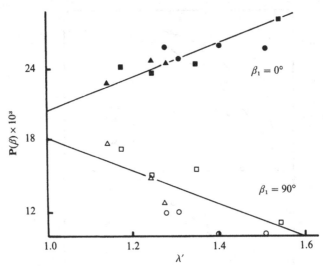

Figure 7.12. $P(\beta)$ at two angles $\beta_1 = 0°$ and $90°$ plotted as a function of λ', the draw ratio of the annealed samples of PTFE. (Reprinted with permission from Brandolini *et al.* (1984). Copyright (1984) American Chemical Society.)

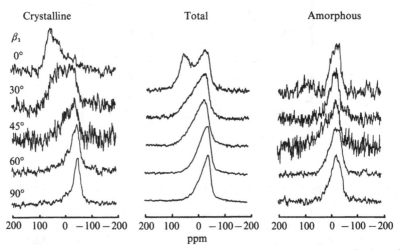

Figure 7.13. Crystalline, amorphous and total chemical shift spectra of oriented PTFE ($\lambda = 2.4$). The gain is $\times 2$ for $\beta_1 = 0°, 30°$ (total spectrum); $\times 1.5$ for $\beta_1 = 0°$ and $\times 2$ for $\beta_1 = 30°, 45°$ (crystalline spectrum). (Reprinted with permission from Kasuboski (1988), Thesis, University of Delaware.)

crystalline contribution was still retained. Crystalline and amorphous spectra for a number of sample orientations in B_0 (fig. 7.13) reveal a weak angular dependence in the amorphous signal. The first moments of the

Table 7.5. *Second moments of the distribution in partially ordered PTFE*
($\lambda = 2.4$). *Data taken from Kasuboski (1988)*

Region	Order parameter S	$\langle \cos^2 \beta \rangle$
Crystalline	0.25 ± 0.03	0.50 ± 0.06
Amorphous	0.05 ± 0.03	0.37 ± 0.22
Total	0.17 ± 0.03	0.45 ± 0.08

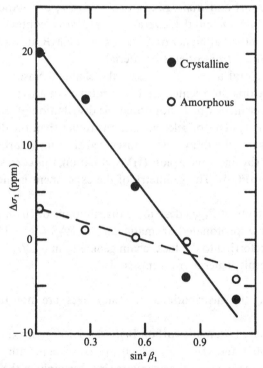

Figure 7.14. The first moment of the spectrum, $\Delta\sigma_1$, relative to the isotropic position for a random sample, recorded as a function of $\sin^2\beta_1$ in uniaxially drawn PTFE ($\lambda = 2.4$). Crystalline regions (●) fitted to $\Delta\sigma_1 = 20.7 - 28.8 \sin^2\beta_1$ ($R = 0.99$); amorphous regions (○) fitted to $\Delta\sigma_1 = 3.4 - 6.4 \sin^2\beta_1$ ($R = 0.94$). (Reprinted with permission from Kasuboski (1988), Thesis, University of Delaware.)

resolved spectra are plotted as a function of $\sin^2\beta_1$ in fig. 7.14 which confirms the linear dependence of $\Delta\sigma_1$ on $\sin^2\beta_1$ and allows order parameters to be determined for each region using eqs 7.9 and 7.12. These are listed in table 7.5.

7.6 Orientation in poly(ethylene terephthalate)

Ward's review (1985) of orientation in PET offers another striking example of the benefits of a multifaceted approach in investigating the structural reorientation induced by uniaxial deformation. The collated results of diverse techniques reveal a number of interesting features: the development of orientation in PET is not pseudo-affine but, rather, mimics that of a stretched network; unlike PE and PTFE, the distributions of chain axes are no longer axially symmetric due to a preferential orientation of the benzene rings towards the plane of the oriented film; the crystalline and amorphous regions are all-*trans* and *trans/gauche*, respectively; orientation induces conformational change in the overall *trans/gauche* content and crystallisation is promoted at higher draw ratios; and NMR anisotropy is due essentially to wholly *trans* oriented material.

That the structural units – in this case the chains themselves – are not transversely isotropic in oriented PET is confirmed in a two-dimensional MAS NMR experiment which monitored the distribution of chain axes in the principal axis systems of selected carbons rather than the distribution of chains relative to the draw axis as discussed thus far (Harbison *et al.*, 1987). In the Harbison–Vogt–Spiess (HVS) notation, this corresponds to a determination of $P(\chi\phi)$. The geometry of the experiment is shown in fig. 7.15.

The contour plots of $P(\chi\phi)$ describing distributions of chain axes in the protonated and unprotonated aromatic carbon PAS (fig. 7.16) are best understood by referring to the local chain geometry in fig. 7.17. A number of interesting results emerge from this study.

- As predicted, the distributions of chain axes are not transversely isotropic.
- In the protonated aromatic carbon PAS (fig. 7.16(a)), maxima occur at $(\chi\phi) = (36°, 90°)$ and $(84°, 90°)$ which are consistent with chain axes distributed in the plane of the aromatic ring, certainly in the crystalline regions and most probably in the amorphous regions as well. The assignment of the principal chemical shift components of the protonated aromatic carbon is also borne out in the χ-dependence of $P(\chi\phi)$.
- NMR data are fully consistent with X-ray measurements which establish the angle between the chain axis and the symmetry axis of the aromatic ring to be 24°. Note that this determination does not depend on prerequisite X-ray information and as such the study represents an example of NMR crystallography (Tycko and Dabbagh, 1991).

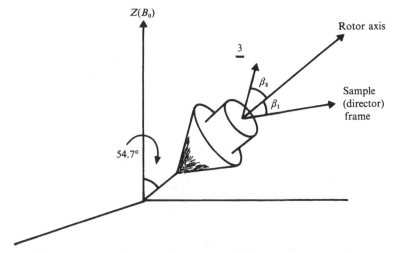

Figure 7.15. Geometry of the two-dimensional MAS experiment described in the text. The angles are as specified in the original paper of Harbison and coworkers (1987).

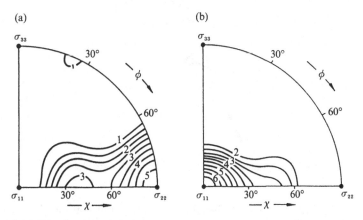

Figure 7.16. Distribution of chain axes $P(\chi\phi)$ in (a) the protonated aromatic carbon PAS and (b) the unprotonated aromatic carbon PAS in PET. (Reprinted with permission from Harbison *et al.* (1987). Copyright (1985) American Institute of Physics.)

- To be consistent, there should be a maximum at $\chi = 24°$ in $P(\chi\phi)$ in the unprotonated carbon PAS (fig. 7.16(b)). Instead, a broad intense peak is observed at $\chi = 0°$ which must dominantly reflect the amorphous

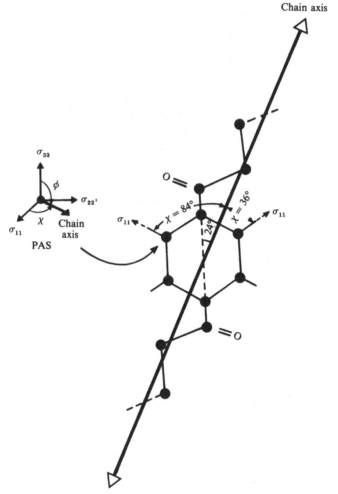

Figure 7.17. Local geometry of the ethylene terephthalate chain in oriented PET.

contribution: recall that the plots represent the combined crystalline and amorphous order parameters. The amorphous contribution also accounts for the broad ridge between the two maxima in fig. 7.16(a). A wholly definitive interpretation must await an experiment which selectively suppresses one or other of the dominant contributions to the spectra.

This again reiterates the point that the strength of NMR in probing local order lies in its greater ability to probe more moderate distributions as opposed to more highly ordered systems for which other techniques such as X-ray analysis are more appropriate.

7.7 Polymer liquid crystals

Liquid crystals (LCs) find many applications in electronics, principally as display devices. These applications exploit their electrical, magnetic and optical properties which stem from the facile molecular alignment and long-range order which can be conferred on the constituent mesogenic groups in a controlled way by the application of electric or magnetic fields. Polymerisation of such mesogenic groups or their incorporation into polymer backbone chains combines the attributes of LCs with the structural integrity and processibility of polymers. Spacer molecules are often interposed between the mesogenic side group and the polymer backbone chain to decouple their respective molecular motions. The response time of polymer liquid crystals (PLCs) to the externally applied field is appreciably slower than their low-molecular-weight counterparts, which can be a disadvantage from an applications point of view, but it allows the underlying molecular reorganisation and dynamics to be examined on a slower timescale using the full array of diagnostic NMR techniques described thus far.

Consider the main-chain liquid crystal polymer

for which spectra shown in fig. 7.18(a) were recorded as a function of time in a magnetic field of 4.7 T (Moore and Stupp, 1987). The change in lineshape reflects the progressive macroscopic alignment in the liquid state which is retained upon solidification. The second moment M_2 of the broadline component is proportional to the square of the order parameter S (eq 7.4) and fig. 7.18(b) illustrates the progression towards perfect alignment ($S = 1$) with time. Determination of M_2 therefore allows the kinetics of reorientation to be followed as a function of temperature, molecular weight, thermal history and so on. Moore and Stupp assigned the narrow central peak to less ordered and therefore more mobile interphase regions which constituted 6–20 % of the polymer; the analogy with the interface region in partially crystalline polymers is evident. On

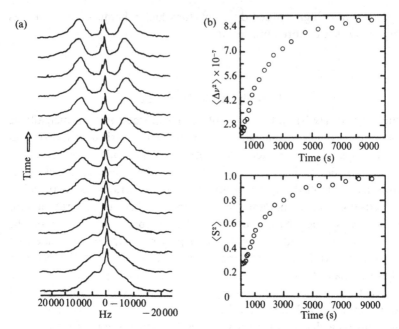

Figure 7.18. (a) ^{1}H NMR spectra in the fluid state of a mesogenic polyester (see text) with increasing time in the magnetic field of 4.7 T ($T = 185$ °C). (b) Evolution of M_2 for the broadline component and the macroscopic order parameter S^2 as a function of time (Reprinted with permission from Moore and Stupp (1987). Copyright (1987) American Chemical Society.)

the basis of their annealing studies, they deduced that chain segments in the boundary regions tended to limit the rate of chain reorientation in the magnetic field.

In a more sophisticated study, Blümich and coworkers (1987) analysed the sidebands of two-dimensional MAS spectra (cf Section 7.3.4) for characteristic carbons in different segments of a liquid crystal side-chain polymer to determine their respective orientation distribution functions. The results are shown in fig. 7.19. Note the development of preferential order in their polyacrylate sample with progression away from the backbone chain where preferential order is negligible, towards the mesogenic side group where orientation is appreciable.

For a more detailed account of this rapidly developing field, the reader is referred to reviews by Spiess (1985a, b) and Khetrapal and Becker (1991).

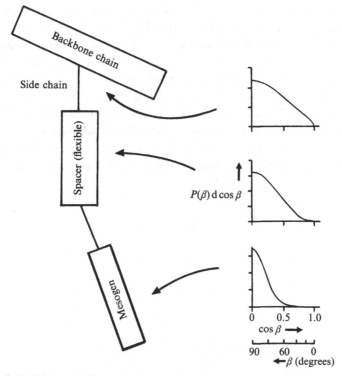

Figure 7.19. Schematic diagram of a side-chain liquid crystal polymer. Orientation distribution functions are portrayed for appropriate carbon atoms in specific segments of a polyacrylate sample. (Adapted from Blümich *et al.* (1987) with permission of the Berichte der Bunsengesellschaft ©.)

7.8 Secondary local probes of orientation

A number of elegant experiments have been devised which utilise secondary probes such as deuterated swelling agents, paramagnetic species or fluorescent chromophores that respond to local order in a host polymer. Deloche and Samulski (1981) used deuterated benzene to examine network deformation in *cis*-polyisoprene. The observed splitting of the deuterium spectrum of the swelling agent, $\Delta \nu_Q$, is linearly dependent on $(\lambda^2 - \lambda^{-1})$ in keeping with the predictions of their lattice model for the deformed network. This model evolved from the observed dependence of $\Delta \nu_Q$ on draw ratio, temperature, degree of swelling and network cross link density, reflecting, in turn, the sensitivity of the ^{2}H NMR spectrum to the orientational order conferred on the solvent probe as it diffuses through the deformed matrix.

Complementary studies on poly(dimethylsiloxane) (PDMS) swollen with C_6D_6 and partially deuterated PDMS swollen with C_6H_6 clarify the individual roles of solvent and network (Toriumi *et al.*, 1985). Again, Δv_Q for both deuterated solvent and deuterated network depend linearly on $(\lambda^2 - \lambda^{-1})$ for low strains ($\lambda < 2$). That computed order parameters for network and solvent exhibit similar strain dependence confirms the view that the orientational diffusion of solvent is strongly coupled to network chain ordering. By the same token, measurements on deuterated 1,4-polybutadiene networks point to a non-affine deformation mechanism whereby short chains are stretched to a greater extent than long chains. There is also evidence of preferred orientation near network junctions (Gronski *et al.*, 1984) (cf Section 8.1).

8

Selected topics

In this chapter, a number of examples are chosen to illustrate further the diversity of application of NMR in the field of solid polymers. The treatment is necessarily cursory since each topic forms a major area of research in its own right. The particular selection also serves to illustrate the point that useful information can be obtained with relatively unsophisticated and routine NMR procedures without the need for absolute recourse to state-of-the-art facilities.

8.1 Network systems

This discussion on network and interpenetrating network (IPN) systems affords the opportunity to comment further on the molecular motional regime between the extremes of solid and liquid behaviour, of which the interfacial effects discussed in Section 6.3 are examples. Generally, topological constraints such as entanglements and cross links induce significant deviations from liquid-like behaviour which are usually reflected in complex NMR relaxation decay (Kimmich, 1988). Motion is inevitably highly anisotropic and this leads to a residual dipolar coupling that typically defines the shape of the transverse relaxation function. Conversely, the shape of the function contains valuable information on the underlying complex motions. These effects were implicit in early measurements of the influence of molecular weight on NMR relaxation (McCall *et al.*, 1962) where the Rouse-like dynamics for low-molecular-weight polymers developed into the more complex behaviour of entangled chains as the molecular weight increased. Early intuitive models envisaged two temporal regimes which typically characterised sol and gel behaviour in cross-linked systems (Folland and Charlesby, 1978).

A consolidated viewpoint has evolved from these initial deliberations in

269

which networks of cross links or entanglements create a 'macrostructure' characterised by much larger space and time scales than those encountered at the microscopic level. Scaling concepts feature prominently in models that describe this situation (de Gennes, 1979). In a series of seminal papers, Cohen-Addad and coworkers (1983, 1984) developed a scale-invariant model by averaging NMR behaviour over local dimensions. This approach delineated the superimposed long timescale dynamics from short timescale events at the molecular level. Brereton (1989, 1990, 1993) and with his coworkers (1991) expanded upon these concepts to generate a general hierarchical model which embodies both short-timescale, Rouse-like, behaviour and longer timescale phenomena such as reptation. These models have been used to examine a number of important practical systems such as deformed cross-linked composites (Brereton, 1990) and swollen gels (Cohen-Addad, 1991). Filled elastomers are another important category of networked systems which serve to illustrate these concepts.

8.1.1 Filled elastomers

The mechanical properties of elastomers are enhanced when a suitable filler is added to them. Typically, a correlation has been drawn between the dynamics of the polymer–filler interface at the molecular level and the tensile strength of the filled composite. The interaction between filler and polymer is both physical and chemical in nature and its overall effect is to alter quite dramatically the structure and molecular motions of the host polymer. Two examples are considered which illustrate the role of NMR in unravelling events on the different structural and temporal scales.

8.1.1.1 Siloxane–silica systems

Poly(dimethyl siloxanes) (PDMS) filled with silica (Aerosil) particles are interesting, first because of the level of improvement achieved in mechanical properties with the introduction of active filler (Warrick *et al.*, 1979) and, second, because they are reasonably well characterised and are therefore useful model systems. Three microregions, distinguished by their characteristic molecular mobility, have been identified: In the tightly adsorbed layer, which is somewhat less than 1 nm thick, the motion of chains is severely restricted, to a degree that depends on filler type. Slow adsorption–desorption processes and fast chain librations have been postulated by Litvinov (1988) who estimates the residence time of chain units to fall within the range 10^{-4}–10^{-3} s. Adsorption of water by the filler or replacement of hydroxyl groups on the surface of the Aerosil particles with hydrophobic $Si(CH_3)_3$ groups reduces the strength of the polymer–filler interaction. In the next structural layer, motions are similar to those of the unfilled polymer but chain translation is limited by polymer–filler adsorptive links and by topological constraints which include physical network junctions. More remote material, having fewer contacts and therefore fewer constraints to motion, behaves much like the unfilled polymer.

NMR relaxation is sensitive to the structural and temporal changes which accompany the introduction of filler. Typically, $T_1(^1H: 88 \text{ MHz})$ minima in the unfilled polymer, associated respectively with CH_3 rotation and the glass transition at 98 K and 193 K, are complemented by a third minimum in heavily loaded PDMS at 295 K which reflects motional events that are directly associated with the presence of filler particles. As expected, there is a broad distribution of motions in the composite. $T_1(^2H)$ measurements above 255 K allow accurate quantification of even small amounts of tightly bound polymer (about 3%) by monitoring lineshape changes during inversion recovery (Litvinov *et al.* 1989).

A further investigative dimension is introduced when filled PDMS is uniaxially stretched (Litvinov and Spiess, 1992). The observed increase in ^2H linewidth is attributed to a decrease in the number of chain loops in the adsorption layer arising from chain slippage as the composite is oriented. This is facilitated by the adsorption–desorption processes which occur above T_g (Litvinov, 1988).

The immobilised interfacial component in the filled network under strain persists above the melting temperature of the filled but unstrained material. It is argued that the motionally constrained chains at the interface act as nucleation centres for stress-induced crystallisation.

When stretched, the single deuterium line for the unoriented composite develops quadrupolar splitting, $\Delta\nu_Q$, which has been analysed in terms of orientation anisotropy induced in the stretching process, as described in Section 7.8. The linear dependence of $\Delta\nu_Q$ on $(\lambda^2 - \lambda^{-1})$ for both filled and unfilled PDMS implies affine deformation. Observed differences in the response of the filled and unfilled networks (fig. 8.1) imply that the number of chain units between effective cross links is less in the filled network and/or that the microscopic strain λ differs from the macroscopic strain L/L_0. The following relationship between the two strains has been proposed (Queslel and Mark, 1989; Meinecke, 1991)

$$\lambda = (L/L_0)(1 + 2.5\phi + 14.1\phi^2) \tag{8.1}$$

where ϕ is the volume fraction of the spherical filler particles. This

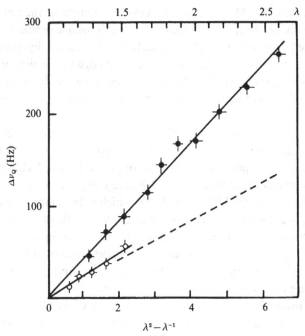

Figure 8.1. Quadrupolar splitting Δv_Q as a function of $(\lambda^2 - \lambda^{-1})$ for unfilled ($\bigcirc$) and filled ($\bullet$) PDMS networks ($T = 305$ K). The dashed line denotes Δv_Q versus $(\lambda^2 - \lambda^{-1})$ for the filled system corrected by eq 8.1 (Reprinted from Litvinov and Spiess, *Makromol. Chemie*, **193**, 1181 (1992) by permission of the publishers Huthig & Wepf Verlag, Basel.)

correction to L/L_0 brings the data for the filled network into coincidence with those of the unfilled network, as shown by the dashed line in fig. 8.1. It would therefore appear that the filler particles do not necessarily constitute a source of permanent cross links since a description invoking an affine term and a corrected draw ratio (eq 8.1) adequately accounts for the data (Litvinov and Spiess, 1992).

It is interesting to note that no appreciable increase in the effective cross-link density has been detected when ²H-labelled *cis*-1,4-polybutadiene (PB) is filled with carbon black (Gronski *et al.*, 1991). This system is considered further in the following section.

8.1.1.2 Carbon black–elastomer systems

That some of the characteristics of silica–PDMS composites can be generalised to other systems is borne out in carbon black-filled elastomers. Early models, based on thermal and dynamic modulus measurements (Rehner Jr, 1965; Smit, 1966), envisaged a shell structure with an adsorbed

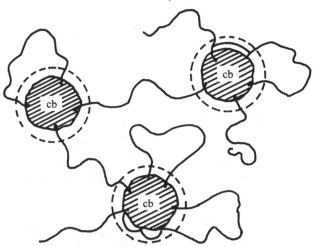

Figure 8.2. Model for carbon black-filled rubber. (Reprinted with permission from O'Brien *et al.* (1976). Copyright (1976) American Chemical Society.)

layer of polymer about 2 nm thick around the carbon black particles. As expected, T_g for this layer is higher than the value for the bulk material. Although contested by others (Kraus *et al.*, 1970), this model was supported and refined in subsequent NMR studies (Roe *et al.*, 1970; Kaufman *et al.*, 1971; O'Brien *et al.*, 1976) which again detected a tightly bonded layer surrounding the filler particles, an outer annular region of more loosely bound rubber and a third component with the overall character of the unfilled polymer. The former two components constitute the insoluble gel whereas the latter can be extracted with a good solvent.

The model shown in fig. 8.2 embodies the principal structural features suggested by NMR, namely, chain folds, cilia, tie molecules between filler particles, chain entanglements and occluded rubber trapped within the structure of carbon black aggregates (O'Brien *et al.*, 1976; Kentgens *et al.*, 1987). Note the structural similarities of the polymer closest to the filler with the crystalline/amorphous interface described in Section 6.3. Recall, too, that the first few units of an anchored chain are fairly rigid (cf eq 6.3) and these define the bound rubber layer in the immediate vicinity of the filler particles. The porous, highly ordered, surface topography of carbon black particles, which is also a feature of the model, has since been confirmed by Donnet and Custodero (1992).

NMR linewidths respond most noticeably to the introduction of carbon black: specifically, T_2 decreases whereas T_1 and $T_{1\rho}$ are only slightly affected. The nature of the interaction between filler and host polymer was

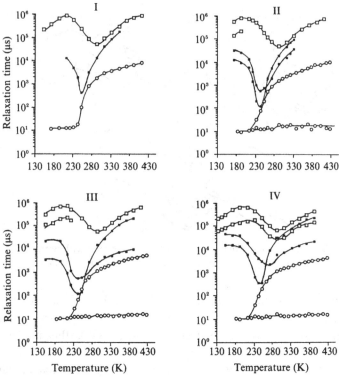

Figure 8.3. Proton T_1 ($\square$), $T_{1\rho}$ ($\blacksquare$) and T_2 ($\bigcirc$) for a range of natural rubber filled with carbon black. (I), natural rubber as received; (II), original masterbatch containing a total rubber content equal to 200 % of the carbon black content; (III), sample of masterbatch extracted in boiling hexane (342 K); (IV), sample extracted in boiling toluene (383 K). (Reprinted with permission from Kenny *et al.* (1991). Copyright (1991) American Chemical Society.)

explored by examining the insoluble gel which survived extraction procedures of progressively increasing severity (Kenny *et al.*, 1991). Typical data are shown in fig. 8.3 for natural rubber filled with 50 phr (parts per hundred rubber) of a HAF grade carbon black. Note the following conclusions.

• The *positions* of T_1 and $T_{1\rho}$ minima and T_2 transitions do not change appreciably from one sample to another implying, at best, a modest dependence of T_g on sample treatment. This is in accord with the generally observed insensitivity of techniques such as DSC and DMTA to the immobilisation of small amounts of tightly bound rubber in the filled composite which is responsible for the persistent short T_2 component above T_g in the filled composite. The collated relaxation data

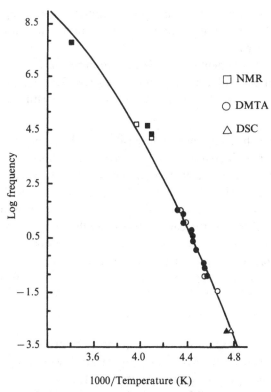

Figure 8.4. Plot of log v_c versus $1000/T$ for the glass transition in sample I (unfilled points) and sample II (filled points) as designated in the caption to fig. 8.3. (Reprinted with permission from Kenny *et al.* (1991). Copyright (1991) American Chemical Society.)

(fig. 8.4) for samples (I) and (II) confirm that T_g for the filled and unfilled material differ by no more than about 1 K.

- The high-temperature T_{2L} plateau progressively decreases as the extraction procedure becomes increasingly severe. Noting that $T_2 = 10^3$ μs and 10 μs above and below T_g respectively, the analysis in Section 3.6 may be used to compute the available solid angle for motions above the glass transition: the excluded angle β is given by $\sin^2\beta\cos\beta \approx 10^{-2}$, from which $\beta \approx 6°$. Thus, most but not all orientations are available to the rubber molecules in this region; hence the term *loosely bound* material.

- The modest transition in T_{2S} near 285 K which coincides with the completion of the initial rapid rise in T_{2L} associated with the glass transition signals cooperative motion between the tightly and loosely bound rubber molecules in the composite. The extent to which T_{2S}

persists to temperatures which can be as much as 150 K higher than in the neat polymer is a measure of the degree to which T_g for the tightly bound polymer increases.

● A variety of experimental observations confirm the $T_{1\rho}$ crossover point at 260 K. In particular, inspection of the FIDs during $T_{1\rho}$ decay shows that $T_{1\rho s}$ is associated with T_{2L} below 260 K and with T_{2s} above this temperature.

Consider now the component T_2 intensity data shown in fig. 8.5(a). First, there is a general increase in the relative proportion of immobilised polymer or bound layer as extraction becomes more severe. Second, the decrease in the intensity of the short T_2 component, $I(T_{2s})$, with increasing temperature is consistent with the notion of a range of bonding energies between the rubber and the carbon black: as the temperature increases, rubber molecules attain sufficient energy to overcome their bonding energies and break away from the filler particles. When plotted as $\log I$ versus inverse temperature (fig. 8.5(b)), a linear relationship is observed which can be represented by

$$I(s) = I_0(s) \exp{(395 \pm 50/T)} \qquad (8.2)$$

indicating that the mechanism which describes the progressive removal of rubber molecules with increasing temperature would appear to be Arrhénius in character within the accuracy of the data.

The Pliskin–Tokita (1972) expression provides the average thickness ΔR_0 of the bound layer.

$$\mathrm{BR} = \Delta R_0 f\{\phi \rho A/(1-\phi)\} + G \qquad (8.3)$$

BR is the fraction of apparently bound rubber (the volume of the insoluble polymer divided by the volume of polymer in the composite), f is the fraction of total carbon black surface area exposed to the polymer, ϕ is the volume fraction of carbon black in the composite, ρ is the density and A the specific surface area of the carbon black, and G is the fraction of insoluble gel in the unfilled material. The plot of BR versus $\phi \rho A/(1-\phi)$ is linear for a wide range of carbon black–elastomer composites and the slope of the line provides ΔR_0. The ability to separate loosely bound and tightly bound components, represented respectively by $I(T_{2L})$ and $I(T_{2s})$ (allowing, of course, for a contribution from rigid proton-containing additives such as stabilisers), permits the corresponding layer thicknesses for the loosely bound and tightly bound polymer to be determined.

Figure 8.6 portrays the relationship between BR and $\phi \rho A/(1-\phi)$ with ϕ and A as disposable parameters, respectively. The host polymer in this case

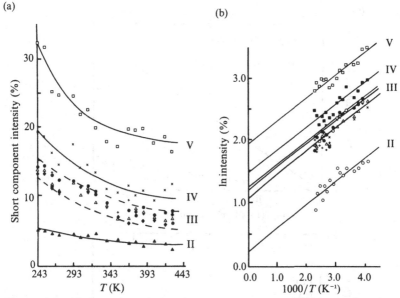

Figure 8.5. (a) Intensity of the short T_2 component $I(T_{2s})$ as a function of temperature. (b) Plot of $\ln[I(T_{2s})]$ versus inverse temperature. Sample designations are specified in the caption to fig. 8.3. Additionally, V denotes the material extracted in boiling mixed xylenes (418 K). (Reprinted with permission from Kenny *et al.* (1991). Copyright (1991) American Chemical Society.)

is *cis*-polybutadiene (O'Brien *et al.*, 1976). In fig 8.6(a), the Pliskin–Tokita formula is reasonably obeyed when carbon black loading ϕ is varied. The inner layer is about 20% of the total layer thickness. When the specific surface area A is varied, however, the plots are no longer linear due to a lower than expected bound layer content for the finer carbon blacks (higher A). This reflects the tendency of the finer carbon blacks to agglomerate, thereby reducing the available carbon black surface area. The sensitivity of $I(T_{2s})$ to agglomeration has been used to quantify the degree of dispersion of carbon black in the composite (Wardell *et al.*, 1981, 1982).

8.2 Water in polymers

Some polymer systems, of which ionomers and hydrogels are examples, can only fulfil their intended purpose when hydrated. In other instances, the presence of water, even in trace amounts, can adversely affect performance: enhanced electrical loss in coaxial cables and reduced effectiveness of certain adhesives are typical. Consequently, the need for a full understanding of the complex manner in which water behaves in host polymers is exceedingly important.

(a)

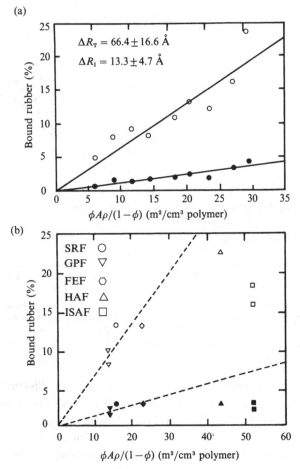

Figure 8.6. (a) Pliskin–Tokita plot of bound rubber as a function of $\phi A\rho/(1-\phi)$ with $A = 29.3$ m^2 g^{-1} and ϕ as a variable. (b) Plot with A as variable and $\phi = 50$ phr. The dashed lines are reproduced from (a). (Reprinted with permission from O'Brien *et al.* (1976). Copyright (1976) American Chemical Society.)

Water behaves anomalously when adsorbed into matrices as diverse as muscle, clay and polymers in the sense that the sharp first-order transitions which characterise pure bulk water are appreciably modified (Rowland, 1980). Many important applications of polymer–water systems rely on the intricate interaction between water molecules and their surroundings. These systems, while important in their own right, also serve as useful models to describe the peculiar behaviour of water in more complex media.

The anomalous behaviour of water is ascribed to several physical and chemical sources. Stillinger (1980) cites finite size effects at boundaries and interfaces which curtail the number of degrees of freedom, thereby disrupting the natural order in the bulk phase. Clustering, too, can lead to supercooling due principally to fewer nucleating seeds in clusters of limited size. Water of hydration and hydrogen bonding are typical of chemical interactions which bind water to specific hydrophilic sites. Regarding the various interactions, Rowland and Kuntz (1980) proposed the following ranking in order of decreasing strength: ion–ion > water–polar = polar–polar = water–water > water–hydrophobic.

Many factors complicate the interpretation of data from hydrated polymers or hydrogels: water may act as plasticiser or antiplasticiser depending on temperature and concentration (cf Section 6.9); the structural organisation of adsorbed water is sensitive to polymer mobility; hydration can induce conformational changes; effects of cross linking can be significant; and additives such as salts play an important role. More generally, water is influenced by both equilibrium and non-equilibrium factors, the latter displaying strong dependence on concentration and thermal history (Pouchly *et al.*, 1979) and hysteresis effects are often observed under temperature cycling. That events in one temperature regime need not extrapolate in any obvious manner to another regime, is evident in NMR experiments which probe water behaviour over a wide temperature range.

The different types of water that coexist in polymers are variously described as *bound* and *free*, *freezable* and *non-freezable* and *associated* water. Estimates of their relative proportions can depend both on the technique used and on thermal history. Thus, DSC measurements reveal a two-step water glass-to-liquid transition in hydrated poly(2-hydroxyethyl methacrylate): a weak endothermic step at 132 ± 4 K followed by a second endothermic step at 162 ± 2 K, which is contiguous with the first (Hofer *et al.*, 1990). NMR detects the onset of the water glass transition at the higher temperature of about 170 K, reflecting the sensitivity of this measurement to higher motional frequencies. Specifically, NMR distinguishes between water that is still mobile at quite low temperatures and normal water that freezes at 273 K. Hydration typically proceeds as a stepwise progression from tightly bound water at low levels of hydration to increasing amounts of free, bulk-like, water at saturation levels (Rupley *et al.*, 1980; Smyth *et al.*, 1988; Wilson and Turner, 1988).

While NMR is richly informative in its own right, its use in conjunction with other techniques has significant added benefit. The fact that different methods respond in uniquely different ways, can be turned to advantage by collating data from a variety of experiments as confirmed in NMR and

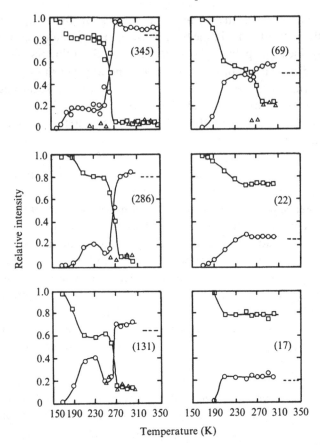

Figure 8.7. Component T_2 intensity data for P(NVP/MMA) with different water contents: T_{2L} (O); T_{2i} (△); T_{2s} (□). The dashed lines denote the calculated water proton fractions. Numbers in parenthesis denote wt% water relative to the dry polymer. (Reprinted with permission from Quinn *et al.* (1988). Copyright (1988) American Chemical Society.)

DSC measurements on the commercially important hydrogel poly(N-vinyl-2-pyrrolidone/methyl methacrylate) P(NVP/MMA) (Quinn *et al.*, 1988). These studies show that the NMR response is not a simple superposition of the individual water and polymer behaviours. Proton relaxation data for the polymer hydrated with D_2O, for example, indicate a broad distribution of cooperative molecular motions for which the non-uniform plasticisation of the polymer matrix by the water is at least partly responsible.

Linewidth data, and more particularly the intensities of resolved T_2 components, are especially revealing (fig. 8.7). Consider first the response

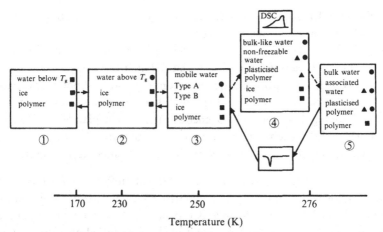

Figure 8.8. Specification of five thermal equilibrium states in hydrated P(NV-P/MMA). The dashed and solid arrows denote increasing and decreasing temperature cycles. The symbols denote the T_2 associated with each component as follows: T_{2s} (■), T_{2i} (▲), T_{2L} (●) (see text). (Reprinted with permission from Quinn *et al.* (1988). Copyright (1988) American Chemical Society.)

of the least hydrated sample, S(17). The increase in I_L beginning at 190 K heralds the onset of motion for the tightly bound water which behaves much like a glass passing through its glass transition (Boyle *et al.*, 1982, Pineri and Eisenberg, 1987). I_L attains a temperature-independent intensity of $I_L = 0.22$, which is equal to the calculated fraction of water protons in the sample within experimental error. It is notable that I_L does not increase near 273 K as would be expected from the melting of bulk-like water, if present. At higher water concentrations, I_L peaks around 230 K and then decreases to form a minimum at about 250 K. An intermediate component is detected in this temperature region. I_L then rises sharply as 273 K is approached, signifying the presence of appreciable amounts of bulk water in the more heavily hydrated samples. Five thermal equilibrium states can be deduced from these and parallel DSC measurements (fig. 8.8):

State 1 (< 170 K): Below 170 K, rigid polymer coexists with ice and bound, glass-like water below its glass-transition temperature T_g.

State 2 (170–230 K): Above T_g, the 76±10 wt % of tightly bound water becomes mobile by 230 K.

State 3 (about 250 K): At about 250 K there is evidence of polymer plasticisation and two bound or non-freezable water components with distinguishably different mobilities are resolved in the amounts of 50±10 wt % (type A) and 28±15 wt % (type B), respectively. The amount of more mobile (type A) water agrees remarkably well with the non-freezable water fraction $(47 \pm 7 \text{ wt}\%)$ determined from endothermic DSC measurements.

State 4 (250–276 K): Bulk-like water, non-freezable water, ice, plasticised polymer and non-plasticised polymer coexist.

State 5 (> 276 K): Both plasticised and non-plasticised polymer coexist with normal bulk water and less mobile or *associated* water which is exchanging between heterogeneous sites. As explained by Resing (1972), these exchange interactions account for the T_2 plateau or shallow T_2 minimum observed in the higher temperature region.

In certain hydrogel applications such as the manufacture of soft contact lenses, it is also important to understand the effects of hydration with saline solution. ^{23}Na NMR affords an added probe of the local environment as revealed in ^{23}Na linewidths and chemical shifts recorded as a function of solution content and modest temperature change (fig. 8.9) (McBrierty *et al.*, 1992). A single symmetric line is observed from which the linewidth Δv and chemical shift σ begin to increase as the degree of hydration falls below the non-freezable water content as determined by DSC. At high levels of hydration, Na$^+$ ions can exchange between relatively free and bound states as described by eq 8.3.

$$\left.\begin{array}{l} \Delta v_{\text{obs}} = P_{\text{b}} \Delta v_{\text{b}} + P_{\text{f}} \Delta v_{\text{f}} \\ \sigma_{\text{obs}} = P_{\text{b}} \sigma_{\text{b}} + P_{\text{f}} \sigma_{\text{f}} \end{array}\right\} \tag{8.4}$$

f and b denote free and bound states and P is the mole fraction in each state. It is evident that Δv_{obs} and σ_{obs} will increase as the water content decreases towards the non-freezable level since Δv and σ are higher in bound states which have longer motional correlation times τ_{c}. More generally, the asymmetry of the electric field gradient experienced by the ^{23}Na nucleus will increase as the hydration sphere around the Na$^+$ ions begins to deplete, leading to the observed increase in Δv and σ. However Δv for the highest level of hydration is still significantly greater than Δv for aqueous electrolytes of comparable salt concentration, indicating that the Na$^+$ ion still senses the polymer matrix even when hydration levels are high (Quinn *et al.*, 1990).

The dynamics of hydration and the nature of the water dispersion achieved are influenced by a number of factors such as temperature, pressure, polymer affinity for water and water concentration (or relative humidity). Crofton and coworkers (1982) examined water uptake in cellulose acetate (CA) as a function of time and water concentration. Degree of acetylation is also important in this instance since replacement

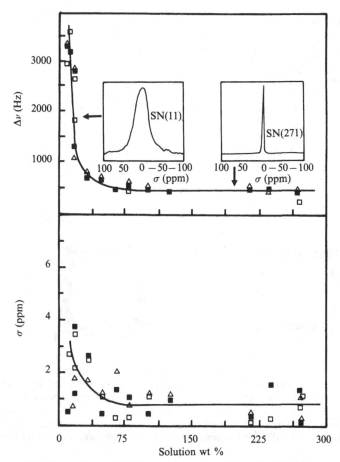

Figure 8.9. Dependence of [23]Na resonance linewidth at half height Δv and chemical shift σ as a function of saline solution wt% in P(NVP/MMA). ($\square$), 293 K, ($\blacksquare$), 303 K, ($\triangle$), 313 K. (Reprinted with permission from McBrierty *et al.* (1992). Copyright (1992) American Chemical Society.)

of OH groups with acetyl groups reduces the overall affinity of CA for water. At short times, the observed Fickian response yields diffusion coefficients in the ranges $1-3 \times 10^{-11} \text{ m}^2 \text{ s}^{-1}$ and $4-13 \times 10^{-11} \text{ m}^2 \text{ s}^{-1}$, depending on degree of acetylation, for relative humidities of 53% and 64%, respectively.

Studies on the ingress of water into nylon-66 (Mansfield *et al.*, 1992) explore the dynamics of water uptake and the actual dispersion of water achieved. This study offers an excellent example of the rapid advances and concomitant practical utility of magnetic resonance imaging (Mansfield

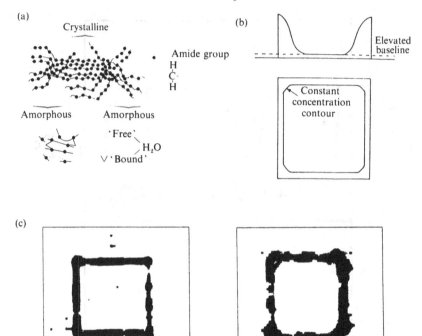

Figure 8.10. (a) Schematic diagram of water uptake in nylon-66. (b) Envisaged water-uptake contour in a section of nylon block together with the projection of water-uptake concentration normal to the block face. (c) Images of water ingress after immersion in water at 100 °C for 4 min (left) and 40 min (right). (Reproduced from Mansfield *et al.* (1992) with permission of the publishers, Academic Press Ltd. ©).

and Hahn, 1992). Diffusion coefficients and activation energies determined from the NMR data support a two-phase exchange model. High pressure data further indicate that water is preferentially take up at the amide sites in the amorphous regions, ultimately achieving three water molecules per site. As shown in fig. 8.10, there is excellent agreement between the anticipated and actual water uptake and dispersion.

8.3 Ionomers

Ionomers, formed by incorporating small concentrations of neutralisable ionic moieties into nonpolar chains, exhibit mechanical and transport properties that differ greatly from those of the neat polymer. The Nafion® membranes developed by the DuPont Company possess remarkable cation-exchange properties of importance, for example, in separators used in electrolytic cells. The polymer is a perfluorosulphonate resin in which hydrophilic side chains terminated with $-SO_3H$ groups are periodically attached to the hydrophobic backbone molecules. Nafion membranes can absorb substantial amounts of water and the acid form may be readily neutralised.

A comprehensive range of experiments points to the formation of ion pairs in the first instance developing at higher concentrations (50–100 ion pairs) into clusters 1–3 nm in diameter and 2–10 nm apart (Pineri and Eisenberg, 1987). A more recent model developed by Eisenberg and coworkers (1990) exploits the concept of motionally constrained chains with a concomitantly higher T_g in the immediate vicinity of clusters and multiplets, described earlier in Section 6.3. At high ionic concentrations these local areas of restricted mobility overlap to form discrete phases with measurable characteristic glass transition temperatures. The extent to which ion aggregation or clustering occurs depends on many factors such as prepolymer molecular weight, chain elasticity and flexibility, the extent of entanglements, the polarity of the medium, the location of the ion on the chain, the nature and concentration of cations and the degree of neutralisation. In essence they are network systems where the clusters act as dynamic cross links. The structure and functionality of the membrane depends sensitively on water content with the aqueous phase assuming a central role in the cation exchange process.

Although the hydrated membrane is exceedingly complex, analysis is greatly simplified due to the ability of NMR (i) to delineate backbone behaviour in Nafions *via* ^{19}F NMR, (ii) to examine selected cations such as [23]Na which exhibit their own characteristic spectra, (iii) to monitor the response of the aqueous phase with ^{1}H resonance and (iv) to exploit the site specificity of ^{13}C NMR. The flexibility of this approach has clarified many of the structural complexities of Nafions (Boyle *et al.*, 1983a, b). Water which preferentially hydrates the hydrophilic moieties in the membrane exhibits the general characteristics described in the foregoing section, the predominant feature being the glass-like character of bound water at low temperatures with a T_g in the region of 170 K.

In an effort to understand more clearly the structural intricacies of these systems, attention has been directed to a class of membranes where the ionic species are incorporated at both ends of a chain. These halatotelechelic polymers (HTPs) are generally considered to be less complicated systems for exploring structure–property relationships. Experiments such as EXAFS (Register *et al.*, 1988), ESR (Schlick and

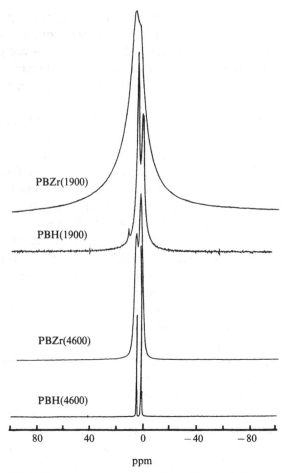

Figure 8.11. 400 MHz proton spectra for four PB telechelic samples showing progressive line broadening with neutralisation and decreasing molecular weight. (Reprinted with permission from McBrierty *et al.* (1993). Copyright (1993) American Chemical Society.)

Alonso-Amigo 1989), excimer fluorescence (Granville *et al.*, 1988) and NMR (Komoroski and Mauritz, 1978) have been used to investigate events in the vicinity of the cluster. From these studies, it is clear that the cross links involve both chemical and physical attachments. The degree to which backbone polymer is constrained depends on the cohesive strength of the cluster and on the molecular weight of the backbone polymer. It is envisaged that the cluster is in a state of dynamic equilibrium.

The site-specific character of high-resolution solid-state ^{1}H and ^{13}C NMR has yielded detailed perspective on HTPs at the molecular level

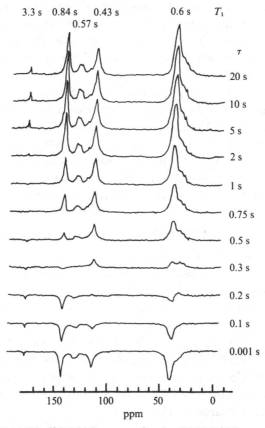

Figure 8.12. 75.4 MHz ^{13}C MAS spectra for dry PBH (1900) recorded with the inversion recovery sequence for various delay times τ. T_1 values for the resolved resonances are as indicated. (Reprinted with permission from McBrierty *et al.* (1993). Copyright (1993) American Chemical Society.)

(McBrierty *et al.*, 1993). By way of example consider the carboxy-terminated polybutadiene telechelic, PB(COOH)$_2$ and its neutralised version PB(COOZr)$_2$ denoted, respectively, PBH and PBZr. In sample designations, the molecular weight is added in parenthesis. The proton spectrum in fig 1.1(b) reveals the CH$_2$, CH and COOH components at progressively higher chemical shifts. Resolution of the carboxylate proton signal allows direct examination of the end groups in the acid telechelic and their role in multiplet and cluster formation. Figure 8.11 compares proton linewidths for PBH and PBZr for two different molecular weights. To interpret these data it is helpful to visualise the clusters as cross links in a state of dynamic equilibrium which act as terminal points for the PB chains. A crude analogy may be drawn with the model in fig. 8.2. Overall the chains will be more constrained in the lower molecular weight sample, which is the reverse of what would be expected in the absence of clustering. The linewidth also broadens upon neutralisation, reflecting greater rigidity and stronger cross links.

Resolution is significantly better in ^{13}C MAS spectra and there is the added advantage that chain end groups can also be studied in the neutralised telechelic. These spectra can monitor the integrity of cluster formation through line broadening and spin–lattice relaxation as a function of temperature. A typical ^{13}C inversion recovery spectrum for PBH(1900) is shown in fig. 8.12, where the measured T_1 values are consistent with the number of attached protons, in accord with eq 2.38.

8.4 Electrically active polymers

Investigations into the electrical properties of polymers constitute a major area of research. They exploit the multifunctionality of polymers and the ability to tailor properties through chemical modification, thermal treatment, copolymerisation and mechanical deformation. Piezo-, pyro- and ferroelectric properties, fast switching response, second harmonic generation and frequency doubling in electro-optics all reflect molecular behaviour to which NMR is sensitive. Here we will focus principally on the piezo- and ferroelectric responses as well as electrical conductivity in the semiconducting regime. NMR has contributed significantly to unravelling the underlying mechanisms in each case.

8.4.1 Conducting polymers

The traditional role of polymers as insulators has been complemented in recent times by a burgeoning interest in their conductive properties. This is due in large measure to many important commercial applications in the areas of synthetic metals, battery and solar cell technology and biosensors (Lovinger, 1983; Skotheim, 1986; Hanack *et al.*, 1991). A brief digression to examine the salient features of conductivity in polymers will place in context the subsequent discussion of the role of NMR in elucidating the fundamental mechanisms involved.

Attention is focussed on conjugated polymers which are semiconducting because of an extended π-electron system along the polymer chain. Conductivity is therefore highly anisotropic and often one-dimensional. This conductivity can be enhanced by chemical or electrochemical doping with n- or p-type dopants which generally are additive rather than

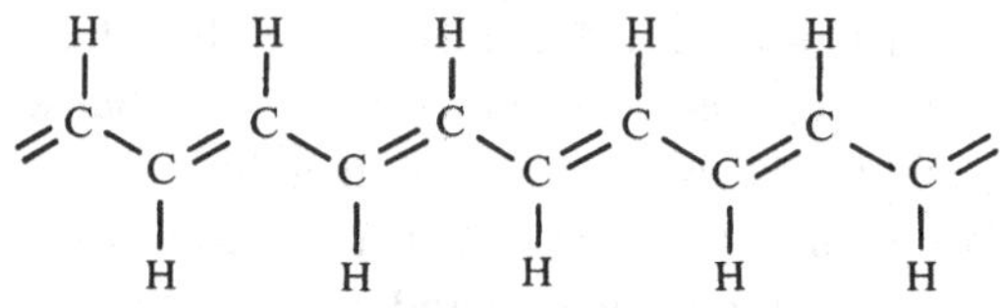

Figure 8.13. Structure of the *trans*-polyacetylene chain.

substitutional (Chiang *et al.*, 1977). As a result, conductivity can be varied over the full range from insulator to semiconductor to metallic conductor by systematic adjustment of dopant levels (Etemad *et al.*, 1982). Specific conductivities per unit mass which are a factor of two higher than copper have been achieved (Naarman, 1987; Basescu *et al.*, 1987).

Of the models developed to describe conductivity in polymers, the soliton model (and variants thereof) (Rice, 1979; Su *et al.*, 1979, 1980, 1983; Takayama *et al.*, 1980), the metallic drop model (Tomkiewicz *et al.*, 1980, 1982) and models that describe the incorporation of dopant species into the host polymer (Baughman *et al.*, 1978; Street and Clarke, 1981; Karasz *et al.*, 1982) are the most prominent. These are dealt with in detail in the original references and in the review literature; here, a cursory treatment will suffice.

Some insight can be gleaned in the context of *trans*-polyacetylene (PA) (fig. 8.13) which structurally is the simplest conjugated polymer. The one-dimensional character of the rigid planar molecule and the twofold ground-state degeneracy associated with an interchange of single and double bonds can be described in terms of conformational excitations or *solitons*. Solitons in PA, each extending over about 14 carbon atoms along the chain, are localised in the sense that they separate topologically distinct parts of the chain. Their propagation describes a delocalisation of the π-electron system along the chain. This, of course, implies that there is no *inter*chain charge transfer, which is true, for example, for most poly-diacetylenes because of their large *inter*chain separation (0.6–0.7 nm) and this renders them truly one-dimensional semiconductors (Feast and Friend, 1990). It is not so for PA which is more accurately described as a three-dimensional semiconductor because of *inter*chain effects. Hirsch and Grabowski (1984) argue that the soliton is only marginally stable in PA. In fact, the situation is more complicated for the greater number of conjugated polymers prepared thus far, for which a description in terms of charge transport *via* lattice distortions or *polarons* has been proposed (Vardeny *et al.*, 1986; Bradley *et al.*, 1989).

The metallic droplet model, in contrast, views the doped material as an inhomogeneous structure with dopant-rich metallic clusters in a sea of undoped polymer. Changes in conductivity are visualised in terms of percolation thresholds beyond which the metallic regions communicate efficiently. This model is less idealised in the sense that it reflects more accurately the non-ideal morphology and structural heterogeneity which typify polymers.

The appropriateness of any one model is dictated not alone by a predilection for one theoretical approach over another but, more tangibly, by a range of factors such as morphology, defects, isomerism, dopant distribution and concentration. In PA, for example, it is proposed that there is a phase transition from conductivity *via* solitons to metallic state conduction when the dopant concentration exceeds 6% (Heeger, 1986). This, in fact, is the concentration at which solitons embracing 14 carbon sites overlap.

As with many other areas of polymer investigation, it is control of morphology which underpins the successful translation of molecular phenomena into useful macroscopic behaviour. For conducting polymers, this goal is thwarted by problems of processibility, control of isomerism and defects, infusibility and environmental stability. The considerable efforts of synthetic chemists to address these issues have been increasingly successful.

It will be evident from the foregoing discussion that NMR assumes an important role because of its sensitivity to local electronic and structural properties both in solution and in the solid state. That different isomers can be identified is evident from table A1.2 which typically lists distinguishably different ^{13}C chemical shifts for the *cis* and *trans* isomers of PA. Diffusion effects can be monitored and the behaviour of dopant nuclei which are amenable to NMR (for example, ^{19}F, ^{23}Na and ^{13}C) can be selectively examined. A review of the formative literature, focussing principally on PA, illustrates this diversity of application and the way in which a basic understanding of conductivity in polymers emerges (McBrierty, 1983).

8.4.1.1 *Doped and undoped polyacetylene*

Early proton investigations by Nechtschein and coworkers (1980) on *trans*-PA and *trans*-PA $(AsF_5)_{0.1}$ showed that nuclear relaxation in both materials was dominated by electronic spins. The ratio of their diffusion coefficients parallel and transverse to the chain, $D_\parallel/D_\perp$ exceeded 10^6, which supported the initial view of one-dimensional conduction in PA, as described by the soliton model (Peo *et al.*, 1981; Kume *et al.*, 1982).

Differences in the NMR and subsequent ESR (Clark and Glover, 1982) estimates of $D_\parallel$ pointed to both fixed (oxygen-trapped) and mobile spins in *trans*-PA. The applicability of the soliton model, however, was contested by Tomkiewicz *et al.* (1982) who, as discussed above, favoured the magnetic droplet model.

^{13}C NMR exploited the sensitivity of ^{13}C chemical shifts to the behaviour of the π-electrons and monitored the distribution of dopant in the polymer along with its effect on the *cis/trans* content (Clarke and Scott, 1982; Resing *et al.*, 1982a, b). ^{13}C enriched studies by Masin and coworkers (1981) invoked an alternative mechanism of three-dimensional spin diffusion to dilute magnetic impurities such as residual catalyst, oxygen-pinned solitons or topological solitons to explain the observed initial $t^{\frac{1}{2}}$ behaviour of ^{13}C magnetisation (cf Section 3.8.5). This apparently ruled out rapid electronic spin or soliton diffusion as the dominant mechanism in this system, although subsequent work by this group on *trans*-PA doped with iodine in the intermediate regime between the insulator and metallic phase supported the soliton model. The onset of molecular motion in PA with increasing temperature is routinely detected by NMR in these studies.

8.4.2 Piezoelectric polymers

Our current understanding of piezoelectricity stems from two schools of thought, one invoking dipole reorientation and the other, charge injection and trapping (Kepler and Anderson, 1980). A combination of factors which include sample morphology, thermal treatment and poling method determines the magnitude of the piezoelectric coefficient (d_{31}) achieved in piezoelectric polymers. The β-polymorph of poly(vinylidene fluoride) (PVF$_2$) is one such example which is either obtained by uniaxially drawing α-PVF$_2$ or by copolymerisation with poly(trifluoroethylene) (PTrFE). Poling is usually achieved by the application of a strong electric field at a suitably elevated temperature which tends to align the dipole moments along the field direction. Alternatively, a corona discharge may be used. About 20 % of the polarisation remains when the field is removed after the sample has cooled down. The need for poling at elevated temperatures is readily understood by recalling the T_2 crystalline α-transition in PVF$_2$ at 120 °C which signifies the onset of rotational freedom of chains about their axes in the crystalline regions, as required in the dipole reorientation process (cf Section 5.4.5).

The drawn and poled β-PVF$_2$ film is highly spatially anisotropic with *c*-axis alignment along the draw axis and *b*-axes alignment along the poling

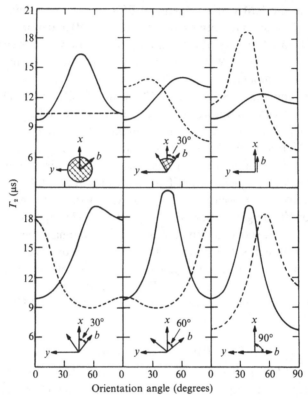

Figure 8.14. T_2 behaviour for the ideal b-axis distributions shown. The solid and dashed curves respectively denote T_2 as a function of draw axis (Z) and poling direction (X) alignment in the laboratory magnetic field $\mathbf{B}_0$. (Reprinted with permission from Douglass *et al.* (1982). Copyright (1982) American Institute of Physics.)

field direction which is normal to the film surface. Fortuitously, in PVF$_2$ the molecular dipole moments are collinear with the b-axes. As described in the foregoing chapter, NMR linewidths are sensitive to these orientation distributions (Douglass *et al.*, 1982; Ishii *et al.*, 1982). It is important to characterise b-axes distributions since transient infra-red and X-ray experiments (Ohigashi, 1976; Naegele and Yoon, 1978) reveal a direct correlation between reorientation of b-axes and the development of piezoelectricity in the polymer. The sensitivity of NMR linewidths to a number of ideal b-axis distributions is shown in fig. 8.14 for different draw axis (solid lines) and poling axis (dashed lines) orientations in the magnetic field B_0 (Douglass *et al.*, 1982). Experimental data for a number of poled and unpoled β-PVF$_2$ samples are portrayed in fig. 8.15. In these

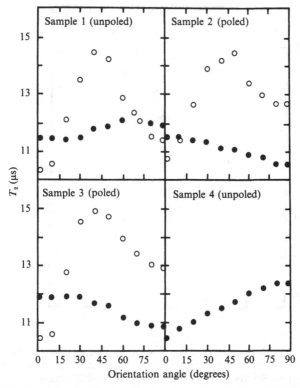

Figure 8.15. Experimental T_2 data for four samples of oriented PVF$_2$ as a function of draw axis (Z) alignment ($\bigcirc$) and poling axis (X) alignment ($\bullet$) in the laboratory magnetic field $\mathbf{B}_0$. (Reprinted with permission from Douglass *et al.* (1982). Copyright (1982) American Institute of Physics.)

measurements, a $T_{1\rho}$ spin-locking sequence preceding the solid-echo sequence suppressed the amorphous contribution to the FID. Equation 3.24 was used to analyse the residual crystalline FID. Results first show clear differences in T_2 anisotropy in B_0 between poled and unpoled polymer; second, the distributions are more complicated than the ideal distributions portrayed in fig. 8.14; and, third, $(n\pi/3)$-fold rather than $n\pi$-fold dipole reorientation is favoured.

Analysis of the experimental T_2 data yields the distribution functions $\langle\cos 2\delta\rangle$ and $\langle\cos 4\delta\rangle$ which characterise the equilibrium remnant polarisation: δ is the angle between a typical *b*-axis and the poling direction. Figure 8.16 shows the correlation obtained between $\langle\cos 2\delta\rangle$ and the piezoelectric coefficient d_{31}. Improved alignment of dipoles towards the direction of the poling field and a correspondingly higher d_{31} is evident as

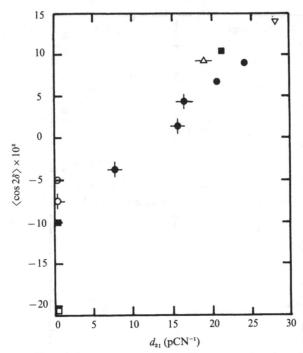

Figure 8.16. Plot of the piezoelectric d_{31} coefficient as a function of $\langle \cos 2\delta \rangle$ for a number of poled materials (see text). X-ray data are denoted (■). (Reprinted with permission from McBrierty *et al.* (1982a). Copyright (1982) American Institute of Physics.)

the degree of poling increases. While NMR cannot provide odd moments of the distribution directly, the change in the sign of $\langle \cos 2\delta \rangle$ reflects their presence. The magnitudes of $\langle \cos 2\delta \rangle$ and $\langle \cos 4\delta \rangle$ ultimately achieved show that the dipoles are far from perfectly aligned along the direction in which the poling field was applied. This confirms that significant relaxation away from the poling direction has occurred when the electric field is removed, which is in keeping with earlier X-ray observations. Note the agreement between X-ray and NMR results in fig. 8.16. Interestingly, the observation that dipole reorientation proceeded reasonably smoothly over the complete time evolution of d_{31} was at odds with the initial rapid and then slower development of macroscopic polarisation observed in other measurements (Drey-Aharon *et al.*, 1980).

8.4.3 Ferroelectric polymers

Of those polymers that exhibit ferroelectric effects (Mathur *et al.*, 1984; Tasaka *et al.*, 1984), PVF_2 and its copolymers are the most widely studied (Lovinger, 1983; Tasaka and Miyata, 1985). The hysteresis curve of electrical displacement reveals ferroelectric switching on a microsecond timescale at ambient temperature (Furukawa *et al.*, 1981). However, the ferroelectric phase in PVF_2 is illusive since it terminates by melting rather than by a well-defined Curie transition. $P(VF_2/TrFE)$ copolymers, on the other hand, exhibit a transition at temperatures which are well below the melting point and, as such, are better systems for examining the ferroelectric phenomenon. The temperature of the first-order Curie transition in $P(VF_2/TrFE)$ increases with VF_2 content and extrapolates to a Curie temperature of 205 °C for neat PVF_2, which indeed is well above its melting point (Lovinger *et al*, 1983).

Figure 8.17(a) presents proton T_1, $T_{1\rho}$, T_2 and collated correlation frequency versus inverse temperature data for the 52/48 mol% $P(VF_2/TrFE)$ material (McBrierty *et al.*, 1982b). Focussing upon the so-called '70° transition', it is immediately evident that there is a discontinuity at this temperature which is most pronounced in the linewidth data. Below 70°, the results, although complex, broadly reflect a superposition of responses for the constituent polymer components. The long and short T_2 components change dramatically at the transition, indicating that both crystalline and amorphous regions participate. The intensities of the long T_2 and short $T_{1\rho}$ components tend towards zero as the transition temperature is approached. The locus of infinite slope on the transition map (fig. 8.17(b)) reflects the first-order nature of the transition. The change in curvature of the β-relaxation as the ferroelectric transition is traversed is consistent with the notion of motional retardation. Alternatively, one may simply be observing structural change from WLF behaviour, which is characteristic of a glass below the transition, to an Arrhénius response above (cf Section 1.5) which is in keeping with the perceived crystal reorganisation which takes place (Lovinger *et al.*, 1983).

The '70° transition' in the copolymer may be interpreted in terms of the onset of either order–disorder fluctuations or motions of metastable glass-like material. Unlike the response of ferroelectric single crystals where T_1 exhibits a characteristic cusp-like maximum at the transition (Blinc and Zumer, 1968), spin lattice relaxation in the copolymer is almost certainly controlled by relaxation *via* spin diffusion to efficiently relaxing amorphous regions which act as sinks (cf Section 3.8). Any treatment of ferroelectric

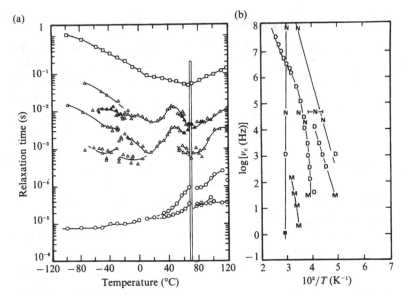

Figure 8.17. (a) Proton T_1 (60 MHz) ($\square$), $T_{1\rho}$ ($B_1 = 10$ G) ($\blacktriangle$) and T_2 ($\bigcirc$) data for 52/48 mole % P(VF$_2$/TrFE) copolymer. The filled triangles denote unresolved components. (b) Transition map describing relaxation in PVF$_2$/TrFE: (N), NMR; (D), dielectric; (M), mechanical; ($\blacksquare$) DSC data. (Reprinted with permission from McBrierty *et al.* (1982b). Copyright (1992) American Chemical Society.)

phenomena must address the fundamental question of the stabilisation or destabilisation of the ordered state and how it is achieved. The effects of drawing and poling are revealing in this respect. First, the ferroelectric transition becomes sharper and shifts to higher temperatures (80–85 °C), in keeping with X-ray results and, second, there is clear evidence of improved crystallographic packing in the ordered regions: typically, T_2(poled) $<$ T_2(unpoled) which reveals a more compact and stable structure in the poled material (McBrierty *et al.*, 1984).

In summary, the '70° transition' involves a complex interplay between conformational and molecular motional events involving both crystalline and amorphous regions. The contribution of NMR, in concert with other techniques, notably X-ray analysis, has been significant in identifying the molecular processes involved.

Appendix 1 Data on common polymers

Table A1.1. *Structures and pertinent data for some common polymers*

Polymer	Chain unit	$T_m(°C)$[a]	$T_g(°C)$[a]
Polyethylene (PE)	$(-CH_2CH_2-)_n$		
high pressure, branched		(115)	—
linear		135	(−125)
Poly(ethylene oxide)	$(-CH_2CH_2O-)_n$		
(PEO)		66	−67
Poly(oxymethylene)	$(-OCH_2-)_n$		
(POM)		195	−85
Poly(ethylene terephthalate) (PET)	$(-\overset{O}{\overset{\|}{C}}-\bigcirc-\overset{O}{\overset{\|}{C}}-O-CH_2CH_2-O-)_n$	265	69
Polypropene (PP)	$(-\underset{CH_3}{CH}-CH_2-)_n$		
isotactic		165	−10
atactic		—	−20
Poly(propene oxide) (PPO)	$(-\underset{CH_3}{CH}-CH_2-O-)_n$	75	−75
Poly(methyl acrylate) (PMA)	$(-\underset{COOCH_3}{CH}CH_2-)_n$	—	6
Poly(ethyl acrylate) (PEA)	$(-\underset{COOC_2H_5}{CH}-CH_2-)_n$	—	−24
Poly(methyl methacrylate) (PMMA)	$(-\overset{CH_3}{\underset{COOCH_3}{C}}-CH_2-)_n$		
isotactic		160	≃45
syndiotactic		200	115
atactic		—	105
Polystyrene (PS)	$(-\underset{\bigcirc}{CH}-CH_2-)_n$		
atactic		—	100
isotactic		240	100

Appendix 1

Table A1.1. *Cont.*

Polymer	Chain unit	$T_m(°C)^{(a)}$	$T_g(°C)^{(a)}$
Poly(styrene oxide) (PSO)	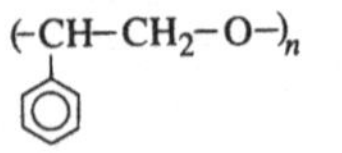 $(-CH-CH_2-O-)_n$ with phenyl	149	37
Poly (phenylene sulphide (PPS)	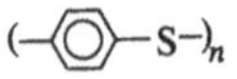 $(-C_6H_4-S-)_n$	—	—
Polysulphone	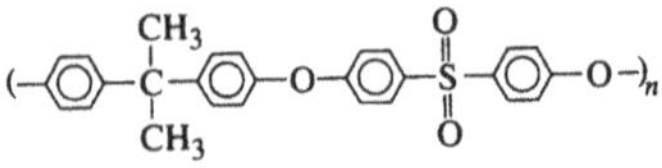 $(-C_6H_4-C(CH_3)_2-C_6H_4-O-C_6H_4-SO_2-C_6H_4-O-)_n$	—	219
Polycarbonate (PC)	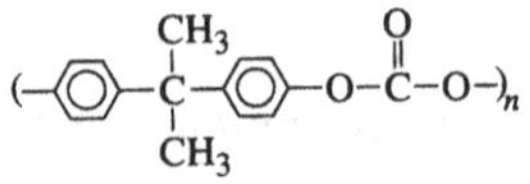 $(-C_6H_4-C(CH_3)_2-C_6H_4-O-CO-O-)_n$	267	149
Polyacrylonitrile (PAN)	$(-CH_2-CH(CN)-)_n$	—	$\simeq 105$
Polybutene-1 (PB-1) isotactic	$(-CH(CH_2CH_3)-CH_2-)_n$	142	-20
Polyisobutene (PIB)	$(-C(CH_3)_2-CH_2-)_n$	—	-73
Polybutadiene (PB)	$(-CH(CH=CH_2)-CH_2-)_n$		
1,2-isotactic		125	-4
1,2-syndiotactic		154	—
cis-1,4	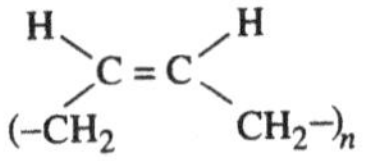 $(-CH_2-)CH=CH(-CH_2-)_n$ cis	6	-108
trans-1,4	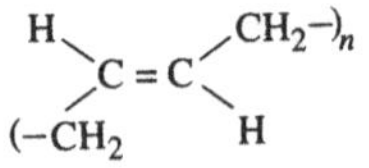 $(-CH_2-)CH=CH(-CH_2-)_n$ trans	148	-18
Polyisoprene (PIP) *cis*-1,4 (natural rubber)	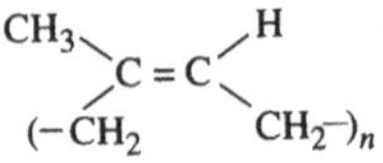 $(-CH_2-)C(CH_3)=CH(-CH_2-)_n$ cis	—	-73
trans-1,4 (gutta percha)	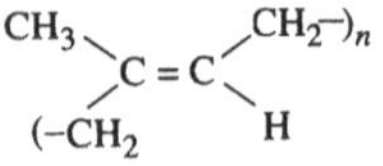 $(-CH_2-)C(CH_3)=CH(-CH_2-)_n$ trans	—	-58
Poly(vinyl acetate) (PVAc)	$(-CH(OCOCH_3)-CH_2-)_n$	—	28
Poly(vinyl alcohol) (PVA)	$(-CH(OH)-CH_2-)_n$	258	85
Poly(vinyl (chloride) (PVC)	$(-CH(Cl)-CH_2-)_n$	—	81

Table A1.1. *Cont.*

Polymer	Chain unit	$T_m(°C)^{(a)}$	$T_g(°C)^{(a)}$
Poly(vinyl fluoride) (PVF)	$(-CH-CH_2-)_n$ with F	200	
Poly(vinylidene fluoride) (PVDF)	$(-CF_2-CH_2-)_n$	171	-45
Poly(vinyl pyrrolidone) (PVPr)	$(-CH-CH_2-)n$	—	—
Poly(dimethyl siloxane) (PDMS)	CH_3 / $(-Si-O-)_n$ / CH_3	—	-123
Polychloroprene (PCP)	$(-CH_2-C=CHCH_2-)_n$ with Cl	—	-48
Polychlorotri-fluoroethylene (PCTFE)	$(-CF_2CF-)_n$ with Cl	218	45
Polytetrafluoroethylene (PTFE)	$(-CF_2-CF_2-)_n$	327	—
Nylon-6	$(-(CH_2)_5CONH-)_n$	—	75 (dry)
Nylon-66 (polyhexamethylene adipamide)	$(-NH(CH_2)_6NHCO(CH_2)_4CO-)_n$	265	49
Nylon-11	$(-(CH_2)_{10}CONH-)_n$	—	46
Nylon-12	$(-(CH_2)_{11}CONH-)_n$	—	37
Cellulose		—	208
Cellulose acetate (unplasticised)		(210)	—

(a) T_m is the melting temperature; T_g is the glass transition temperature. The accuracy of temperatures in brackets is doubtful. (F. A. Bovey, private communication).

Table A1.2. *Chemical shift data for polymers. (Compiled from Duncan (1990) with additional data from the literature)*

Polymer	Formula	Temperature (K)	δ_{11}	δ_{22} (ppm)	δ_{33}	δ_{iso}	Notes	References
Carbon[a]								
Polyethylene (PE)	$(-CH_2-)_n$	RT	50	37	13	33.3		Urbina and (Waugh 1974)
			52	39	14	35		Opella and Waugh (1977)
			50.1	36.7	12.7	33.2		Nakai et al. (1988b)
						33.6	ortho; drawn	VanderHart and
						35.0	mono; drawn	Khoury (1984)
			51.4	38.9	12.9	34.4	drawn	VanderHart (1979)
Polyacetylene (PA)	$(-CH=CH-)_n$	RT	220	142	51	137	trans	Resing et al. (1982a)
			234	146	34	138		Mehring et al. (1983)
			218	138	22	126		Terao et al. (1984)
			217	143	45	135		Manenschijn et al. (1984)
			220	142	48	137		Nakai et al. (1988a)
			221	137	20	126	cis	Resing et al. (1982a)
			228	139	17	128		Mehring et al. (1983)
			219	144	47	137		Terao et al. (1984)
			219	138	25	127		Nakai et al. (1988a)
Polycarbonate bisphenol A		113	210	141	18	123		O'Gara et al. (1985)
		< 133	207	142	22	124		Roy et al. (1986)
Polyoxy-methylene (POM)	$(-OC^*H_2-)_n$	RT	114	87	64	88		Veeman (1984)
			111	86	67	88		Maas et al. (1987)
			107	86	72	88		Kurosu et al. (1987)
			108	87	70	88		Kurosu et al. (1988)
			101	86	84	90		
			108	84	66	86		
Poly(ethylene oxide) (PEO)	$(-C^*H_2C^*H_2O-)_n$	133	91	83	33	69		Fleming et al. (1980)
Polybutadiene (PB)	$(-CH_2C^*H=C^*HCH_2-)_n$	123	236	115	35	129	cis 1,4	Fleming et al. (1980)
Poly(methyl methacrylate) (PMMA)	$(-C^*H_2CCH_3COOCH_3-)_n$	RT	79	48	29	52		Edzes (1983)
	$(-CH_2C^*CH_3COOCH_3-)_n$		53	43	37	44		
	$(-CH_2CCH_3C^*OOCH_3-)_n$		268	150	112	177		
	$(-CH_2CCH_3COOC^*H_3-)_n$		80	63	10	51		
Poly(ethylene terephthalate) (PET)	$(-OCOC_6H_4COOC^*H_2C^*H_2)_n$	RT	80	80	28	63		Murphy et al. (1982)
	$(-OC^*OC_6H_4C^*OOCH_2CH_2-)_n$		250	122	122	165		
	$(-OCOC^*_6H_4COOCH_2CH_2-)_n$		226	153	15	131		

Table A1.2. *Cont.*

Polymer	Formula	Temperature (K)	Chemical shift parameters				Notes	References
			δ_{11}	δ_{22}	δ_{33} (ppm)	δ_{iso}		
Poly(vinylidene fluoride) (PVDF)	$(-C^*H_2CF_2-)_n$ $(-CH_2C^*F_2-)_n$	297	56 131	49 120	30 111	45 121		Fleming *et al.* (1980)
Proton[b] Polyethylene (PE)	$(-CH_2-)_n$	77	3.5	3.5	-2.4	0.5		Burum and Rhim (1979)
Fluorine[c] Poly(tetrafluoro ethylene) (PTFE)	$(-CF_2-)_n$	81	-43	-144	-182	-123		Mehring *et al.* (1971)
		153	-51	-142	-173	-122		
		77	-27	-149	-205	-127		Mansfield *et al.* (1973)
		77	-53	-143	-203	-133		Garroway *et al.* (1975)
		145	-41	-141	-178	-120		Vega and English (1979)
		HT	-49	-156	-156	-120		Vega and English (1980)
Nitrogen[d] Nylon-6 (^{15}N)	$(-CO(CH_2)_5N^*H-)_n$	RT	-170	-290	-345	-268	Semi-crystalline	Powell and Mathais (1990)
Silicon[e] Poly(diethyl siloxane) (PDS)	$(-Si^*(CH_2CH_3)_2O-)_n$	180	1	1	-53	-17	β_1 form	Litvinov *et al.* (1989)
		268	-7	-7	-48		β_2 form	
		298	-10	-10	-29		mesophase, α_m	

[a] Relative to TMS on the δ scale such that $C_6H_6 = 128.7$ ppm and $CS_2 = 192.8$ ppm.

[b] Relative to TMS on the δ scale such that $C_6H_6 = +7.3$ ppm.

[c] Relative to $CFCl_3$ on the δ scale such that $C_6F_6 = -163$ ppm and $F_2 = +429$ ppm.

[d] Relative to CH_3NO_2 on the δ scale such that $NH_3 = -380$ ppm.

[e] Relative to TMS on the δ scale such that SiO_2 (quartz) $= -114$ ppm and $SiF_4 = -113.4$ ppm. In all cases, upfield is negative.

Springer and Xu (1991) have considered magnetic susceptibility corrections in detail.

Table A1.3. ^1H and ^{13}C chemical shifts for polymers

CARBON-13 CHEMICAL SHIFTS (ppm from TMS)

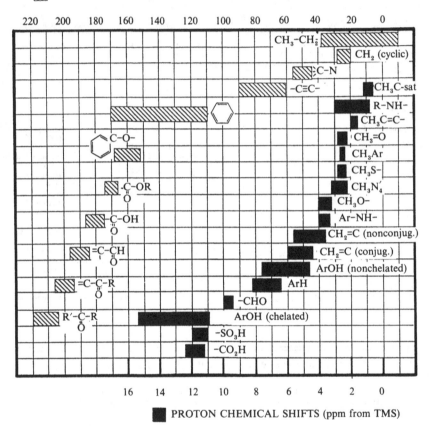

PROTON CHEMICAL SHIFTS (ppm from TMS)

Appendix 2 The rotation operator

Euler rotations in terms of the operator $D(\alpha\beta\gamma)$ have been treated in detail by several authors (Rose, 1957; Edmonds, 1974; Wolf, 1969). The matrix elements of $D(\alpha\beta\gamma)$ are defined as

$$\mathscr{D}_{mn}^{(l)}(\alpha\beta\gamma) = e^{-im\alpha}d_{mn}^{(l)}(\beta)\,e^{-in\gamma} \qquad (A2.1)$$

Note that $d_{mn}^{(l)}(\beta) = 1$ for $l = 0$, and

$$d_{mn}^{(l)}(\beta) = (-1)^{m+n}d_{-m-n}^{(l)}(\beta) = (-1)^{m+n}d_{nm}^{(l)}(\beta) \qquad (A2.2)$$

Expressions for $d_{nm}^{(2)}(\beta)$ are listed in table A2.1.

The complex conjugate of $\mathscr{D}_{mn}^{(l)}(\alpha\beta\gamma)$ is written

$$\mathscr{D}_{mn}^{(l)*}(\alpha\beta\gamma) = (-1)^{m-n}\,\mathscr{D}_{-m-n}^{(l)}(\alpha\beta\gamma) \qquad (A2.3)$$

The orthogonality property is

$$\frac{1}{8\pi^2}\int_0^{2\pi}\int_{-1}^{1}\int_0^{2\pi}\mathscr{D}_{mn}^{(l)}(\alpha\beta\gamma)\,\mathscr{D}_{m'n'}^{(l')*}(\alpha\beta\gamma)\,\mathrm{d}\alpha\,\mathrm{d}(\cos\beta)\,\mathrm{d}\gamma = \delta_{ll'}\,\delta_{mm'}\,\delta_{nn'}\cdot\frac{1}{2l+1} \quad (A2.4)$$

Note the following useful identities:

$$\mathscr{D}_{0n}^{(l)}(\alpha\beta\gamma) = (-1)^n\left(\frac{4\pi}{2l+1}\right)^{\frac{1}{2}}Y_{ln}^{*}(\beta\gamma) \qquad (A2.5)$$

$$\mathscr{D}_{m0}^{(l)}(\alpha\beta\gamma) = \left(\frac{4\pi}{2l+1}\right)^{\frac{1}{2}}Y_{lm}^{*}(\beta\alpha) \qquad (A2.6)$$

$$\mathscr{D}_{00}^{(l)}(\alpha\beta\gamma) = P_l(\cos\beta) \qquad (A2.7)$$

$$\mathscr{D}_{mn}^{(l)}(0) = \delta_{mn}, \text{ the Dirac delta function} \qquad (A2.8)$$

The spherical harmonics $Y_{lm}(\theta\phi)$ required in this text are listed in table A2.2.

The product of two matrices may be expanded in terms of Wigner 3-j symbols as follows:

$$\mathscr{D}_{mn}^{(l)}(\alpha\beta\gamma)\,\mathscr{D}_{m'n'}^{(l')}(\alpha\beta\gamma) = \sum_{l''m''n''}(2l''+1)\begin{pmatrix} l & l' & l'' \\ m & m' & m'' \end{pmatrix}\cdot\begin{pmatrix} l & l' & l'' \\ n & n' & n'' \end{pmatrix}\cdot\mathscr{D}_{m''n''}^{(l'')*}(\alpha\beta\gamma)$$

$$(A2.9)$$

Table A2.1. *Analytical expressions for $d_{mn}^{(2)}(\beta)$*

$d_{00}^{(2)}(\beta)$	$\frac{1}{2}(3\cos^2\beta - 1)$
$d_{10}^{(2)}(\beta)$	$-(\frac{3}{2})^{\frac{1}{2}}\sin\beta\cos\beta$
$d_{1\pm 1}^{(2)}(\beta)$	$\pm\frac{1}{2}(2\cos^2\beta \pm \cos\beta - 1)$
$d_{20}^{(2)}(\beta)$	$(\frac{3}{8})^{\frac{1}{2}}\sin^2\beta$
$d_{2\pm 1}^{(2)}(\beta)$	$-\frac{1}{2}\sin\beta(1\pm\cos\beta)$
$d_{2\pm 2}^{(2)}(\beta)$	$\frac{1}{4}(1\pm\cos\beta)^2$

Table A2.2. *Spherical harmonics for $l = 0, 2, 4$*

l	m	$Y_{lm}(\theta\phi)$
0	0	$\frac{1}{2}\left(\frac{1}{\pi}\right)^{\frac{1}{2}}$
2	0	$\frac{1}{4}\left(\frac{5}{\pi}\right)^{\frac{1}{2}}(3\cos^2\theta - 1)$
2	± 1	$\mp\frac{1}{2}\left(\frac{15}{2\pi}\right)^{\frac{1}{2}}\cos\theta\sin\theta\,e^{\pm i\phi}$
2	± 2	$\frac{1}{4}\left(\frac{15}{2\pi}\right)^{\frac{1}{2}}\sin^2\theta\,e^{\pm 2i\phi}$
4	0	$\frac{3}{16}\left(\frac{1}{\pi}\right)^{\frac{1}{2}}(35\cos^4\theta - 30\cos^2\theta + 3)$
4	± 1	$\mp\frac{3}{8}\left(\frac{5}{\pi}\right)^{\frac{1}{2}}\sin\theta\cos\theta\,(7\cos^2\theta - 3)\,e^{\pm i\phi}$
4	± 2	$\frac{15}{8}\left(\frac{1}{10\pi}\right)^{\frac{1}{2}}\sin^2\theta\,(7\cos^2\theta - 1)\,e^{\pm 2i\phi}$
4	± 3	$\mp\frac{15}{8}\left(\frac{7}{5\pi}\right)^{\frac{1}{2}}\cos\theta\sin^3\theta\,e^{\pm 3i\phi}$
4	± 4	$\frac{15}{16}\left(\frac{7}{10\pi}\right)^{\frac{1}{2}}\cos^4\theta\,e^{\pm 4i\phi}$

Note also that

$$\sum_m \mathscr{D}_{mn}^{(l)*}(\alpha\beta\gamma)\,\mathscr{D}_{mn'}^{(l)}(\alpha\beta\gamma) = \delta_{nn'} \tag{A2.10}$$

$$\sum_n \mathscr{D}_{mn}^{(l)*}(\alpha\beta\gamma)\,\mathscr{D}_{m'n}^{(l)}(\alpha\beta\gamma) = \delta_{mm'} \tag{A2.11}$$

Table A2.3. *Values of* $\begin{pmatrix} j_1 & j_2 & j_3 \\ m & -m & 0 \end{pmatrix}$

j_1	j_2	j_3\\m	0	1	2	3	4
0	0	0	1	—	—	—	—
2	2	0	$1/\sqrt{5}$	$-1/\sqrt{5}$	$1/\sqrt{5}$	—	—
2	2	2	$-2/\sqrt{70}$	$1/\sqrt{70}$	$2/\sqrt{70}$	—	—
2	2	4	$2/\sqrt{70}$	$4/3\sqrt{70}$	$1/3\sqrt{70}$	—	—
2	4	2	$2/\sqrt{70}$	$-1/\sqrt{21}$	$1/\sqrt{42}$	—	—
2	4	4	$-10/3\sqrt{385}$	$1/\sqrt{462}$	$3/\sqrt{231}$	—	—
2	4	6	$5/\sqrt{715}$	$4/\sqrt{858}$	$1/\sqrt{429}$	—	—
4	4	0	$1/3$	$-1/3$	$1/3$	$-1/3$	$1/3$
4	4	2	$-10/3\sqrt{385}$	$17/6\sqrt{385}$	$-4/3\sqrt{385}$	$-7/6\sqrt{385}$	$14/3\sqrt{385}$
4	4	4	$6/\sqrt{2002}$	$-3/\sqrt{2002}$	$-11/3\sqrt{2002}$	$7/\sqrt{2002}$	$14/3\sqrt{2002}$
4	4	6	$-10/3\sqrt{715}$	$-1/6\sqrt{715}$	$11/3\sqrt{715}$	$17/6\sqrt{715}$	$2/3\sqrt{715}$
4	4	8	$70/3\sqrt{24310}$	$280/15\sqrt{24310}$	$140/15\sqrt{24310}$	$40/15\sqrt{24310}$	$1/3\sqrt{24310}$

3-j symbols have a number of useful properties. For example

$$\sum_{l''m''} (2l''+1) \begin{pmatrix} l & l' & l'' \\ m & m' & m'' \end{pmatrix} \begin{pmatrix} l & l' & l'' \\ n & n' & n'' \end{pmatrix} = \delta_{mn} \delta_{m'n'} \qquad (A2.12)$$

$$\sum_{mm'} \begin{pmatrix} l & l' & l'' \\ m & m' & m'' \end{pmatrix} \begin{pmatrix} l & l' & l''' \\ m & m' & m''' \end{pmatrix} = \delta_{l''l''} \delta_{m''m'''} \cdot \frac{1}{(2l''+1)} \qquad (A2.13)$$

$$\begin{pmatrix} l & l' & l'' \\ m & m' & m'' \end{pmatrix} = \begin{pmatrix} l' & l'' & l \\ m' & m'' & m \end{pmatrix} = \begin{pmatrix} l'' & l & l' \\ m'' & m & m' \end{pmatrix} \qquad (A2.14)$$

$$(-1)^{l+l'+l''} \begin{pmatrix} l & l' & l'' \\ m & m' & m'' \end{pmatrix} = \begin{pmatrix} l' & l & l'' \\ m' & m & m'' \end{pmatrix} = \begin{pmatrix} l & l'' & l' \\ m & m'' & m' \end{pmatrix} = \begin{pmatrix} l'' & l' & l \\ m'' & m' & m \end{pmatrix} \qquad (A2.15)$$

$$\begin{pmatrix} l & l' & l'' \\ m & m' & m'' \end{pmatrix} = (-1)^{l+l'+l''} \begin{pmatrix} l & l' & l'' \\ -m & -m' & -m'' \end{pmatrix} \qquad (A2.16)$$

$$|m| \leqslant l; \qquad |m'| \leqslant l'; \qquad |m''| \leqslant l''$$
$$|l+l'| \leqslant l''; \qquad m+m'+m'' = 0 \qquad (A2.17)$$

The numerical values of relevant 3-j symbols are listed in table A2.3.

Appendix 3 Rotation of tensors between coordinate frames

Consider fig. 3.2 where the *new* PAS is generated from the *old* laboratory frame (XYZ) by Euler rotations $\Omega_0 \equiv \alpha_0 \beta_0 \gamma_0$.

Spherical harmonics in the two frames are related as follows:

$$\left.\begin{array}{l} Y_{lm}(\theta'\phi') = \sum_n \mathscr{D}^{(l)}_{nm}(\alpha_0 \beta_0 \gamma_0)\, Y_{ln}(\theta\phi) \\ \text{(new)} \qquad\qquad n \qquad\qquad\qquad \text{(old)} \\[2mm] Y_{ln}(\theta\phi) = \sum_m \mathscr{D}^{(l)*}_{nm}(\alpha_0 \beta_0 \gamma_0)\, Y_{lm}(\theta'\phi') \\ \text{(old)} \qquad\qquad m \qquad\qquad\qquad \text{(new)} \end{array}\right\} \tag{A3.1}$$

The angles $(\theta'\phi')$ and $(\theta\phi)$ are polar angles in the *new* PAS and *old* laboratory system (XYZ), respectively. The coupling tensor $R_{j,m}$ in the Hamiltonian rotates in similar fashion (cf Appendix 5).

Successive rotations may be effected *via* intermediate coordinate frames if required. Consider, for example, the orientation of the PAS relative to the laboratory frame (XYZ). Referring to fig. 3.1, the direct orientation Ω_0 may be expressed in terms of intermediate steps, Ω_1, Ω and Ω_2 as follows:

$$\mathscr{D}^{(l)}_{mm'}(\Omega_0) = \sum_{m_1 m_2} \mathscr{D}^{(l)}_{mm_1}(\Omega_1)\, \mathscr{D}^{(l)}_{m_1 m_2}(\Omega)\, \mathscr{D}^{(l)}_{m_2 m'}(\Omega_2) \tag{A3.2}$$

The identity $\mathscr{D}^{(l)}_{mn}(0) = \delta_{mn}$ (eq A2.8) allows for the suppression of one or more of the intermediate steps.

Appendix 4 Spatial distribution of structural units in a polymer

Consider the most general orientation of a structural unit whose direction is specified by the molecular coordinate frame (xyz) relative to the sample coordinate frame $(X_0 Y_0 Z_0)$ as in fig. 7.1(a). Z_0 is invariably the draw direction in the polymer. Let $P(\Omega)$ be the function that describes the orientational distribution of these structural units relative to $(X_0 Y_0 Z_0)$; $P(\Omega) d\Omega$ is then the fraction of units with orientations between Ω and $\Omega + d\Omega$. For our purposes, $P(\Omega)$ may be expanded in terms of Wigner matrices as follows (McBrierty, 1974b):

$$P(\Omega) = \sum_{lmn} P_{lmn} \mathscr{D}_{mn}^{(l)}(\Omega) \tag{A4.1}$$

where P_{lmn} are moments of the distribution. $P(\Omega)$ is normalised thus

$$\int_0^{2\pi} \int_{-1}^1 \int_0^{2\pi} P(\Omega) \, d\alpha \, d(\cos \beta) \, d\gamma = 1 \tag{A4.2}$$

In subsequent analysis, it will be evident that averaged functions, represented throughout with angle brackets, are required where, typically

$$\langle \mathscr{D}_{mn}^{(l)*}(\Omega) \rangle = \int_0^{2\pi} \int_{-1}^1 \int_0^{2\pi} P(\Omega) \, \mathscr{D}_{mn}^{(l)*}(\Omega) \, d\alpha \, d(\cos \beta) \, d\gamma \tag{A4.3}$$

Substituting A4.1 into A4.3 and recalling the normalisation properties of the matrices (A2.4) it follows that

$$P_{lmn} = \frac{2l+1}{8\pi^2} \langle \mathscr{D}_{mn}^{(l)*}(\Omega) \rangle \tag{A4.4}$$

These expressions which characterise the distribution in its most general form greatly simplify when symmetry is invoked, as illustrated in Chapter 7. For example,

$$P_{l00} = \frac{2l+1}{8\pi^2} \langle P_l(\cos \beta) \rangle \tag{A4.5}$$

Appendix 5 The internal Hamiltonian

In the nomenclature and definitions of McBrierty (1974b), the general form of the internal Hamiltonians, in irreducible tensor notation as described by Haeberlen (1976) and Hentschel and coworkers (1978) is

$$\mathcal{H} = C \sum_{j=0,2} \sum_{m=-j}^{+j} (-1)^m T_{j,m} R_{j,-m} \qquad (A5.1)$$

$T_{j,m}$ and $R_{j,-m}$ are, respectively, spin and coupling tensors specified in the laboratory system (XYZ). The coefficient C is defined in table A5.1 (Hentschel et al., 1978). Specification of tensors in the molecular coordinate system or principal axis system (fig. 3.2) is an essential step in the application of eq A5.1. Using the rotation procedures described in Appendix 3, the coupling tensor R_{j-m} in (XYZ) is related to the coupling tensor $\rho_{j,m'}$ in the PAS as follows.

$$R_{j,-m} = \sum_{m'} \mathscr{D}^{(j)*}_{-mm'}(\Omega_0) \rho_{j,m'} \qquad (A5.2)$$

in which case

$$\mathcal{H} = C \sum_{j=0,2} \sum_{mm'} (-1)^m T_{j,m} \rho_{j,m'} \mathscr{D}^{(j)*}_{-mm'}(\Omega_0) \qquad (A5.3)$$

The elements of the rotation matrix $\mathscr{D}(\Omega_0)$ (Appendix 2) contain all the angular information required to specify the orientation of the PAS relative to the laboratory system (XYZ) via intermediate coordinate frames if required.

The tensor quantities in eq A5.3 are defined in table A5.1. In high magnetic fields where the Larmor frequency ω_0 is dominant, only secular terms are important and, accordingly, summations in eq A5.3 are limited to $j = 0, 2$; $m = 0$; $m' = 0, \pm 2$, terms, in which case

$$\mathcal{H}_{secular} = C \sum_{j=0,2} \sum_{m'=-j}^{j} T_{j,0} \rho_{j,m'} \mathscr{D}^{(j)*}_{0m'}(\Omega_0)$$

$$= C \sum_{j=0,2} \sum_{m'} \left(\frac{4\pi}{2j+1}\right)^{\frac{1}{2}} (-1)^{m'} T_{j,0} \rho_{j,m'} Y_{jm'}(\beta_0, \gamma_0) \qquad (A5.4)$$

The most general situation likely to be encountered in polymers involves the

Table A5.1. *NMR spin and coupling tensors in eq A3.6 for B_0 along the z axis of the laboratory system*

Interaction	C	$\rho_{0,0}$	$\rho_{2,0}$	$\rho_{2,\pm2}$	$T_{0,0}$	$T_{2,0}$	$T_{2,\pm1}$	$T_{2,\pm2}$
Anisotropic chemical shift	$\gamma\hbar$	$\bar\sigma^{(a)}$	$\left(\tfrac{2}{3}\right)^{\frac12}\Delta\sigma^{(b)}$	$-\tfrac13\Delta\sigma\eta^{(c)}$	$B_0 I_0^{(d)}$	$\left(\tfrac{2}{3}\right)^{\frac12}B_0 I_0$	$\left(\tfrac{1}{2}\right)^{\frac12}B_0 I_{\pm1}$	$\cdots$
Dipolar	$-2\gamma_1\gamma_2\hbar^2$	0	$\left(\tfrac{3}{2}\right)^{\frac12}r_k^{-3(e)}$	0	$\cdots$	$\left(\tfrac{1}{6}\right)^{\frac12}(3I_0^1 I_0^2 - \mathbf{I}^1\cdot\mathbf{I}^2)$	$\left(\tfrac{1}{2}\right)^{\frac12}(I_{\pm1}^1 I_0^2 + I_0^1 I_{\pm1}^2)$	$I_{\pm1}^1 I_{\pm1}^2$
Quadrupolar	$\dfrac{eQ}{2I(2I-1)}$	0	$\left(\tfrac{3}{2}\right)^{\frac12}eq$	$-\tfrac12 eq\eta_Q$	$\cdots$	$\left(\tfrac{1}{6}\right)^{\frac12}[3I_0^2 - I(I+1)]$	$\left(\tfrac{1}{2}\right)^{\frac12}(I_{\pm1}I_0 + I_0 I_{\pm1})$	$I_{\pm1}^2$

(a) $\bar\sigma = \frac13(\sigma_{11}+\sigma_{22}+\sigma_{33})$: $\sigma_{33} \geqslant \sigma_{22} \geqslant \sigma_{11}$.

(b) $\Delta\sigma = \sigma_{33} - \frac12(\sigma_{22}+\sigma_{11})$.

(c) $\eta = (\sigma_{22}-\sigma_{11})/(\sigma_{33}-\bar\sigma)$.

(d) $I_0 = I_z$; $I_{\pm1} = \mp\frac{1}{\sqrt2}[I_x \pm iI_y]$.

(e) r_k is the internuclear vector joining spins 1 and 2.

transformation of tensor components between the PAS (123) and the laboratory system (XYZ) *via* the sample coordinate frame $(X_0 Y_0 Z_0)$ (fig. 3.1) for which

$$\mathscr{D}^{(J)*}_{0m'}(\Omega_0) = \sum_{m_1} \mathscr{D}^{(J)*}_{0m_1}(\Omega_1) \mathscr{D}^{(J)*}_{m_1 m'}(\Omega) \tag{A5.5}$$

which is obtained directly from A3.2 with $\Omega_2 \equiv 0$, which brings the PAS (123) into coincidence with the molecular coordinate frame (xyz). Substitution of eq A5.5 into eq A5.4 leads to the expression

$$\begin{aligned}
\mathscr{H}_{\text{secular}} = \; & CT_{0,0}\rho_{0,0} \\
& + CT_{2,0}\{\rho_{2,0} \sum_{m_1} \mathscr{D}^{(2)*}_{0m_1}(\Omega_1) \mathscr{D}^{(2)*}_{m_1 0}(\Omega) \\
& + \rho_{2,2} \sum_{m_1} \mathscr{D}^{(2)*}_{0m_1}(\Omega_1) [\mathscr{D}^{(2)*}_{m_1-2}(\Omega) + \mathscr{D}^{(2)*}_{m_1 2}(\Omega)]\}
\end{aligned} \tag{A5.6}$$

This is the most general form of the secular internal Hamiltonian of immediate interest.

Appendix 6 Spectral lineshapes

The following analysis parallels closely the treatment of Hentschel *et al.* (1978) and extends earlier studies on lineshapes, in terms of moments (McBrierty, 1974b), to include dipolar, chemical shift and quadrupolar interactions.

The lineshape is the weighted sum of single-crystal spectra taking due account of their orientation with respect to the laboratory coordinate frame (XYZ). Define the transition probability and frequency between states M and N for a typical spin k to be W_{MN}^k and ω_{MN}^k, respectively. The lineshape $S(\omega)$ may then be written

$$S(\omega) = \sum_k \sum_{MN} \int P(\Omega_0)\, W_{MN}^k(\Omega_0)\, \delta[\omega - \omega_{MN}^k(\Omega_0)]\, d\Omega_0 \qquad (A6.1)$$

where the Dirac delta function denotes the resonance condition and Ω_0 is the orientation of the kth spin relative to the laboratory system (XYZ). Equation A6.1 may be simplified first by noting that the fraction of spins whose orientations lie between Ω_0 and $\Omega_0 + d\Omega_0$ is equal to the fraction of molecules with orientations between Ω and $\Omega + d\Omega$ which is $P(\Omega)\, d\Omega$; second, that in high magnetic fields W_{MN}^k is independent of Ω_0 and equal to unity for allowed transitions; and, third, that the orientational distribution function $P(\Omega)$ may be expressed in terms of the angles Ω_0 by successive transformations described in fig. 3.1. Thus

$$\mathscr{D}_{nm}^{(l)}(\Omega) = \sum_{n_1 n_2} \mathscr{D}_{mn_1}^{(l)*}(\Omega_2)\, \mathscr{D}_{n_2 n_1}^{(l)}(\Omega_0)\, \mathscr{D}_{n_2 n}^{(l)*}(\Omega_1) \qquad (A6.2)$$

For the illustrative case of axial symmetry where the structural units are randomly disposed about the sample polar axis Z_0 ($m = 0$) and the units are themselves transversely isotropic about their own polar z axis ($n = 0$),

$$P(\Omega) = \sum_l P_{l00} \sum_{nm} \mathscr{D}_{0n}^{(l)*}(\Omega_2)\, \mathscr{D}_{mn}^{(l)}(\Omega_0)\, \mathscr{D}_{m0}^{(l)*}(\Omega_1) \qquad (A6.3)$$

in which case

$$S(\omega) = \sum_l P_{l00} \sum_{mn} \mathscr{D}_{0n}^{(l)*}(\Omega_2)\, \mathscr{D}_{m0}^{(l)*}(\Omega_1)$$

$$\cdot \int_0^{2\pi} \int_{-1}^{1} \int_0^{2\pi} \mathscr{D}_{mn}^{(l)}(\Omega_0)\, \delta[\omega - \omega(\beta_0\, \gamma_0)]\, d\alpha_0\, d(\cos\beta_0)\, d\gamma_0 \qquad (A6.4)$$

312

Hentschel *et al.* (1978) have evaluated the integral, guided by the earlier treatment of Bloembergen and Rowland (1953). It is assumed for simplicity that $\omega_0 = 0$ and $\omega(\beta_0 \gamma_0)$ is given by eq 3.4. Note the integration over the angle α_0 is zero unless $m = 0$ in which case

$$S(\omega) = \sum_{ln} P_{l00} \left(\frac{4\pi}{2l+1} \right) Y_{ln}(\beta_2 \gamma_2) P_l(\cos \beta_1)$$

$$\cdot \int_0^{2\pi} \int_{-1}^{1} \int_0^{2\pi} Y^*_{ln}(\beta_0 \gamma_0) \, \delta[\omega - \omega(\beta_0 \gamma_0)] \, d\alpha_0 \, d(\cos \beta_0) \, d\gamma_0 \quad (A6.5)$$

This is the lineshape for a polymer exhibiting axial symmetry. If, in addition, the spin coupling tensors are axially symmetric ($\eta = 0$), $S(\omega)$ assumes a much simpler form where $n = 0$ in eq A6.5. As shown by Hentschel *et al.* (1978), eq A6.5 reduces to

$$S(\omega) = 8\pi^2 \sum_l P_{l00} P_l(\cos \beta_2) P_l(\cos \beta_1) S_l(\omega) \quad (A6.6)$$

where $S_l(\omega)$ are 'sub-spectra' defined as follows:

$$S_l(\omega) = [2\mathfrak{d}\sqrt{3}]^{-1} [\omega/\mathfrak{d} + 1]^{-\frac{1}{2}} P_l[x(\omega)] \quad (A6.7)$$

and $x(\omega) = [(\omega/\mathfrak{d}+1)/3]^{\frac{1}{2}}$. The parameter $\mathfrak{d}$ is defined in table 3.1. Expressions for planar and conical distributions are listed in the review paper by Spiess (1982).

It must be recalled that these considerations apply only to the case of high magnetic field. Hentschel *et al.* (1978) have extended this analysis to low fields for $I = 1$ and $\eta = 0$.

Harbison and coworkers (1987) likewise express two-dimensional MAS spectra $S(\Omega_1, \Omega_2)$ in terms of sub-spectra $S_L(\Omega_1, \Omega_2)$ as in eq A6.6 in their analysis of spinning sidebands for oriented polymers. Note in this case that the $(2L+1)$ term contained in P_{L00} is incorporated into $S_L(\Omega_1, \Omega_2)$; the sub-spectra depend on sample orientation β_1 in $\mathbf{B}_0$.

Appendix 7 Analysis of spinning sidebands

Consider the geometry of fig. 7.15 where the molecular coordinate frame (xyz) in fig. 3.1 is identified as the rotor frame. Herzfeld and Berger (1980) derived an expression for the spectral frequency ω. Retaining their angle assignments,

$$\omega = \mathfrak{w}_0 + \tfrac{2}{3}[A_2 \cos 2(\psi + \gamma) + B_2 \sin 2(\psi + \gamma) + \sqrt{2}(A_1 \cos(\psi + \gamma) + B_1 \sin(\psi + \gamma))] \tag{A7.1}$$

where

$$\left. \begin{aligned}
A_1 &= -(\mathfrak{d}/2) \sin 2\beta (3 + \eta \cos 2\alpha) \\
B_1 &= \mathfrak{d}\eta \sin \beta \sin 2\alpha \\
A_2 &= (\mathfrak{d}/2)[3 \sin^2 \beta + \eta \cos 2\alpha (\sin^2 \beta - 2)] \\
B_2 &= \mathfrak{d}\eta \cos \beta \sin 2\alpha
\end{aligned} \right\} \tag{A7.2}$$

The FID for the sample is

$$g(t) = \frac{1}{8\pi^2} \int_0^{2\pi} \int_0^{\pi} \int_0^{2\pi} \exp\left[i\Theta(\alpha, \beta, \gamma, t)\right] d\alpha \, d(\cos \beta) \, d\gamma \tag{A7.3}$$

where

$$\Theta(\alpha, \beta, \gamma, t) = \int_0^t \omega(\alpha, \beta, \gamma, t) \, dt \tag{A7.4}$$

Evaluation of eq A7.3 exploits the following properties of Bessel functions of the first kind:

$$\exp(iz \sin \phi) = \sum_{k=-\infty}^{\infty} \exp(ik\phi) J_k(z) \tag{A7.5a}$$

and

$$J_k(z) = \frac{1}{2\pi} \int_0^{2\pi} \exp[-i(k\theta - z \sin \theta)] \, d\theta \tag{A7.5b}$$

The Fourier transform of the FID comprises a central resonance at $\bar{\sigma}$ flanked by

314

a series of sidebands at intervals of $N\omega_r$. The relative intensity of the Nth sideband is

$$I_N = \frac{1}{4\pi} \int_0^\pi \int_0^{2\pi} |F|^2 \, d\alpha \, d(\cos\beta) \qquad (A7.6)$$

where

$$F = \frac{1}{2\pi} \int_0^{2\pi} e^{-iN\theta} \cdot e^{iF(\alpha, \beta, \theta)} \, d\theta \qquad (A7.7)$$

and

$$F(\alpha, \beta, \theta) = \frac{-1}{3\omega_r} [A_2 \sin 2\theta + B_2 \cos 2\theta + 2\sqrt{2}(A_1 \sin\theta + B_1 \cos\theta)] \qquad (A7.8)$$

Inspection of A7.2 and A7.6–A7.8 shows that the line intensities depend only on σ_{11}, σ_{22} and σ_{33} for a chosen $\gamma B_0/\omega_r$.

Harbison, Voigt and Speiss (1986, 1987) derived an expression for the sideband intensity for each resonance in a uniaxially ordered polymer. They analysed the two-dimensional MAS experiment (cf Chapter 4) where data acquisition is synchronised with rotor position. The geometry of the experiment is portrayed in fig. 7.15. The expression for the sideband intensity in the two-dimensional MAS experiment in their notation is

$$I_{M,N}(\beta_2) = \frac{1}{8\pi^3} e^{-iM\gamma_{20}} \int_0^{2\pi} K_{M,N}(\alpha_2, \beta_2, \bar{\sigma}) \, d\alpha_2 \qquad (A7.9)$$

where $\bar{\sigma}$ is the shielding tensor in the rotor frame and

$$K_{M,N}(\alpha_2 \beta_2 \bar{\sigma}) = \int_0^{2\pi} \int_0^{2\pi} e^{i(N-M)\gamma} \cdot e^{-iN\vartheta} F^*(\alpha_2, \beta_2, \gamma, \bar{\sigma}) \cdot F(\alpha_2, \beta_2, \vartheta, \bar{\sigma}) \, d\gamma \, d\vartheta$$

$$\qquad (A7.10)$$

Note that γ and ϑ are integration variables. $\alpha_{20} = \alpha_2 + \omega_r t_1$ allows for the fact that α_2 may not be zero at $t_1 = 0$. F has the same functional form as in A7.8 but the expressions for A_1, A_2, B_1 and B_2 are further complicated by the fact that the shielding tensor may not be diagonal in the rotor frame of reference. The appropriate expressions are listed in the original publication (Harbison *et al.*, 1987). Identification of the molecular frame (xyz) as the rotor frame in fig. 3.1 facilitates intercomparison of the Euler angle definitions used here with those in the Herzfeld–Berger and Harbison–Vogt–Spiess analyses.

Bibliography

Chapter 1

Blümich, B. and Spiess, H. W. (1988). *Agnew, Chem. Int. Ed. Engl.* **27**, 1655.
de Gennes, P.-G. (1990). *Introduction to Polymer Dynamics*, Cambridge University Press, Cambridge.
Ernst, R. R., Bodenhausen, G. and Wokaun, A. (1987). *Principles of Nuclear Magnetic Resonance in One and Two Dimensions*, Clarendon Press, Oxford.
Fyfe, C. A. (1983). *Solid State NMR for Chemists*, CFC Press, Ontario.
Haeberlen, U. (1976). *High Resolution NMR in Solids – Selective Averaging*, Waugh, J. S. (ed.), Supplement 1, Adv. Magnetic Res. Academic Press, New York.
Hedvig, P. (1977). *Dielectric Spectroscopy of Polymers*, Adam Hilger, Bristol.
Kausch, H. H. and Zachmann, H. G. (eds.) (1985). *Characterisation of Polymers in the Solid State, I: Part A. NMR and Other Spectroscopic Methods*, Adv. Polym. Sci, **66**, Springer-Verlag, Berlin.
Komoroski, R. A. (1986). *High Resolution NMR Spectroscopy of Synthetic Polymers in Bulk*, VCH Publishers Inc., Deerfield Beach, Florida.
McBrierty, V. J. and Douglass, D. C. (1980). *Phys. Repts.*, **63(2)**, 61.
McBrierty, V. J. and Douglass, D. C. (1981). *Macromolecular Rev.*, **16**, 295.
McBrierty, V. J. (1983). *Magnetic Res. Rev.*, **8**, 165.
McCall, D. W. (1969). Relaxation in solid polymers, in *Molecular Dynamics and Structure*, Carter, R. S. and Rush, J. J. (eds.), National Bureau of Standards, Special Publn 301, Washington, pp. 475–537.
Roy, A. K. and Inglefield, P. T. (1990). Solid State NMR Studies of Local Motion in Polymers, in *Progress in NMR Spectroscopy*, **22**, 569–603, Pergamon Press, Oxford.
Tonelli, A. E. (1989). *NMR Spectroscopy and Polymer Microstructure: the Conformational Connection*, VCH Publishers Inc., Deerfield Beach, Florida.
Ward, I. M. (1971). *Mechanical Behaviour of Polymers*, Interscience, New York.
Ward, I. M. (ed.) (1982). *Developments in Oriented Polymers*, Applied Science, London.
Ward, I. M. (1983). *Mechanical Properties of Solid Polymers*, 2nd edition, J. Wiley and Sons, London.
Young, R. J. (1981). *Introduction to Polymers*, Chapman and Hall, London.

Chapter 2

Abragam, A. (1961). *The Principles of Nuclear Magnetism*, Clarendon Press, Oxford.

Bloembergen, N., Purcell, E. M. and Pound, R. V. (1948). *Phys. Rev.*, **73**, 679.

Carrington, A. and McLachlan, A. D. (1967). *Introduction to Magnetic Resonance*, Harper and Row, London.

Farrar, T. C. and Becker, E. D. (1971). *Pulse and Fourier Transform NMR*, Academic Press, New York.

Goldman, M. (1970). *Spin Temperature and Nuclear Magnetic Resonance in Solids*, Clarendon Press, Oxford.

Haeberlen, U. (1976). *High Resolution NMR in Solids – Selective Averaging*, Waugh, J.S. (ed.), Supplement 1, Adv. Magnetic Res., Academic Press, New York.

Harris, R. K. (1983). *Nuclear Magnetic Resonance Spectroscopy*, Pitman, London.

Kubo, R. and Tomita, K. (1954). *J. Phys. Soc. Japan*, **9**, 888.

Mehring, M. (1976). In *NMR Basic Principles and Progress*, Diehl, P., Fluck, E. and Kosfeld, R. (eds.), **11**.

Slichter, C. P. (1990). *Principles of Magnetic Resonance*, 3rd edition, Springer-Verlag, New York.

Wolf, D. (1979). *Spin Temperature and Nuclear-Spin Relaxation in Matter*, Clarendon Press, Oxford.

Chapter 3

Abragam, A. (1961). *The Principles of Nuclear Magnetism*, Clarendon Press, Oxford.

Andrew, E. R. (1975). *High Resolution NMR in Solids*, Intnl. Rev. Sci., Phys. Chem., Series II, Vol. 4, Magnetic Resonance, pp. 173–20.

Bloembergen, N., Purcell, E. M. and Pound, R. V. (1948). *Phys. Rev.*, **73**, 679.

Carrington, A. and McLachlan, A. D. (1967). *Introduction to Magnetic Resonance*, Harper and Row, London.

Haeberlen, U. (1976). *High Resolution NMR in Solids – Selective Averaging*, Supplement 1, Adv. Magnetic Res., Academic Press, New York.

Kubo, R. and Tomita, K. (1954). *J. Phys. Soc. Japan*, **9**, 888.

Mansfield, P. (1972). Pulsed NMR in solids in *Progress in NMR Spectroscopy*, Vol. 8, Pergamon Press, Oxford, pp. 41–101.

McBrierty, V. J. and Douglass, D. C. (1980). *Nuclear Magnetic Resonance of Solid Polymers*, Phys. Repts., **63**, 61–147, North Holland, Amsterdam.

Mehring, M. (1983). *Principles of High Resolution NMR in Solids*, Springer-Verlag, New York.

Slichter, C. P. (1990). *Principles of Magnetic Resonance*, 3rd edition, Springer-Verlag, New York.

Spiess, H. W. (1978). *NMR Basic Principles and Progress*, Vol. 15, New York, p. 55.

Wolf, D. (1979). *Spin Temperature and Nuclear-Spin Relaxation in Matter*, Clarendon Press, Oxford.

Chapter 4

Blümich, B. and Spiess, H. W. (1988). *Angew. Chem. Int. Ed. Engl.*, **27**, 1655.

Ernst, R. R., Bodenhausen, G. and Wokaun, A. (1987). *Principles of Nuclear Magnetic Resonance in One and Two Dimensions*, Clarendon Press, Oxford.

Farrar, T. C. and Becker, E. D. (1971). *Pulse and Fourier Transform NMR*, Academic Press, New York.

Fleming, W. W., Lyerla, J. R. and Yannoni, C. S. (1984). In *NMR and Macromolecules, Sequence, Dynamic and Domain Structure*, Randall, J. C. (ed.), *ACS Symp. Ser.*, **247**, Washington, pp. 83–94.

Fukushima, E. and Roeder, S. B. W. (1981). *Experimental Pulse NMR: A Nuts and Bolts Approach*, Addison-Wesley, London.

Goldman, M. (1970). *Spin Temperature and Nuclear Magnetic Resonance in Solids*, Clarendon Press, Oxford.

Harris, R. K. (1983).*Nuclear Magnetic Resonance Spectroscopy*, Pitman, London p. 112.

Mehring, M. (1976). In *NMR Basic Principles and Progress*, Diehl, P., Fluck, E. and Kosfeld, R. (eds.), **11**.

Slichter, C. P. (1990). *Principles of Magnetic Resonance*, 3rd edition, Springer-Verlag, New York.

Chapter 5

Blümich, B. and Spiess, H. W. (1988). *Angew. Chem. Int. Ed. Engl.*, **27**, 1655.

Bovey, F. A. (1988). *Nuclear Magnetic Resonance Spectroscopy*, 2nd edition, Academic Press, New York.

Fyfe, C. A. (1983). *Solid State NMR for Chemists*, CFC Press, Ontario.

Lyerla, J. R. and Yannoni, C. S. (1983). *IBM J. Res. and Dev.*, **27**, 302.

McBrierty, V. J. (1989). In *Comprehensive Polymer Science*, Booth, C. and Price, C. (eds.), Pergamon Press, London, pp. 397–428.

Roy, A. K. and Inglefield, P. T. (1990). Solid state NMR studies of local motions in polymers, in *Progress in NMR Spectroscopy*, **22**, 569–603, Pergamon Press, Oxford.

Tonelli, A. E. (1989). *NMR Spectroscopy and Polymer Microstructure: the Conformational Connection*, VCH Publishers Inc., Deerfield Beach, Florida.

Voelkel, R. (1988). *Angew. Chem. Int. Ed. Engl.*, **27**, 1468–83.

Chapter 6

Bovey, F. A. (1988). *Nuclear Magnetic Resonance Spectroscopy*, 2nd edition, Academic Press, New York.

Boyer, R. F. (1977). *Transitions and Relaxations in Amorphous and Semicrystalline Organic Polymers and Copolymers*, J. Wiley and Sons, New York.

Ernst, R. R., Bodenhausen, G. and Wokaun, A. (1987). *Principles of Nuclear Magnetic Resonance in One and Two Dimensions*, Clarendon Press, Oxford.

Fyfe, C. A. (1983). *Solid State NMR for Chemists*, CFC Press, Ontario.

Horsley, R. A. (1958), *Progress in Plastics, 1957*, Iliffe, London, pp. 77–88.

Lyerla, J. R. and Yannoni, C. S. (1983). *IBM J. Res. and Dev.*, **27**, 302.

Mansfield, P. and Hahn, E. L. (1991). *NMR Imaging: Recent Developments and Future Prospects*, Cambridge University Press, Cambridge.

McBrierty, V. J. (1989). In *Comprehensive Polymer Science*, Booth, C. and Price, C. (eds.), Pergamon Press, London, pp. 397–428.

McCall, D. W. (1969). Relaxation in solid polymers, in *Molecular Dynamics and Structure*, Carter, R. S. and Rush, J. J. (eds.), National Bureau of Standards, Special Publn. 301, Washington, pp. 475–537.

Olabisi, O., Robeson, L. M. and Shaw, M. T. (1979). *Polymer–Polymer Miscibility*, Academic Press, New York.

Paul, D. R. and Newman, S. (eds.) (1978). *Polymer Blends*, Academic Press, New York.

Chapter 7

Nomura, S. (1989). In *Comprehensive Polymer Science*, Allen, G. (ed.), Vol. 2, Pergamon Press, Oxford, pp. 459–85.

Ward, I. M. (ed.) (1975). *Structure and Properties of Oriented Polymers*, Applied Science, London.

Ward, I. M. (ed.) (1982). *Developments in Oriented Polymers*, Applied Science, London.

Chapter 8

Allen, P. C., Bott, D. C., Brown, C. S., Connors, L. M., Gray, S., Walker, N. S., Clemenson, P. I. and Feast, W. J. (1989). In *Electronic Properties of Conducting Polymers*, Kusmany, E. H. (ed.), Springer Series on Solid State Sciences, Springer-Verlag, Berlin.

Broadhurst, M. G. and Davis, G. T. (1981). In *Topics in Modern Physics – Electrets*, Sessler, G. M. (ed.), Springer-Verlag, Berlin.

Cohen-Addad, J. P. (1991). *NMR and Fractal Properties of Polymeric Liquids and Gels*, Pergamon Press, London. *in press.*

de Gennes, P.-G. (1979). *Scaling Concepts in Polymer Physics*, Cornell University Press, Ithaca, New York.

Doi, M. and Edwards, S. F. (1986). *The Theory of Polymer Dynamics*, Clarendon Press, Oxford.

Hanack, M., Roth, S. and Schier, H. (eds.) (1991). *Proc. Intnl. Conference on Science and Technology of Synthetic Metals Part 1*, ICSM 90, Tübingen.

Kimmich, R. (1988). *Progr. in NMR Spectra*, **20**, 385.

Klempner, D. and Frisch, K. C. (eds.) (1989, 1990), *Advances in Interpenetrating Polymer Networks*, Vols. I and II, Technomic Press.

Kraus, G. (ed.) (1965). *Reinforcement of Elastomers*, Interscience, New York.

Lovinger, A. J. (1983). *Science*, **220**, 1115.

McBrierty, V. J. (1983). *Magnetic Res. Rev.*, **8**, 165.

Naarman, H. and Theophilou, N. (1988). In *Electroresponsive Molecular and Polymeric Systems*, Skotheim, T. J. (ed.), Marcel Dekker, New York.

Rowland, S. P. (ed.) (1980). *Water in Polymers*, ACS Symp. Ser., **127**, Washington.

Skotheim, T. J. (ed.) (1986). *Handbook of Conducting Polymers*, Marcel Dekker, New York.

'*Specialty Polymers: Present and Future*', Birmingham. (1984). Conference Issue, *Polymer*, **26**, 1281–432.

Appendices

Edmonds, A. R. (1974). *Angular Momentum in Quantum Mechanics*, 3rd edition, Princeton Univ. Press, Princeton.

Rose, M. E. (1957). *Elementary Theory of Angular Momentum*, Wiley and Sons, New York.

Wolf, A. A. (1969). *Amer. J. Phys.*, **37**, 531.

References

Abragam, A. (1961). *The Principles of Nuclear Magnetism*, Clarendon Press, Oxford.

Abragam, A. and Goldman, M. (1982). *Nuclear Magnetism: Order and Disorder*, Clarendon Press, Oxford.

Ahmad, S. and Packer, K. J. (1979a). *Mol. Phys.*, **37**, 47.

Ahmad, S. and Packer, K. J. (1979b). *Mol. Phys.*, **37**, 59.

Ailion, D. and Slichter, C. P. (1965). *Phys. Rev.*, **137**, A235.

Albert, B., Jérôme, R., Teyssié, Ph., Smyth, G. and McBrierty, V. J. (1984). *Macromolecules*, **17**, 2552.

Albert, B., Jérôme, R., Teyssié, Ph., Smyth, G., Boyle, N. G. and McBrierty, V. J. (1985). *Macromolecules*, **18**, 388.

Alla, M., Eckman, R. and Pines, A. (1980). *Chem. Phys. Letters*, **71**, 148.

Anavi, S., Shaw, M. T. and Johnson, J. F. (1979). *Macromolecules*, **12**, 1227.

Anderson, P. W. (1954). *J. Phys. Soc. Japan*, **9**, 316.

Anderson, P. W. (1958). *Phys. Rev.*, **109**, 1492.

Anderson, J. E. and Liu, K.-J. (1971). *Macromolecules*, **4**, 260.

Andrew, E. R., Bradbury, A. and Eades, R. G. (1958). *Nature*, **182**, 1659.

Andrew, E. R. and Newing, R. A. (1958). *Proc. Phys. Soc.*, **72**, 959.

Andrew, E. R. and Jenks, G. J. (1962). *Proc. Phys. Soc.*, **80**, 663.

Andrew, E. R. and Lipofsky, J. (1972). *J. Magnetic Res.*, **8**, 217.

Andrew, E. R. (1975). High resolution NMR in solids, *Intnl. Rev. Sci., Phys. Chem., Series II, Vol. 4, Magnetic Resonance*, pp. 173–20.

Assink, R. A. and Wilkes, G. L. (1977). *Polym. Eng. Sci.*, **17**, 606.

Assink, R. A. (1978). *Macromolecules*, **11**, 1233.

Aujla, R. S., Harris, R. K., Packer, K. J., Parameswaran, M. and Say, B. J. (1982). *Polym. Bull.*, **8**, 253.

Axelson, D. E., Mandelkern, L., Popli, R. and Mathieu, P. (1983). *J. Polym. Sci: Polym. Phys. Ed.*, **21**, 2319.

Bacon, G. E. (1956). *J. Appl. Chem.*, **6**, 477.

Balimann, G. E., Groombridge, C. J., Harris, R. K., Packer, K. J., Say, B. J. and Tanner, S. F. (1981). *Phil. Trans. Roy. Soc. London*, **A299**, 643.

Bank, M., Leffingwell, L. and Thies, C. (1971). *Macromolecules*, **4**, 43.

Basescu, N., Liu, Z. X., Moses, D., Heejer, A. J., Naarman, H. and Theophilou, N. (1987). *Nature*, **327**, 403.

Baughman, R. H., Hzn, S. L., Pez, G. P. and Signorelli, A. J. (1978). *J. Chem. Phys.*, **68**, 5405.

Baum, J., Munowitz, M., Garroway, A. N. and Pines, A. (1985). *J. Chem. Phys.*, **83**, 2015.
Becker, E. D., Ferretti, J. A., Gupta, R. K. and Weiss, G. H. (1980). *J. Magnetic Res.*, **37**, 381.
Beckmann, P. A. (1988). *Phys. Repts.*, **171**, 85.
Belfiore, L. A., Schilling, F. C., Tonelli, A. E., Lovinger, A. J. and Bovey, F. A. (1984). *Macromolecules*, **17**, 2561.
Belfiore, L. A. (1986). *Polymer*, **27**, 80.
Bergmann, K. (1978). *J. Polym. Sci: Polym. Phys. Ed.*, **16**, 1611.
Bergmann, K. (1981). *Polym. Bull.*, **5**, 355.
Bersohn, R. (1952). *J. Chem. Phys.*, **20**, 1505.
Beshah, K., Mark, J. E., Ackerman, J. L. and Himstedt, A. (1986). *J. Polym. Sci: Polym. Phys. Ed.*, **24**, 1207.
Bishop, E. T. and Davidson, S. (1969). *J. Polym. Sci.*, **C26**, 59.
Blinc, R. and Zumer, S. (1968). *Phys. Rev. Letters*, **21**, 1004.
Bloembergen, N. C. (1949). *Physica*, **15**, 388.
Bloembergen, N., Purcell, E. M. and Pound, R. V. (1948). *Phys. Rev.*, **73**, 679.
Bloembergen, N. and Rowland, J. A. (1953). *Acta Met.*, **1**, 731.
Bloembergen, N. (1961). *Nuclear Magnetic Relaxation*, Benjamin, New York.
Blumberg, W. E. (1960). *Phys. Rev.*, **119**, 79.
Blümich, B., Böffel, C., Harbison, G. S., Yang, Y. and Spiess, H. W. (1987). *Ber Bunsenges. Phys. Chem.*, **91**, 1100.
Blümich, B. and Spiess, H. W. (1988). *Angew. Chem. Int. Ed. Engl.*, **27**, 1655.
Bonart, R. and Muller, E. H. (1974). *J. Macromol. Sci: Phys.*, **B10**, 177 and 345.
Bonart, R., Morbitzer, L. and Muller, E. H. (1974). *J. Macromol. Sci: Phys.*, **B9**, 447.
Booth, A. D. and Packer, K. J. (1987). *Molecular Phys.*, **62**, 811.
Bovey, F. A., Schilling, F. C., McCrackin, F. L. and Wagner, H. L. (1976). *Macromolecules*, **9**, 76.
Bovey, F. A. (1988). *Nuclear Magnetic Resonance Spectroscopy*, 2nd edition, Academic Press, New York.
Boyd, R. H. (1985). *Polymer*, **26**, 323 and 1123.
Boyle, N. G., Coey, J. M. D. and McBrierty, V. J. (1982). *Chem. Phys. Letters*, **86**, 16.
Boyle, N. G., McBrierty, V. J. and Douglass, D. C. (1983a). *Macromolecules*, **16**, 75.
Boyle, N. G., McBrierty, V. J. and Eisenberg, A. (1983b). *Macromolecules*, **16**, 80.
Bradley, D. D. C., Colaneri, N. F. and Friend, R. H. (1989). *Synth. Metals*, **29**, 121.
Brandolini, A. J., Apple, T. M., Dybowski, C. R. and Pembleton, R. G. (1982). *Polymer*, **23**, 39.
Brandolini, A. J., Alvey, M. D. and Dybowski, C. R. (1983). *J. Polym. Sci: Polym. Phys. Ed.*, **21**, 2511. [Typographical errors have been corrected in table 7.2.]
Brandolini, A. J., Rocco, K. J. and Dybowski, C. R. (1984). *Macromolecules*, **17**, 1455.
Brereton, M. G. (1989). *Macromolecules*, **22**, 3667.
Brereton, M. G. (1990). *Macromolecules*, **23**, 1119.
Brereton, M. G. (1993). *Macromolecules*, **26**, 1152.
Brereton, M. G., Ward, I. M., Boden, N. and Wright, P. (1991). *Macromolecules*, **24**, 2068.

Brode, G. L. and Koleske, J. V. (1972). *J. Macromol. Sci: Chem.*, **A6**, 1109.

Bunn, A., Cudby, M. E. A., Harris, R. K., Packer, K. J. and Say, B. J. (1981). *J. Chem. Soc: Chem. Commun.*, p. 15.

Bunn, A., Cudby, M. E. A., Harris, R. K., Packer, K. J. and Say, B. J. (1982). *Polymer*, **23**, 694.

Burum, D. P. and Rhim, W.-K. (1979). *J. Chem. Phys.*, **71**, 944.

Caravatti, P., Deli, J. A., Bodenhausen, G. and Ernst, R. R. (1982). *J. Amer. Chem. Soc.*, **104**, 5506.

Caravatti, P., Neuenschwander, P. and Ernst, R. R. (1985). *Macromolecules*, **18**, 119.

Caravatti, P., Neuenschwander, P. and Ernst, R. R. (1986). *Macromolecules*, **19**, 1889.

Carlson, B. C. and Rushbrooke, G. S. (1950). *Proc. Cambridge Phil. Soc.*, **46**, 626.

Carr, H. Y. (1953). Ph.D. Thesis, Harvard University.

Carr, H. Y. and Purcell, E. M. (1954). *Phys. Rev.*, **94**, 630.

Carrington, A. and McLachlan, A. D. (1967). *Introduction to Magnetic Resonance*, Harper and Row, London.

Cheng, H. N. and Smith, D. A. (1986). *Macromolecules*, **19**, 2065.

Cheung, T. T. P., Gerstein, B. C., Ryan, L. M., Taylor, R. E. and Dybowski, C. R. (1980). *J. Chem. Phys.*, **73**, 6059.

Cheung, T. T. P. and Gerstein, B. C. (1981). *J. Appl. Phys.*, **52**, 5517.

Chiang, C. K., Fincher, C. R., Park, Y. W., Heeger, A. J., Shirakawa, H., Louis, E. J., Gau, S. C. and MacDiarmid, A. G. (1977). *Phys. Rev. Letters*, **39**, 1098.

Chirlian, L. E. and Opella, S. (1989). *Adv. Magnetic Res.*, **14**, 183.

Clark, W. G. and Glover, K. (1982). *Mol. Cryst. Liq. Cryst.*, **86**, 237.

Clarke, T. C. and Scott, J. C. (1982). *Solid State Comm.*, **41**, 389.

Cohen-Addad, J.-P. and Dupeyre, R. (1983). *Polymer*, **24**, 400.

Cohen-Addad, J.-P. and Guillermo, A. (1984). *J. Polym. Sci.*, **22**, 931.

Cohen-Added, J.-P. and Viallat, A. (1986). *Polymer*, **27**, 1855.

Cohen-Addad, J.-P. (1991). *NMR and Fractal Properties of Polymeric Liquids and Gels*, Pergamon Press, London.

Coleman, M. M. and Zarian, J. (1979). *J. Polym. Sci: Polym. Phys. Ed.*, **17**, 837.

Colombo, M. G., Meier, B. H. and Ernst, R. R. (1988). *Chem. Phys. Letters*, **146**, 189.

Colquhoun, I. J. and Packer, K. J. (1987). *British Polym. J.*, **19**, 151.

Connolly, J. J., Inglefield, P. T. and Jones, A. A. (1987). *J. Chem. Phys.*, **86**, 6602.

Connor, C., Naito, A., Takegoshi, K. and McDowell, C. A. (1985). *Chem. Phys. Letters*, **113**, 123.

Connor, T. M. (1963). *Trans. Faraday Soc.*, **60**, 1574.

Connor, T. M. (1970). *J. Polym. Sci.*, A-2, **8**, 191.

Coombes, A. and Keller, A. (1979). *J. Polym. Sci: Polym. Phys. Ed.*, **17**, 1637.

Cory, D. G., de Boer, J. C. and Veeman, W. S. (1989). *Macromolecules*, **22**, 1618.

Cosgrove, T., Crowley, T. L., Vincent, B., Barnett, K. G. and Tadros, T. F. (1981). *Faraday Symp., Chem. Soc.*, **16**, 101.

Couchman, P. R. (1980). *Macromolecules*, **13**, 1272.

Crofton, D. J., Moncrieff, D. and Pethrick, R. A. (1982). *Polymer*, **23**, 1605.

Cruz, C. A., Barlow, J. W. and Paul, D. R. (1979). *Macromolecules*, **12**, 726.

Cudby, M. E. A., Packer, K. J. and Hendra, P. J. (1984). *Polym. Comm.*, **25**, 303.
Cudby, M. E. A., Harris, R. K., Metcalfe, K., Packer, K. J., Smith, P. W. R. and Bunn, A. (1985). *Polymer*, **26**, 169.
Cunningham, A. (1974). Ph.D. Thesis, Leeds University.
Davis, D. and Slichter, W. P. (1973). *Macromolecules*, **6**, 728.
de Gennes, P.-G. (1958). *J. Phys. Chem. Solids*, **7**, 345.
de Gennes, P.-G. (1971). *J. Chem. Phys.*, **55**, 572.
de Gennes, P.-G. (1976). *Macromolecules*, **9**, 587.
de Gennes, P.-G. (1979). *Scaling Concepts in Polymer Physics*, Cornell University Press, Ithaca, New York.
de Gennes, P.-G. (1980). *Macromolecules*, **13**, 1069.
Deloche, B. and Samulski, E. T. (1981). *Macromolecules*, **14**, 575.
Demco, D., Tegenfeldt, J. and Waugh, J. S. (1975). *Phys. Rev.*, **B11**, 4133.
Di Paola-Baranyi, G. and Degré, P. (1981). *Macromolecules*, **14**, 1456.
Dixon, W. T. (1981). *J. Magnetic Res.*, **44**, 220.
Dixon, W. T. (1982). *J. Chem. Phys.*, **77**, 1800.
Doddrell, D., Glushko, V. and Allerhand, A. (1972). *J. Chem. Phys.*, **56**, 3683.
Doi, M. (1975). *J. Phys. A: Math. Gen.*, **8**, 417.
Doolittle, A. K. (1951). *J. Appl. Phys.*, **22**, 1471.
Doolittle, A. K. (1952). *J. Appl. Phys.*, **23**, 236.
Donnet, J. B. and Custodero, E. (1992). *Carbon*, **30**, 813.
Doskocilova, D., Schneider, B., Jakes, J., Schmidt, P., Baldrian, J., Hernandez-Fuentes, I. and Caceres Alonso, M. (1986). *Polymer*, **27**, 1658.
Douglass, D. C. and Jones, G. P. (1966). *J. Chem. Phys.*, **45**, 956.
Douglass, D. C. and McBrierty, V. J. (1971). *J. Chem. Phys.*, **54**, 4085.
Douglass, D. C., McBrierty, V. J. and Weber, T. A. (1976). *J. Chem. Phys.*, **64**, 1533.
Douglass, D. C., McBrierty, V. J. and Weber, T. A. (1977). *Macromolecules*, **10**, 178.
Douglass, D. C. and McBrierty, V. J. (1978). *Macromolecules*, **11**, 766.
Douglass, D. C. and McBrierty, V. J. (1979). *Polym. Eng. Sci.*, **19**, 1054.
Douglass, D. C. (1980). In *Polymer Characterisation by ESR and NMR*, Woodward, A. E. and Bovey, F. A. (eds.), *ACS Symp. Ser.* **142**, p. 147.
Douglass, D. C., McBrierty, V. J. and Wang, T. T. (1982). *J. Chem. Phys.*, **77**, 5826.
Douglass, D. C. (1991), private communication.
Doyle, S., Pethrick, R. A., Harris, R. K., Lane, J. M., Packer, K. J. and Heatley, F. (1986). *Polymer*, **27**, 19.
Drey-Aharon, H., Sluckin, T. J., Taylor, P. L. and Hopfinger, A. J. (1980). *Phys. Rev.*, **B21**, 3700.
Dumais, J. J., Jelinski, L., Leung, L. M., Gancarz, I., Galambos, A. and Koberstein, J. T. (1985). *Macromolecules*, **18**, 116.
Duncan, T. M. and Dybowski, C. R. (1981). *Surf. Sci. Repts.*, **1**, 157.
Duncan, T. M. (1990). *A Compilation of Chemical Shift Anisotropies*. The Farragut Press, Madison, Wisconsin.
Duplessix, R., Escoubes, M., Rodmacq, B., Volino, F., Roche, E., Eisenberg, A. and Pineri, M. (1980). In *Water in Polymers*, Rowland, S. P. (ed.), *ACS Symp. Ser.* **127**, Washington, p. 467.
Earl, W. L. and VanderHart, D. L. (1979). *Macromolecules*, **12**, 762.
Edmonds, A. R. (1974). *Angular Momentum in Quantum Mechanics*, 3rd edition, Princeton Univ. Press, Princeton.
Edwards, S. F. (1967). *Proc. Phys. Soc.*, **91**, 513; **92**, 9.

Edzes, H. T. (1983). *Polymer*, **24**, 1425.

Eisenberg, A., Hird, B. and Moore, R. B. (1990). *Macromolecules*, **23**, 4098.

English, A. D. (1984). *Macromolecules*, **17**, 2182.

Ernst, R. R., Bodenhausen, G. and Wokaun, A. (1987). *Principles of Nuclear Magnetic Resonance in One and Two Dimensions*, Clarendon Press, Oxford.

Etemad, S., Heeger, A. J. and MacDiarmid, A. G. (1982). *Annual Rev. Phys. Chem.*, **33**, 443.

Farrar, T. C. and Becker, E. D. (1971). *Pulse and Fourier Transform NMR*, Academic Press, New York.

Feast, W. J. and Friend, R. H. (1990). *J. Mat. Sci.*, **25**, 3796.

Fleming, W. W., Fyfe, C. A., Kendrick, R. D., Lyerla, J. R., Vanni, H. and Yannoni, C. S. (1980). In *Polymer Characterisation by ESR and NMR*, Woodward, A. E. and Bovey, F. A. (eds.), *ACS Symp. Ser.*, **142**, p. 193.

Folland, R. and Charlesby, A. (1978). *J. Polym. Sci: Polym. Letters Ed.*, **16**, 339.

Fox, T. G. (1956). *Bull. Amer. Phys. Soc.*, **1**, 123.

Fox, T. G. and Flory, P. J. (1950). *J. Appl. Phys.*, **21**, 581.

Fox, T. G. and Flory, P. J. (1951). *J. Phys. Chem.*, **55**, 221.

Fox, T. G. and Flory, P. J. (1954). *J. Polym. Sci.*, **14**, 315.

Frank, F. C., Gupta, V. B. and Ward, I. M. (1970). *Phil. Mag.*, **21**, 1127.

Frey, M. H. and Opella, S. J. (1980). *J. Chem. Soc., Chem. Comm.*, p. 474.

Froix, M. F., Williams, D. J. and Goedde, A. O. (1976). *Macromolecules*, **9**, 354.

Frye, J. S. and Maciel, G. E. (1982). *J. Magnetic Res.*, **48**, 125.

Fujimoto, K., Nishi, T. and Kado, R. (1972). *Polymer J.*, **3**, 448.

Fukushima, E. and Roeder, S. B. W. (1981). *Experimental Pulse NMR: A Nuts and Bolts Approach*, Addison-Wesley, London.

Fulcher, G. S. (1925). *J. Amer. Ceramic Soc.*, **77**, 3701.

Furukawa, T., Johnson, G. E., Bair, H. E., Tajitsu, Y., Chiba, A. and Fukada, E. (1981). *Ferroelectrics*, **32**, 61.

Fyfe, C. A., Lyerla, J. R., Volksen, W. and Yannoni, C. S. (1979). *Macromolecules*, **12**, 757.

Fyfe, C. A., Rudin, A. and Tchir, W. (1980). *Macromolecules*, **13**, 1320.

Fyfe, C. A., Lyerla, J. R. and Yannoni, C. S. (1982). *Acc. Chem. Res.*, **15**, 208.

Gabrys, B., Fumitaka, H. and Kitamaru, R. (1987). *Macromolecules*, **20**, 175.

Garroway, A. N., Stalker, D. C. and Mansfield, P. (1975). *Polymer*, **16**, 161.

Garroway, A. N., Moniz, W. B. and Resing, H. A. (1979). *ACS Symp. Ser.*, **103**, 67.

Garroway, A. N., VanderHart, D. L. and Earl, W. L. (1981). *Phil. Trans. Roy. Soc. London*, **A299**, 609.

Garroway, A. N., Ritchey, W. M. and Moniz, W. B. (1982). *Macromolecules*, **15**, 1051.

Gashgari, M. A. and Frank, C. W. (1981). *Macromolecules*, **14**, 1558.

Gerstein, B. C. (1983). *Anal. Chem.*, **55**, 781A, 899.

Gevert, T. U. and Svanson, S. E. (1985). *Polymer*, **26**, 307: *Euro. Polym. J.*, **21**, 401.

Glarum, S. H. (1960). *J. Chem. Phys.*, **33**, 639.

Gobbi, G. C., Silvestri, R., Russell, T. P., Lyerla, J. R., Fleming, W. W. and Nishi, T. (1987). *J. Polym. Sci: Polym. Letters*, **25**, 61.

Goldman, M. and Shen, L. (1966). *Phys. Rev.*, **144**, 321.

Goldman, M. (1970). *Spin Temperature and Nuclear Magnetic Resonance in Solids*, Clarendon Press, Oxford.

Gordon, M. and Taylor, J. S. (1952). *J. Appl. Chem.*, **2**, 493.

Granville, M., Jérôme, R. J., Teyssié, P., de Schryver, F. C. (1988). *Macromolecules*, **21**, 2894.

Gronski, W., Stadler, R. and Jacobi, M. M. (1984). *Macromolecules*, **17**, 741.

Gronski, W., Hoffmann, U., Baumann, K., Simon, G. and Straube, E. (1991). *Abstracts of the 29th Europhysics Conference on Macromolecular Physics: Physics of Polymer Networks*, Alexisbad/Harz (Germany), p. 52.

Groombridge, C. J., Harris, R. K., Packer, K. J., Say, B. J. and Tanner, S. F. (1980). *J. Chem. Soc., Chem. Commun.*, p. 175.

Gullion, T. and Schaefer, J. (1989). *J. Magnetic Res.*, **81**, 196.

Gullion, T., McKay, R. A. and Schmidt, A. (1991). *J. Magnetic Res.*, **94**, 362.

Gunther, E., Blümich, B. and Spiess, H. W. (1990). *Mol. Phys.*, **71**, 477.

Gupta, V. B. and Ward, I. M. (1970). *J. Macromol. Sci: Phys.*, **4**, 453.

Gupta, M. K., Ripmeester, J. A., Carlsson, D. J. and Wiles, D. M. (1983). *J. Polym. Sci: Polym. Letters Ed., **21**, 211.

Hadziioanou, G. and Stein, R. S. (1984). *Macromolecules*, **17**, 567.

Haeberlen, U. (1976). *High Resolution NMR in Solids – Selective Averaging*, Waugh, J. S. (ed.), Supplement 1, Adv. Magnetic Res., Academic Press, New York.

Hagemeyer, A., Brombacher, L., Schmidt-Rohr, K. and Spiess, H. W. (1990). *Chem. Phys. Letters*, **167**, 583.

Hagemeyer, A., van der Putten, D. and Spiess, H. W. (1991). *J. Magnetic Res.*, **92**, 628.

Hahn, E. L. (1950). *Phys. Rev.*, **80**, 580.

Hanack, M., Roth, S. and Schier, H., (eds.) (1991). *Proc. Intnl. Conference on Science and Technology of Synthetic Metals Part 1*, ICSM 90, Tübingen.

Harbison, G. S. and Spiess, H. W. (1986). *Chem. Phys. Letters*, **124**, 128.

Harbison, G. S., Vogt, V.-D. and Spiess, H. W. (1987). *J. Chem. Phys.*, **86**, 1206.

Harris, R. K. (1983a). *Nuclear Magnetic Resonance Spectroscopy*, Pitman, London, p. 112.

Harris, R. K. (1983b). *Nuclear Magnetic Resonance Spectroscopy*, Pitman, London, pp. 107 and 110.

Harris, R. K., Jonsen, P. and Packer, K. J. (1985). *Magnetic Res. in Chem.*, **23**, 565.

Harris, R. K., Kenwright, A. M., Packer, K. J., Stark, B. P. and Everatt, B. (1986). *Magnetic Res. in Chem.*, **25**, 80.

Hartmann, S. R. and Hahn, E. L. (1962). *Phys. Rev.*, **128**, 2042.

Havens, J. R., Ishida, H. and Koenig, J. L. (1981). *Macromolecules*, **14**, 1327.

Havens, J. R. and VanderHart, D. L. (1985). *Macromolecules*, **18**, 1663.

Heatley, F. and Begum, A. (1977). *Makromol. Chem.*, **178**, 1205.

Heatley, F. (1979). *Progress in NMR Spectroscopy*, **13**, 47.

Heatley, F. (1989). In *Comprehensive Polymer Science*, Booth, C. and Price, C. (eds.), Pergamon Press, London, pp. 377–96.

Heeger, A. J. (1986). In *Handbook of Conducting Polymers*, Skotheim, T. J. (ed.), Marcel Dekker, New York.

Helfand, E. and Wassermann, Z. R. (1982). Chapter 4 in *Developments in Block Copolymers-1*, Goodman, I., (ed.), Applied Science, London.

Henrichs, P. M. and Linder, M. (1984). *J. Magnetic Res.*, **58**, 458.

Henrichs, P. M., Hewitt, J. M. and Linder, M. (1984). *J. Magnetic Res.*, **60**, 280.

Henrichs, P. M., Linder, M. and Hewitt, J. M. (1986). *J. Chem. Phys.*, **85**, 7077.

Henrichs, P. M., Tribone, J., Massa, D. J. and Hewitt, J. M. (1988). *Macromolecules*, **21**, 1282.

Hentschel, R., Schlitter, J., Sillescu, H. and Spiess, H. W. (1978). *J. Chem. Phys.*, **68**, 56.

Hentschel, R., Sillescu, H. and Spiess, H. W. (1981). *Polymer*, **22**, 1516.

Hentschel, R., Sillescu, H. and Spiess, H. W. (1984). *Polymer*, **25**, 1078.

Hermans, J. J., Hermans, P. H., Vermaas, D. and Weidinger, A. (1946). *Rec. Trav. Chim.*, **65**, 427.

Herzfeld, J. and Berger, A. E. (1980). *J. Chem. Phys.*, **73**, 6021.

Hirsch, J. E. and Grabowski, M. (1984). *Phys. Rev. Letters*, **52**, 1713.

Hirschinger, J., Miura, H., Gardner, K. H. and English, A. D. (1990). *Macromolecules*, **23**, 2153.

Hirschinger, J., Schaefer, D., Spiess, H. W. and Lovinger, A. J. (1991). *Macromolecules*, **24**, 2428.

Hofer, K., Mayer, E. and Johari, G. P. (1990). *J. Phys. Chem.*, **94**, 2689.

Illers, K. H. (1969). *Kolloid. Z. Z. Polym.*, **231**, 622.

Ishii, F., Sawatari, T. and Odajima, A. (1982). *Jpn. J. Appl. Phys.*, **21**, 251.

Ito, M., Serizawa, H., Tanaka, K., Leung, W. P. and Choy, C. L. (1983). *J. Polym. Sci: Polym. Phys. Ed.*, **21**, 2299.

Ito, M., Kanamoto, T., Tanaka, K. and Porter, R. S. (1985). *J. Polym. Sci: Polym. Phys. Ed.*, **23**, 59.

Jackson, W. J. and Caldwell, J. R. (1967). *J. Appl. Polym. Sci.*, **11**, 211, 227.

Jeener, J. and Broekaert, P. (1967). *Phys. Rev.*, **157**, 232.

Jelinski, L. W., Schilling, F. C. and Bovey, F. A. (1981). *Macromolecules*, **14**, 581.

Jelinski, L. W., Dumais, J. J., Schilling, F. C. and Bovey, F. A. (1982). *ACS Symp. Ser.*, **191**, 345.

Jelinski, L. W., Dumais, J. J. and Engel, A. K. (1983a). *Macromolecules*, **16**, 403.

Jelinski, L. W., Dumais, J. J., Watnick, P. I., Engel, A. K. and Sefcik, M. D. (1983b). *Macromolecules*, **16**, 409.

Jelinski, L. W., Dumais, J. J. and Engel, A. K. (1983c). *Macromolecules*, **16**, 492.

Jelinski, L. W., Dumais, J. J. and Engel, A. K. (1983d). *ACS Org. Coatings and Appl. Polym. Sci. Proc.*, **48**, 102.

Jelinski, L., Dumais, J. J. and Engel, A. K. (1984). *ACS Symp. Ser.*, **247**, 55.

Jones, A. A. (1989). In *Molecular Dynamics in Restricted Geometries*, Klafter, J. and Drake, T. M. (eds.), Wiley, London, p. 247.

Jones, G. P. (1966). *Phys. Rev.*, **148**, 332.

Kaplan, S. and O'Malley, J. J. (1979). *Polym. Prepr. Div. Polym. Chem. Amer. Chem. Soc.*, **20**, 266.

Karasz, F. E., Chein, J. C. W., Shimamura, K. and Hirsch, J. A. (1982). *Mol. Cryst. Liq. Cryst.*, **86**, 223.

Kasuboski, L. A. (1988). Ph.D. Thesis, University of Delaware.

Kaufman, S., Slichter, W. P. and Davis, D. D. (1971). *J. Polym. Sci.*, A-2, **9**, 829.

Kaufmann, S., Wefing, S., Schaefer, D. and Spiess, H. W. (1990). *J. Chem. Phys.*, **93**, 197.

Keller, A., Pedemonte, E. and Willmouth, F. M. (1970). *Nature*, **225**, 538: *Kolloid. Z.*, **238**, 385.

Kelley, F. N. and Bueche, F. (1961). *J. Polym. Sci.*, **50**, 549.

Kenny, J. C., McBrierty, V. J., Rigbi, Z. and Douglass, D. C. (1991). *Macromolecules*, **24**, 436.

Kentgens, A. P. M., de Jong, A. F., de Boer, E. and Veeman, W. S. (1985). *Macromolecules*, **18**, 1045.

Kentgens, A. P. M., Veeman, W. S. and van Bree, J. (1987). *Macromolecules*, **20**, 234.

Kenwright, A. M., Packer, K. J. and Say, B. J. (1986). *J. Magnetic Res.*, **69**, 426.

Kepler, R. G. and Anderson, R. A. (1980). *CRC Crit. Rev. Solid State Mat. Sci.*, **9**, 399.

Khetrapal, C. L. and Becker, E. D. (1991). *Magnetic Res. Rev.*, **16**, 35.

Khutsishvili, G. R. (1956). *Proc. Inst. Phys. Acad. Sci. Georgia (USSR)*, **4**, 3.

Kim, H.-G. (1972). *Macromolecules*, **5**, 594.

Kimmich, R. (1988). *Progr. in NMR Spectra*, **20**, 385.

Kinjo, N. J. and Nakagawa, T. (1973). *Polymer J.* **4**, 143.

Kitamaru, R., Horii, F. and Murayama, K. (1986). *Macromolecules*, **19**, 636.

Klesper, E., Gronski, W. and Barth, V. (1970). *Makromol. Chem.*, **139**, 1.

Koberstein, J. T. and Stein, R. S. (1983). *J. Polym. Sci: Polym. Phys. Ed.*, **21**, 1439.

Kolbert, A. C. and Griffin, R. G. (1991). *J. Magnetic Res.*, **93**, 242.

Komoroski, R. A. and Mauritz, K. A. (1978). *J. Amer. Chem. Soc.*, **100**, 7487.

Komoroski, R. A. (1983). *J. Polym. Sci: Polym. Phys. Ed.*, **21**, 1569.

Komoroski, R. A. (1986). *High-Resolution NMR Spectroscopy of Synthetic Polymers in Bulk: Methods of Stereochemical Analysis*, Volume 7, Wiley, London.

Kratky, O. (1933). *Kolloid. Z.*, **64**, 213.

Kraus, G., Rollmann, K. W. and Gruver, J. T. (1970). *Macromolecules*, **3**, 92.

Krause, S. (1978). *Macromolecules*, **11**, 1288.

Krause, S., Lu, Z. and Iskandar, M. (1982). *Macromolecules*, **15**, 1076.

Kretz, M., Meurer, B., Spegt, P. and Weill, G. (1988). *J. Polym. Sci: Polym. Phys. Ed.*, **26**, 1553.

Krigbaum, W. R. and Roe, R.-J. (1964). *J. Chem. Phys.*, **41**, 737.

Kubo, A. and McDowell, C. A. (1988). *J. Chem. Phys.*, **89**, 63: *J. Chem. Soc. (Faraday 1)*, **84**, 3713.

Kubo, R. and Tomita, K. (1954). *J. Phys. Soc. Japan*, **9**, 888.

Kuhlmann, K. F. and Grant, D. M. (1968). *J. Amer. Chem. Soc.*, **90**, 7355.

Kuhn, W. and Grün, F. (1942). *Kolloid. Z.*, **101**, 248.

Kume, K., Mizuno, K., Mizoguchi, K., Nomura, K., Tanaka, J., Tanaka, M. and Fujimoto, H. (1982). *Mol. Cryst. Liq. Cryst.*, **83**, 49.

Kurosu, H., Yamanobe, T., Komoto, T. and Ando, I. (1987). *Chem. Phys.*, **116**, 391.

Kurosu, H., Komoto, T. and Ando, I. (1988). *J. Mol. Structure*, **176**, 279.

Kwei, T. K., Nishi, T. and Roberts, R. F. (1974). *Macromolecules*, **7**, 667.

Lacher, R. C., Bryant, J. L., Howard, L. N. and Sumners, D. W. (1986). *Macromolecules*, **19**, 2639.

Laupretre, F., Monnerie, L., Barthelemy, L., Vairon, J.-P., Sauzeau, A. and Roussel, D. (1986a). *Polym. Bull.*, **15**, 159.

Laupretre, F., Monnerie, L. and Bloch, B. (1986b). *Anal. Chim. Acta*, **189**, 117.

Leung, L. M. and Koberstein, J. T. (1985). *J. Polym. Sci.: Polym. Phys. Ed.*, **23**, 1883.

Lind, A. C. (1977). *J. Chem. Phys.*, **66**, 3482.

Linder, M., Henrichs, P. M., Hewitt, J. M. and Massa, D. J. (1985). *J. Chem. Phys.*, **82**, 1585.

Lippmaa, E., Alla, M. and Turherm, T. (1976). *Proc. 19th Congress Ampere*, Heidelberg (Groupment Ampere, Heidelberg, 1976), p. 113.

Litvinov, V. M. (1988). *Polym. Sci., USSR*, **30**, 2250.

Litvinov, V. M., Whittaker, A. K., Hagemeyer, A. and Spiess, H. W. (1989). *Colloid Polym. Sci.*, **267**, 681.

Litvinov, V. M. and Spiess, H. W. (1992). *Makromol. Chem.*, **193**, 1181.

Liu, Y., Roy, A. K., Jones, A. A., Inglefield, P. T. and Ogden, P. (1990). *Macromolecules*, **23**, 968.

Lovinger, A. J., Furukawa, T., Davis, G. T. and Broadhurst, M. G. (1983). *Polymer*, **24**, 1225, 1233.

Lovinger, A. J. (1983). *Science*, **220**, 1115.

Lowe, I. J. and Norberg, R. E. (1957). *Phys. Rev.*, **107**, 46.

Lowe, I. J. (1959). *Phys. Rev. Letters*, **2**, 285.

Lowe, I. J. and Gade, S. (1967). *Phys. Rev.*, **156**, 817.

Löwenhaupt, B. and Hellmann, G. P. (1990). *Colloid Polym. Sci.*, **268**, 885.

Lu, F. J., Benedetti, E. and Hsu, S. L. (1983). *Macromolecules*, **16**, 1525.

Lyerla, J. R. and Levy, G. C. (1974). In *Topics in ^{13}C NMR Spectroscopy*, Levy, G. C. (ed.), Wiley, New York, Vol. 1, p. 79.

Maas, W. E. J. R., Kentgens, A. P. M. and Veeman, W. S. (1987). *J. Chem. Phys.*, **87**, 6854.

Maciel, G. E., Szeverenyi, N. M., Early, T. A. and Myers, G. E. (1983). *Macromolecules*, **16**, 598.

Maciel, G. E., Chuang, I. S. and Gollob, L. (1984). *Macromolecules*, **17**, 1081.

Maciel, G. E., Szeverenyi, N. M. and Sardashti, M. (1985). *J. Magnetic Res.*, **64**, 365.

Maier, W. and Saupe, A. (1960). *Z. Naturforsch.*, **15**, 287.

Makaruk, L. and Polanska, H. (1981). *Polym. Bull.*, **4**, 127.

Manenschijn, A., Duijvestijn, M. J., Smidt, J., Wind, R. A., Yannoni, C. S. and Clarke, T. C. (1984). *Chem. Phys. Letters*, **112**, 99.

Mansfield, M. L. (1987). *Macromolecules*, **20**, 1384.

Mansfield, P. and Ware, D. (1966). *Phys. Letters*, **22**, 133.

Mansfield, P. (1970). *Phys. Letters A*, **32**, 485.

Mansfield, P. (1972). Pulsed NMR in solids in *Progress in NMR Spectroscopy*, Vol. 8, Pergamon Press, Oxford, pp. 41–101.

Mansfield, P., Orchard, M. J., Stalker, D. C. and Richards, K. H. B. (1973). *Phys. Rev.*, **B7**, 90.

Mansfield, P. and Hahn, E. L. (1991). *NMR Imaging: Recent Developments and Future Prospects*, Cambridge University Press, Cambridge.

Maricq, M. M. and Waugh, J. S. (1979). *J. Chem. Phys.*, **70**, 3300.

Masin, F., Gusman, G. and Deltour, R. (1981). *Solid State Comm.*, **40**, 415: **40**, 513.

Mathur, S. C., Scheinbeim, J. I. and Newman, B. A. (1984). *J. Appl. Phys.*, **56**, 2419.

Maxwell, J. C. (1892). *Electricity and Magnetism I*, 3rd edition, Clarendon Press, Oxford, p. 440.

McBrierty, V. J. and Ward, I. M. (1968). *Brit. J. Appl. Phys. D*, **1**, 1529.

McBrierty, V. J. and Douglass, D. C. (1970). *J. Magnetic Res.*, **2**, 352.

McBrierty, V. J., McCall, D. W., Douglass, D. C. and Falcone, D. R. (1970). *J. Chem. Phys.*, **52**, 512.

McBrierty, V. J., McCall, D. W., Douglass, D. C. and Falcone, D. R. (1971a). *Macromolecules*, **4**, 584.

McBrierty, V. J., McDonald, I. R. and Ward, I. M. (1971b). *J. Phys. D*, **4**, 88.

McBrierty, V. J., Douglass, D. C. and Falcone, D. R. (1972). *J. Chem. Soc. Faraday II*, **68**, 1051.

McBrierty, V. J. and McDonald, I. R. (1973). *J. Phys. D*, **6**, 131.

McBrierty, V. J. (1974a). *Polymer*, **15**, 503.

McBrierty, V. J. (1974b). *J. Chem. Phys.*, **61**, 872.

McBrierty, V. J., Douglass, D. C. and Weber, T. A. (1976). *J. Polym. Sci: Polym. Phys. Ed.*, **14**, 1271.

McBrierty, V. J. and Douglass, D. C. (1977). *Macromolecules*, **10**, 855.

McBrierty, V. J., Douglass, D. C. and Kwei, T. K. (1978). *Macromolecules*, **11**, 1265.

McBrierty, V. J. (1979). *Faraday Disc., Chem. Soc.*, **68**, 78.

McBrierty, V. J. and Douglass, D. C. (1980). *Physics Reports*, **63**, pp. 61–147.

McBrierty, V. J., Douglass, D. C. and Barham, P. J. (1980). *J. Polym. Sci: Polym. Phys. Ed.*, **18**, 1561.

McBrierty, V. J. and Douglass, D. C. (1981). *Macromolecular Rev.*, **16**, 295.

McBrierty, V. J., Douglass, D. C. and Wang, T. T. (1982a). *Appl. Phys. Letters*, **41**, 1051.

McBrierty, V. J., Douglass, D. C. and Furukawa, T. (1982b). *Macromolecules*, **15**, 1063.

McBrierty, V. J. (1983). *Magnetic Res. Rev.* **8**, 165.

McBrierty, V. J., Douglass, D. C. and Furukawa, T. (1984). *Macromolecules*, **17**, 1136.

McBrierty, V. J., Smyth, G. and Douglass, D. C. (1987). In *Structure and Properties of Ionomers*, NATO ASI Series C, Maths and Phys Sciences, Pineri, M. and Eisenberg, A. (eds.), D. Reidel and Co., Holland, pp. 149–61.

McBrierty, V. J. (1989). In *Comprehensive Polymer Science*, Booth, C. and Price, C. (eds.), Pergamon Press, London, pp. 397–428.

McBrierty, V. J., Quinn, F. X., Keely, C., Wilson, A. C. and Friends, G. D. (1992). *Macromolecules*, **25**, 4281.

McBrierty, V. J., Douglass, D. C., Zhang, X., Quinn, F. X. and Jérôme, R. (1993). *Macromolecules*, **26**, 1734.

McCall, D. W. and Hamming, R. W. (1959). *Acta Cryst.*, **12**, 81.

McCall, D. W., Douglass, D. C. and Anderson, E. W. (1962). *J. Polym. Sci.*, **59**, 301.

McCall, D. W. and Anderson, E. W. (1963). *J. Polym. Sci. A*, **1**, 1175.

McCall, D. W. and Douglass, D. C. (1963). *Polymer*, **4**, 433.

McCall, D. W. and Douglass, D. C. (1965). *Appl. Phys. Lett.*, **7**, 12.

McCall, D. W. (1969). Relaxation in solid polymers. *Molecular Dynamics and Structure*, Carter, R. S. and Rush, J. J. (eds.), National Bureau of Standards, Special Publn 301, Washington, pp. 475–537.

McConnell, J. (1983). *Physica*, **117A**, 251.

Mehring, M., Griffin, R. G. and Waugh, J. S. (1971). *J. Chem. Phys.*, **55**, 746.

Mehring, M. (1976). *NMR Basic Principles and Progress*, Diehl, P., Kluck, E. and Kosfeld, R. (eds.), **11**, 1.

Mehring, M., Weber, H., Müller, W. and Wegner, G. (1983). *Solid State Comm.*, **45**, 1079.

Meiboom, S. and Gill, D. (1958). *Rev. Sci. Insts.*, **29**, 688.

Meinecke, E. (1991). *Rubber Chem. Technol.*, **64**, 269.

Mikeš, F., Morawetz, H. and Dennis, K. S. (1980). *Macromolecules*, **13**, 969.
Mirau, P. A., Tanaka, H. and Bovey, F. A. (1988). *Macromolecules*, **21**, 2929.
Miyamoto, Y., Miyaji, H. and Asai, K. (1980). *J. Polym. Sci: Polym. Phys. Ed.*, **18**, 597.
Moore, J. S. and Stupp, S. I. (1987). *Macromolecules*, **20**, 282.
Morèse-Séguéla, B., St. Jacques, M., Renaud, J. M. and Prud'homme, J. (1980). *Macromolecules*, **13**, 100.
Munowitz, M. G., Griffin, R. G., Bodenhausen, G. and Huang, T. H. (1981). *J. Amer. Chem. Soc.*, **103**, 2529.
Munowitz, M. G. and Griffin, R. G. (1982). *J. Chem. Phys.*, **76**, 2848.
Murphy, P. B., Taki, T., Gerstein, B. C., Henrichs, P. M. and Massa, D. J. (1982). *J. Magnetic Res.*, **49**, 99.
Murthy, N. S., Correale, S. T. and Minor, H. (1991). *Macromolecules*, **24**, 1185.
Naarman, H. (1987). *Proc. ACS Meeting, Denver, Colorado*, April.
Naegele, D. and Yoon, D. Y. (1978). *Appl. Phys. Letters*, **33**, 132.
Naito, K., Johnson, G. E., Allara, D. L. and Kwei, T. K. (1978). *Macromolecules*, **11**, 1260.
Nakai, T., Terao, T. and Shirakawa, H. (1988a). *Chem. Phys. Letters*, **145**, 90.
Nakai, T., Ashida, J. and Terao, T. (1988b). *J. Chem. Phys.*, **88**, 6049.
Nassar, T. R., Paul, D. R. and Barlow, J. W. (1979). *J. Appl. Polym. Sci.*, **23**, 85.
Natansohn, A. and Eisenberg, A. (1987). *Macromolecules*, **20**, 323.
Natarajan, K. M., Samulski, E. T. and Cukier, R. I. (1978). *Nature*, **275**, 527.
Nechtschein, M., Devreux, F., Greene, R. L., Clarke, T. C. and Street, G. B. (1980). *Phys. Rev. Letters*, **44**, 356.
Nieminen, A. K. and Koenig, J. L. (1988). *J. Adhesion Sci. Technol.*, **2**, 407.
Nishi, T. and Wang, T. T. (1975). *Macromolecules*, **8**, 909.
Nomura, S. (1989). In *Comprehensive Polymer Science*, Allen, G. (ed.), Vol. 2, Pergamon Press, Oxford, pp. 459–85.
O'Brien, J. and McBrierty, V. J. (1975). *Proc. Roy. Irish Acad. A*, **75**, 331.
O'Brien, J., Cashell, E. M., Wardell, G. E. and McBrierty, V. J. (1976). *Macromolecules*, **9**, 653.
O'Gara, J. F., Jones, A. A., Hung, C.-C. and Inglefield, P. T. (1985). *Macromolecules*, **18**, 1117.
Ohigashi, H. (1976). *J. Appl. Phys.*, **47**, 949.
Olabisi, O. (1975). *Macromolecules*, **8**, 316.
Olabisi, O., Robeson, L. M. and Shaw, M. T. (1979). *Polymer–Polymer Miscibility*, Academic Press, New York.
Olejniczak, E. T., Vega, S. and Griffin, R. G. (1984). *J. Chem. Phys.*, **81**, 4804.
Olf, H. G. and Peterlin, A. (1970). *J. Polym. Sci.*, **A2**, 771.
Olf, H. G. and Peterlin, A. (1971). *J. Polym. Sci.*, **A-2**, **9**, 1449.
Opella, S. J. and Waugh, J. S. (1977). *J. Chem. Phys.*, **66**, 4919.
Packer, K. J. (1980). *Mol. Phys.*, **39**, 15.
Packer, K. J., Pope, J. M., Yeung, P. R. and Cudby, M. E. A. (1984). *J. Polym. Sci: Polym. Phys. Ed.*, **22**, 589.
Packer, K. J., Poplett, I. J. F. and Taylor, M. J. (1988). *J. Chem. Soc: Faraday Trans. 1*, **84**, 3851.
Pake, G. E. (1956). *Solid State Phys.*, **2**, 1.
Parmer, J. F., Dickinson, L. C., Chien, J. C. W. and Porter, R. S. (1987). *Macromolecules*, **20**, 2308.

Parmer, J. F., Dickinson, L. C., Chien, J. C. W. and Porter, R. S. (1988). *Polym. Prepr.*, **29**, 476.
Patterson, G. D., Nishi, T. and Wang, T. T. (1976). *Macromolecules*, **9**, 603.
Patterson, P. M., Patterson, D. J., Blackwell, J., Koenig, J. L., Jamieson, A. M., Carignan, Y. P. and Turngreen, E. V. (1985). *J. Polym. Sci: Polym. Phys. Ed.*, **23**, 483.
Pearce, E. M., Kwei, T. K. and Min, B. Y. (1984). *J. Macromol. Sci: Chem.*, **A21**, 1181.
Pembleton, R. G., Wilson, R. C. and Gerstein, B. C. (1977). *J. Chem. Phys.*, **66**, 5133.
Peo, M., Forster, H., Menke, K., Hocker, J., Gardner, J. A., Roth, S. and Dransfield, K. (1981). *Solid State Comm.*, **38**, 467.
Pérez, E., VanderHart, D. L., Buckley, C. and Howard, P. R. (1987). *Macromolecules*, **20**, 78.
Pérez, E. and VanderHart, D. L. (1988). *J. Polym. Sci: Polym. Phys. Ed.*, **26**, 1979.
Pietralla, M., Schubach, H. R., Dettenmaier, M. and Heise, B. (1985). *Progr. Colloid. Polym. Sci.*, **71**, 125.
Pineri, M. and Eisenberg, A. (eds.) (1987). *Structure and Properties of Ionomers*, NATO ASI Series C, Maths and Phys Sciences, D. Reidel and Co., Holland, pp. 149–61.
Pines, A., Gibby, M. G. and Waugh, J. S. (1972). *J. Chem. Phys.*, **56**, 1776.
Pines, A., Gibby, M. G. and Waugh, J. S. (1973). *J. Chem. Phys.*, **59**, 569.
Pliskin, I. and Tokita, N. (1972). *J. Appl. Polym. Sci.*, **16**, 473.
Pochan, J. M., Beatty, C. L. and Pochan, D. F. (1979). *Polymer*, **20**, 879.
Pouchly, J., Biros, J. and Benes, S. (1979). *Makromol. Chem.*, **180**, 745.
Powell, D. G. and Mathais, L. J. (1990). *Polym. Commun.*, **31**, 58.
Powles, J. G. and Mansfield, P. (1962). *Phys. Letters*, **2**, 58.
Powles, J. G. and Strange, J. H. (1963). *Proc. Phys. Soc. London*, **82**, 6.
Prabhu, P., Schlinder, A. and Gilbert, R. D. (1978). *Polym. Prepr. Div. Polym. Chem., Amer. Chem. Soc.*, **19**, 642.
Prabhu, P., Schindler, A., Theil, M. H. and Gilbert, R. D. (1980). *J. Polym. Sci: Polym. Letters Ed.*, **18**, 389.
Provotorov, B. N. (1962). *Soviet Physics, JEJP*, **15**, 611. [English translation.]
Queslel, J.-P. and Mark, J. E. (1989). In *Comprehensive Polymer Science*, Allen, G. (ed.), Vol. 2, Pergamon Press, Oxford, p. 271.
Quinn, F. X., Kampff, E., Smyth, G. and McBrierty, V. J. (1988). *Macromolecules*, **21**, 3191.
Quinn, F. X., McBrierty, V. J., Wilson, A. C. and Friends, G. D. (1990). *Macromolecules*, **23**, 4576.
Raleigh, D. P., Olejnczak, E. T., Vega, S. and Griffin, R. G. (1987). *J. Magnetic Res.*, **72**, 238.
Raleigh, D. P., Creuzet, F., Das Gupta, S. K., Levitt, M. H. and Griffin, R. G. (1989). *J. Amer. Chem. Soc.*, **111**, 4502.
Razinskaya, I. N., Vidyakina, L. I., Radbil, T. I. and Shtarkman, B. P. (1972). *Polym. Sci. USSR*, **14**, 1079.
Register, R. A., Foucart, M., Jérôme, R. T., Ding, Y. S. and Cooper, S. L. (1988). *Macromolecules*, **21**, 1009.
Rehner Jr, J. R. (1965). In *Reinforcement of Elastomers*, Kraus, G. (ed.), Interscience, New York, pp. 153–5.
Resing, H. A. (1965). *J. Chem. Phys.*, **43**, 669.
Resing, H. A. (1968). *Adv. Mol. Relax. Processes*, **1**, 109.

Resing, H. A. (1972). *Adv. Mol. Relax. Processes*, **3**, 199.

Resing, H. A., Weber, D. C., Anderson, M., Miller, G. R., Moran, M., Poranski Jr, C. F. and Mattix, L. (1982a). *Polymer. Prepr.*, **23**, 101.

Resing, H. A., Garroway, A. N., Weber, D. C., Ferraris, J. and Slotfeldt-Ellingsen, D. (1982b). *Pure Appl. Chem.*, **54**, 595.

Rhim, W. K., Elleman, D. D. and Vaughan, R. W. (1973). *J. Chem. Phys.*, **59**, 3740.

Rhim, W. K., Elleman, D. D., Schrieber, L. B. and Vaughan, R. W. (1974). *J. Chem. Phys.*, **60**, 1595.

Rice, M. J. (1979). *Phys. Letters*, **71A**, 152.

Richardson, I. D. and Ward, I. M. (1970). *J. Phys. D: Appl. Phys.*, **3**, 643.

Rim, P. B. and Orler, E. B. (1987). *Macromolecules*, **20**, 433.

Robard, A., Patterson, D. and Delmas, G. (1977). *Macromolecules*, **10**, 706.

Roe, R.-J. and Krigbaum, W. R. (1964a). *J. Appl. Phys.*, **35**, 2215.

Roe, R.-J. and Krigbaum, W. R. (1964b). *J. Chem. Phys.*, **40**, 2608.

Roe, R.-J. (1965). *J. Appl. Phys.*, **36**, 2024.

Roe, R.-J. (1966). *J. Appl. Phys.*, **37**, 2069.

Roe, R.-J. (1970). *J. Polym. Sci.*, A-2, **8**, 1187.

Roe, R.-J., Davis, D. D. and Kwei, T. K. (1970). *Bull. Amer. Phys. Soc.*, **15**, 308.

Roe, R.-J., Fishkis, M. and Chang, J. C. (1981). *Macromolecules*, **14**, 1091.

Rose, M. E. (1957). *Elementary Theory of Angular Momentum*, Wiley and Sons, New York.

Rossler, E., Sillescu, H. and Spiess, H. W. (1985). *Polymer*, **26**, 203.

Rothwell, W. P. and Waugh, J. S. (1981). *J. Chem. Phys.*, **74**, 2721.

Rowland, S. P. (ed.) (1980). *Water in Polymers*, ACS Symp. Ser., **127**, Washington.

Rowland, S. P. and Kuntz, I. D. (1980). In *Water in Polymers*, Rowland, S. P. (ed.), *ACS Symp. Ser.*, **127**, Washington, p. 1.

Roy, A. K., Jones, A. A. and Inglefield, P. T. (1985). *J. Magnetic Res.*, **64**, 441.

Roy, A. K., Jones, A. A. and Inglefield, P. T. (1986). *Macromolecules*, **19**, 1356.

Roy, A. K. and Inglefield, P. T. (1990). *Prog. NMR Spect.*, **22**, 569.

Rupley, J. A., Yang, P.-H. and Tollin, G. (1980). In *Water in Polymers*, Rowland, S. P. (ed.), *ACS Symp. Ser.*, **127**, Washington, p. 111.

Schaefer, J. (1973). *Macromolecules*, **6**, 882.

Schaefer, J. (1974). In *Topics in ^{13}C NMR Spectroscopy*, Levy, G. C. (ed.), Wiley, New York, Vol. 1, p. 150.

Schaefer, J. and Stejskal, E. O. (1976). *J. Amer. Chem. Soc.*, **98**, 1031.

Schaefer, J., Stejskal, E. O. and Buchdahl, R. (1977). *Macromolecules*, **10**, 384.

Schaefer, J., McKay, R. A. and Stejskal, E. O. (1979). *J. Magnetic Res.*, **34**, 443.

Schaefer, J., Stejskal, E. O., Steger, T. R., Sefcik, M. D. and McKay, R. A. (1980). *Macromolecules*, **13**, 1121.

Schaefer, J., Sefcik, M. D., Stejskal, E. O. and McKay, R. A. (1981). *Macromolecules*, **14**, 188.

Schaefer, J., McKay, R. A., Stejskal, E. O. and Dixon, W. T. (1983). *J. Magnetic Res.*, **52**, 123.

Schaefer, J., Sefcik, M. D., Stejskal, E. O. and McKay, R. A. (1984a). *Macromolecules*, **17**, 1118.

Schaefer, J., Sefcik, M. D., Stejskal, E. O., McKay, R. A., Dixon, W. T. and Cais, R. E. (1984b). *Macromolecules*, **17**, 1107.

Schaefer, J., Stejskal, E. O., McKay, R. A. and Dixon, W. T. (1984c). *Macromolecules*, **17**, 1479.

Schaefer, J., Stejskal, E. O., Garbow, J. R. and McKay, R. A. (1984d). *J. Magnetic Res.*, **59**, 150.

Schaefer, J., Stejskal, E. O., Perchak, D., Skolnick, J. and Yaris, R. (1985). *Macromolecules*, **18**, 368.

Schajor, W., Pislewski, N., Zimmerman, H. and Haeberlen, U. (1980). *Chem. Phys. Letters*, **76**, 409.

Schilling, F. C., Bovey, F. A., Tseng, S. and Woodward, A. E. (1983). *Macromolecules*, **16**, 808.

Schlick, S. and Alonso-Amigo, M. G. (1989). *Macromolecules*, **22**, 2634.

Schmidt-Rohr, K., Clauss, J., Blümich, B. and Spiess, H. W. (1990). *Magnetic Res. in Chem.*, **28**, S3.

Schmidt-Rohr, K. and Spiess, H. W. (1991a). *Phys. Rev. Letters*, **66**, 3020.

Schmidt-Rohr, K. and Spiess, H. W. (1991b), *Macromolecules*, **24**, 5288.

Schnauss, W., Fujara, F., Hartmann, K. and Sillescu, H. (1990). *Chem. Phys. Letters*, **166**, 381.

Schultz, A. R. and Young, A. L. (1980). *Macromolecules*, **13**, 663.

Schurer, J. W., de Boer, A. and Challa, G. (1975). *Polymer*, **16**, 201.

Sefcik, M. D., Stejskal, E. O., McKay, R. A. and Schaefer, J. (1979). *Macromolecules*, **12**, 423.

Sefcik, M. D., Schaefer, J., Stejskal, E. O. and McKay, R. A. (1980). *Macromolecules*, **13**, 1132.

Sefcik, M. D., Schaefer, J., May, F. L., Raucher, D. and Dub, S. M. (1983). *J. Polym. Sci: Polym. Phys. Ed.*, **21**, 1041.

Semerak, S. N. and Frank, C. W. (1981). *Macromolecules*, **14**, 443.

Sidles, J. A. (1992). *Phys. Rev. Letters*, **68**, 1124.

Sillars, R. W. (1939). *Proc. Roy. Soc. London*, *A* **169**, 66.

Sillescu, H. (1971). *J. Chem. Phys.*, **54**, 2110.

Sillescu, H. (1982). *Pure Appl. Chem.*, **54**, 619.

Skotheim, T. J. (ed.) (1986). *Handbook of Conducting Polymers*, Marcel Dekker, New York.

Slichter, C. P. and Ailion, D. (1964). *Phys. Rev.*, **135A**, 1099.

Slichter, C. P. (1990). *Principles of Magnetic Resonance*, 3rd edition, Springer-Verlag, New York.

Smit, P. P. A. (1966). *Rheol. Acta*, **5**, 277.

Smith, J. B., Manuel, A. J. and Ward, I. M. (1975). *Polymer*, **16**, 57.

Smith, P. W. R. (1986). Ph.D. Thesis, University of East Anglia.

Smyth, G., Quinn, F. X. and McBrierty, V. J. (1988). *Macromolecules*, **21**, 3198.

Solomon, I. (1955). *Phys. Rev.*, **99**, 559.

Spiess, H. W. (1980). *J. Chem. Phys.*, **72**, 6755.

Spiess, H. W. and Sillescu, H. (1981). *J. Magnetic Res.*, **42**, 381.

Spiess, H. W. (1982). In *Developments in Oriented Polymers* 1, Ward, I. M. (ed.), Applied Science Publns, London, p. 47.

Spiess, H. W. (1983). *Colloid Polym. Sci.*, **261**, 193.

Spiess, H. W. (1985a). *Adv. Polym. Sci.*, **66**, 23.

Spiess, H. W. (1985b). In *Characterisation of Polymers in the Solid State I: Part A: NMR and other Spectroscopic Methods*, Kausch, H. H. and Zachmann, H. G. (eds.), Springer-Verlag, Berlin, pp. 23–58.

Spiess, H. W. (1991). *Chem. Rev.*, **91**, 1321.

Springer Jr, C. S. and Xu, Y. (1991). *Aspects of Bulk Magnetic Susceptibility in in vivo MRI and MRS*, Euro. Workshop on Magnetic Resonance in Medicine, Rinck, P. A. (ed.), Blonay, Switzerland.

Stejskal, E. O., Schaefer, J. and Waugh, J. S. (1977). *J. Magnetic Res.*, **28**, 105.

Stejskal, E. O., Schaefer, J., Sefcik, M. D. and McKay, R. A. (1981). *Macromolecules*, **14**, 275.

Stillinger, F. H. (1980). In *Water in Polymers*, Rowland, S. P. (ed.), *ACS Symp. Ser.* **127**, Washington, p. 11.

Street, G. B. and Clarke, T. C. (1981). *IBM J. Res. Develop.*, **25**, 51.

Su, W. P., Schrieffer, J. R. and Heeger, A. J. (1979). *Phys. Rev. Letters*, **42**, 1698.

Su, W. P., Schrieffer, J. R. and Heeger, A. J. (1980). *Phys. Rev.*, **B22**, 2099.

Su, W. P., Schrieffer, J. R. and Heeger, A. J. (1983). *Phys. Rev.*, **B28**, 1138. [Erratum.]

Suter, D. and Ernst, R. R. (1985). *Phys. Rev.*, **B32**, 5608.

Suwelack, D., Rothwell, W. P. and Waugh, J. S. (1980). *J. Chem. Phys.*, **73**, 2559.

Takayama, H., Lin-Liu, T. R. and Maki, K. (1980). *Phys. Rev.*, **B21**, 2388.

Takegoshi, K. and McDowell, C. A. (1986). *J. Magnetic Res.*, **66**, 14.

Takegoshi, K. and Hikichi, K. (1991). *J. Chem. Phys.*, **94**, 3200.

Tamman, G. and Hesse, W. (1926). *Z. Anorg. Allgem. Chem.*, **156**, 245.

Tanaka, H. and Nishi, T. (1985). *J. Chem. Phys.*, **82**, 4326.

Tanaka, H. and Nishi, T. (1986). *Phys. Rev.*, **B33**, 32.

Tanaka, H., Hayashi, T. and Nishi, T. (1986). *J. Appl. Phys.*, **59**, 653 and 3627.

Tasaka, S., Miyasato, K., Yoshikawa, M., Miyata, S. and Ko, M. (1984). *Ferroelectrics*, **57**, 267.

Tasaka, S. and Miyata, S. (1985). *J. Appl. Phys.*, **57**, 906.

Taylor, P. C., Baugher, J. F. and Kriz, H. M. (1975). *Chem. Rev.*, **75(2)**, 203.

Tegenfeldt, J. and Haeberlen, U. (1979). *J. Magnetic Res.*, **36**, 453.

Tékély, P., Laupretre, F. and Monnerie, L. (1985). *Polymer*, **26**, 1081.

Terao, T., Maeda, S. and Saika, A. (1983). *Macromolecules*, **16**, 1535.

Terao, T., Maeda, S., Yamabe, T., Akagi, K. and Shirakawa, H. (1984). *Chem. Phys. Letters*, **103**, 347.

Tomkiewicz, Y., Mortensen, K., Thewaldt, M. L. W., Clarke, T. C. and Street, G. B. (1980). *Phys. Rev. Letters*, **45**, 490.

Tomkiewicz, Y., Shiren, N. S., Schultz, T. D., Thomann, H., Dalton, L. R., Zettl, A., Gruner, G. and Clarke, T. C. (1982). *Mol. Cryst. Liq. Cryst.*, **83**, 17.

Tonelli, A. E. (1989). *NMR Spectroscopy and Polymer Microstructure: the Conformational Connection*, VCH Publishers Inc., Deerfield Beach, Florida.

Torchia, D. A. (1978). *J. Magnetic Res.*, **30**, 613.

Toriumi, H., Deloche, B., Herz, J. and Samulski, E. T. (1985). *Macromolecules*, **18**, 304.

Tse, D. and Hartmann, S. R. (1968). *Phys. Rev. Letters*, **21**, 511.

Tycko, R. and Dabbagh, G. (1991). *J. Amer. Chem. Soc.*, **113**, 3592.

Urbina, J. and Waugh, J. S. (1974). *Proc. Nat. Acad. Sci.*, **71**, 5062.

Van Bogart, J. W. C., Gibson, P. E. and Cooper, S. L. (1983). *J. Polym. Sci: Polym. Phys. Ed.*, **21**, 65.

VanderHart, D. L., Gutowsky, H. S. and Farrar, T. C. (1967). *J. Amer. Chem. Soc.*, **89**, 5056.

VanderHart, D. L. (1979). *Macromolecules*, **12**, 1232.

VanderHart, D. L. and Garroway, A. N. (1979). *J. Chem. Phys.*, **71**, 2773.

VanderHart, D. L. and Khoury, F. (1984). *Polymer*, **25**, 1589.

VanderHart, D. L. and Atalla, R. H. (1984). *Macromolecules*, **17**, 1465.

VanderHart, D. L. and Pérez, E. (1986). *Macromolecules*, **19**, 1902.

VanderHart, D. L. (1987). *J. Magnetic Res.*, **72**, 13.

VanderHart, D. L., Manders, W. F., Stein, R. S. and Herman, W. (1987). *Macromolecules*, **20**, 1724.

Vanderschueren, J., Ladang, M. and Heuschen, J. M. (1980). *Macromolecules*, **13**, 973.

Vanderschueren, J., Janssens, A., Ladang, M. and Niezette, J. (1982). *Polymer*, **23**, 395.

Van Vleck, J. H. (1948). *Phys. Rev.*, **74**, 1168.

Vardeny, Z., Ehrenfreund, E., Brafman, O., Nowak, M., Schaffer, H., Heeger, A. J. and Wudl, F. (1986). *Phys. Rev. Letters*, **56**, 671.

Veeman, W. S. (1984). *Progr. NMR Spectr.*, **16**, 193.

Veeman, W. S. and Cory, D. G. (1989). *J. Magnetic Res.*, **84**, 392.

Vega, A. J. and English, A. D. (1979). *Macromolecules*, **12**, 353.

Vega, A. J. and English, A. D. (1980). *Macromolecules*, **13**, 1635. [There is a typographical error in eq B 6.]

Veregin, R. P., Fyfe, C. A. and Marchessault, R. H. (1986). *Macromolecules*, **19**, 2379.

Voelkel, R. and Sillescu, H. (1979). *J. Magnetic Res.*, **34**, 559.

Vogel, H. (1921). *Physik. Z.*, **22**, 645.

Volkoff, G. M. (1953). *Can. J. Phys.*, **31**, 820.

Vorenkamp, E. J., ten Brinke, G., Meijer, J. G., Jager, H. and Challa, G. (1985). *Polymer*, **26**, 1725.

Wagner, K. W. (1914). *Arch. Elektrotech. (Berlin)*, **2**, 371.

Ward, I. M. (1962). *Proc. Phys. Soc.*, **80**, 1176.

Ward, I. M. (ed.) (1982). *Developments in Oriented Polymers*, Applied Science Publishers, London.

Ward, I. M. (1985). In *Characterisation of Polymers in the Solid State I: Part A: NMR and other Spectroscopic Methods*, Kausch, H. H. and Zachmann, H. G. (eds.), Springer-Verlag, Berlin, pp. 81–116.

Wardell, G. E., McBrierty, V. J. and Douglass, D. C. (1974). *J. Appl. Phys.*, **45**, 3441.

Wardell, G. E., Douglass, D. C. and McBrierty, V. J. (1976). *Polymer*, **17**, 41.

Wardell, G. E. and McBrierty, V. J. (1981). US Patent no. 4301411. [(1983), UK Patent 2043262B.]

Wardell, G. E., McBrierty, V. J. and Marsland, V. (1982). *Rubber Chem. Tech.*, **55**, 1095.

Warner, M., Higgins, J. S. and Carter, A. J. (1983). *Macromolecules*, **16**, 1931.

Warrick, E. L, Pierre, O. R. and Polmanteer, K. E. (1979). *Rubber Chem. Tech.*, **52**, 437.

Waugh, J. S., Huber, L. M. and Haeberlen, U. (1968). *Phys. Rev. Letters*, **20**, 180.

Wecker, S. M., Davidson, T. and Baker, D. W. (1972). *J. Appl. Phys.*, **43**, 4344.

Wecker, S. M., Cohen, J. B. and Davidson, T. (1974). *J. Appl. Phys.*, **45**, 4453.

Wefing, S., Kaufmann, S. and Spiess, H. W. (1988). *J. Chem. Phys.*, **89**, 1234.

Wefing, S. and Spiess, H. W. (1988). *J. Chem. Phys.*, **89**, 1219.

Wehrle, M., Hellmann, G. P. and Spiess, H. W. (1987). *Colloid Polym. Sci.*, **265**, 815.

Weiss, G. H., Gupta, R. K., Ferretti, J. A. and Becker, E. D. (1980). *J. Magnetic Res.*, **37**, 369.

Wetton, R. E., Moore, J. D. and Ingram, P. (1973). *Polymer*, **14**, 161.

Wetton, R. E., MacKnight, W. J., Fried, J. R. and Karasz, F. E. (1978). *Macromolecules*, **11**, 158.

Williams, E. A., Donahue, P. E. and Cargioli, J. D. (1981). *Macromolecules*, **14**, 1016.

Williams, G. and Watts, S. B. (1970). *Trans. Faraday Soc.*, **66**, 80.

Williams, M. L., Landel, R. F. and Ferry, J. D. (1955). *J. Amer. Chem. Soc.*, **77**, 3701.

Wilkes, C. E. (1977). *J. Polym. Sci: Polym. Symp.*, **60**, 161.

Wilson, C. W. and Pake, G. E. (1953). *J. Polym. Sci.*, **10**, 503.

Wilson, T. W. and Turner, D. T. (1988). *Macromolecules*, **21**, 1184.

Windle, A. H. (1982). In *Developments in Oriented Polymers 1*, Ward, I. M. (ed.), Applied Science, London, p. 1.

Wokaun, A. and Ernst, R. R. (1977). *J. Chem. Phys.*, **67**, 1752.

Wolf, A. A. (1969). *Amer. J. Phys.*, **37**, 531.

Wolf, D. (1979). *Spin Temperature and Nuclear-Spin Relaxation in Matter*, Clarendon Press, Oxford.

Wong, A. C., Garroway, A. N. and Ritchey, W. M. (1981). *Macromolecules*, **14**, 832.

Woodward, A. E. (1989). *Atlas of Polymer Morphology*, Hanser-Verlag, München.

Yang, H., Shibayama, M., Stein, R. S., Shimizu, N. and Hashimoto, T. (1986). *Macromolecules*, **19**, 1667.

Yang, Y., Hagemeyer, A., Blümich, B. and Spiess, H. W. (1988). *Chem. Phys. Letters*, **150**, 1.

Yang, Y., Hagemeyer, A., Zemke, K. and Spiess, H. W. (1990). *J. Chem. Phys.*, **93**, 7740.

Zhang, C., Wang, P., Jones, A. A., Inglefield, P. T. and Kambour, R. P. (1991). *Macromolecules*, **24**, 338.

Zhang, X., Qui, L., Wang, D. and Wang, Y. (1988). *Chinese J. Polym. Sci.*, **6**, 159.

Zhang, X., Takegoshi, K. and Hikichi, K. (1991a). *Macromolecules*, **24**, 5756.

Zhang, X., Takegoshi, K. and Hikichi, K. (1991b). *Polymer J.*, **23**, 79 and 87.

Zwijnenberg, A. and Pennings, A. J. (1976). *Colloid Polym. Sci.*, **254**, 868: *J. Polym. Sci: Polym. Letters Ed.*, **14**, 339.

Author index

337

Subject index